MONOGRAPHS ON
STATISTICS AND APPLIED PROBABILITY

General Editors

D.R. Cox, D.V. Hinkley, D. Rubin and B.W. Silverman

1 Stochastic Population Models in Ecology and Epidemiology *M.S. Bartlett* (1960)

2 Queues *D.R. Cox and W.L. Smith* (1961)

3 Monte Carlo Methods *J.M. Hammersley and D.C. Handscomb* (1964)

4 The Statistical Analysis of Series of Events *D.R. Cox and P.A.W. Lewis* (1966)

5 Population Genetics *W.J. Ewens* (1969)

6 Probability, Statistics and Time *M.S. Bartlett* (1975)

7 Statistical Inference *S.D. Silvey* (1975)

8 The Analysis of Contingency Tables *B.S. Everitt* (1977)

9 Multivariate Analysis in Behavioural Research *A.E. Maxwell* (1977)

10 Stochastic Abundance Models *S. Engen* (1978)

11 Some Basic Theory for Statistical Inference *E.J.G. Pitman* (1979)

12 Point Processes *D.R. Cox and V. Isham* (1980)

13 Identification of Outliers *D.M. Hawkins* (1980)

14 Optimal Design *S.D. Silvey* (1980)

15 Finite Mixture Distributions *B.S. Everitt and D.J. Hand* (1981)

16 Classification *A.D. Gordon* (1981)

17 Distribution-free Statistical Methods *J.S. Maritz* (1981)

18 Residuals and Influence in Regression *R.D. Cook and S. Weisberg* (1982)

19 Applications of Queueing Theory *G.F. Newell* (1982)

20 Risk Theory, 3rd edition *R.E. Beard, T. Pentikäinen and E. Pesonen* (1984)

21 Analysis of Survival Data *D.R. Cox and D. Oakes* (1984)

22 An Introduction to Latent Variable Models *B.S. Everitt* (1984)

23 Bandit Problems *D.A. Berry and B. Fristedt* (1985)

24 Stochastic Modelling and Control *M.H.A. Davis and R. Vinter* (1985)

25 The Statistical Analysis of Compositional Data *J. Aitchison* (1986)

26 Density Estimation for Statistics and Data Analysis *B.W. Silverman* (1986)

27 Regression Analysis with Applications *G.B. Wetherill* (1986)

28 Sequential Methods in Statistics, 3rd edition *G.B. Wetherill* (1986)

29 Tensor Methods in Statistics *P. McCullagh* (1987)

30 Transformation and Weighting in Regression *R.J. Carroll and D. Ruppert* (1988)

31 Asymptotic Techniques for Use in Statistics *O.E. Barndorff-Nielsen and D.R. Cox* (1989)

32 Analysis of Binary Data, 2nd edition *D.R. Cox and E.J. Snell* (1989)

33 Analysis of Infectious Disease Data *N.G. Becker* (1989)

34 Design and Analysis of Cross-Over Trials *B. Jones and M.G. Kenward* (1989)

35 Empirical Bayes Methods, 2nd edition *J.S. Maritz and T. Lwin* (1989)

36 Symmetric Multivariate and Related Distributions *K.-T. Fang, S. Kotz and K. Ng* (1989)

37 Generalized Linear Models, 2nd edition *P. McCullagh and J.A. Nelder* (1989)

(Full details concerning this series are available from the publishers)

Empirical Bayes Methods

SECOND EDITION

J.S. MARITZ

Professor of Statistics, La Trobe University

and

T. LWIN

Senior Research Scientist, CSIRO

LONDON NEW YORK
CHAPMAN AND HALL

First published in 1970 by
Methuen London Ltd
Second edition published in 1989 by
Chapman and Hall Ltd
11 New Fetter Lane, London EC4P 4EE
Published in the USA by
Chapman and Hall
29 West 35th Street, New York NY 10001

Typeset in Times 10/12 by
Thomson Press (India) Limited, New Delhi
Printed in Great Britain by St Edmundsbury Press Ltd
Bury St Edmunds, Suffolk

ISBN 0 412 27760 3

British Library Cataloguing in Publication Data

Maritz, J.S.
Empirical Bayes methods
1. Statistical analysis. Bayesian theories
I. Title II. Lwin, T. *1944 –* III. Series
519.5'42

ISBN 0 412 27760 3

Library of Congress Cataloging in Publication Data

Maritz, J.S.
Empirical Bayes methods/J.S. Maritz and T. Lwin
p. cm.—(Monographs on statistics and applied probability)
Bibliography: p.
Includes index.
ISBN 0 412 27760 3 (U.S.)
1. Bayesian statistical decision theory. I. Lwin, T., 1944 – .
II. Title. III. Series.
QA279.5.M864 1989
519.5'42—dc20 89-31954
CIP

Contents

Acknowledgements

T. Lwin thanks the CSIRO Division of Mathematics and Statistics for support in the preparation of material for this book. J.S. Maritz thanks the Department of Statistics, La Trobe University.

Preface

Neyman (1962) referred to the empirical Bayes approach as a *breakthrough* in the theory of statistical decision making, and in the time since the publication of the first edition of *Empirical Bayes Methods* there have certainly been many contributions to this theory. A measure of the importance of a theory is its impact on the practice of Statistics. The empirical Bayes approach has not revolutionized the practice of Statistics, but there can be little argument that it has had a telling influence on the thinking of many statisticians, and on their practice in certain areas of application. One of the objects in preparing this new edition was to collect and present several practical examples of the application of empirical Bayes ideas and techniques so as to give an indication of the sorts of problems in which they may be useful. It is worth pointing out that *meta analysis* is now regarded as a highly desirable undertaking, especially in the social sciences and in medicine. It has clear connections with empirical Bayes.

The empirical Bayes approach can be thought of as a way of looking at data arising in a sequence of similar experiments. It has competitors, and the relationships between them has received a good deal of attention in recent publications by many authors. A discussion of alternatives to empirical Bayes is given in Chapter 7.

Some topics in empirical Bayes theory have been given scant attention in this book, notably the question of rates of convergence of risks of certain methods. This is not to deny their theoretical interest. We have concentrated on topics which appear to have more immediate practical relevance. For example, an examination of applications suggests that linear Bayes and empirical Bayes methods are important. Various studies, including those of robustness by several authors, indicate that the use of parametric priors is more readily defensible than might have been suggested in the first edition of this book.

Briefly in summary, the main changes to *Empirical Bayes Methods* are: inclusion of more details of published accounts of applications, more emphasis on linear EB methods, an account of some competitors of EB, a chapter on interval estimation, more material on multiparameter problems.

Notation and Abbreviations

$A_n \xrightarrow{P} A$: A_n tends to A in probability

$X \stackrel{d}{=} F$: the distribution of random variable X is F

$X \stackrel{d}{=} N(\mu, \sigma^2)$: the distribution of X is normal with mean μ and variance σ^2

$X \stackrel{d}{=} Bin(n, \theta)$: the distribution of X is Binomial with index n and probability parameter θ

m.g.f.: moment generating function
c.f.: characteristic function
c.d.f.: cumulative distribution function
p.d.f.: probability density function
r.v.: random variable
p.d.: probability distribution
EB: empirical Bayes
a.o.:asymptotically optimal
FB: full Bayesian
CD: compound decision

CHAPTER 1

Introduction to Bayes and empirical Bayes methods

1.1 The problem, Bayes conventional and empirical Bayes methods

Empirical Bayes (EB) and related techniques come into play when data are generated by repeated execution of the same type of random experiment. The individual experiments are often called component experiments, or simply components. It is convenient when considering the data from a particular component to think of it as the current component which has been preceded by the other components. Empirical Bayes methods provide a way in which such historical data can be used in the assessment of the current results. This temporal view of the data sequence is a convenience and does not play an active role in EB analysis.

An early example of an EB nature is given by von Mises (1943); see also Chapter 8, section 8.3.1. In examining the quality of a batch of water for possible bacterial contamination $m = 5$ samples of a given volume are taken. A sample registers a positive result if it contains at least one bacterium. Interest centres on the probability, θ, of a positive result. Typically there are many repetitions of this experiment with different batches, and the probability θ can be regarded as varying randomly between experiments according to a prior distribution $G(\theta)$. For a given θ the probability of x positive results in $m = 5$ samples is

$$p(x \mid \theta) = \binom{5}{x} \theta^x (1 - \theta)^{5-x},$$

and in repetitions of the same procedure with different batches the marginal distribution of the number of positive results in five samples is the mixed binomial distribution

$$p_G(x) = \int \binom{5}{x} \theta^x (1-\theta)^{5-x}\, dG(\theta).$$

If the distribution G is known, a Bayesian analysis of the current experiment can be performed. For example, a Bayes point estimate of θ can be calculated as a possible competitor for the classical maximum likelihood estimate. When the prior, or mixing, distribution G is not known it is possible to estimate it by using the observed marginal distribution of the x values. The essence of the EB method in this case is that all calculations of a Bayes nature are performed after replacing G by its estimate. In the example discussed by von Mises there are $N = 3420$ observations from the marginal population characterized by $p_G(x)$, and particular attention is paid to the estimation of G.

Much of the work in EB methods over the past two decades or so has been stimulated by Robbins in papers beginning with the reference Robbins (1955) where the terminology 'empirical Bayes' was introduced. However, as the examples in Chapter 8 show, it has become clear that the applicability of EB ideas is much wider than might have been suggested by the earlier writings. It may be noted especially that EB ideas are applicable in many problems involving mixtures of distributions.

Generally, and more formally, we shall be concerned with problems arising in the following manner: an observation x is made on a random variable X whose distribution depends on the parameter λ. Our task is to make a decision $\delta(x)$ about the value of λ. Typically the decision may be the calculation of a point estimate of λ, or it may be a choice between two hypothetical values of λ. The dependence of the decision on x is indicated by using the symbol $\delta(x)$, which is said to represent the decision function. In practice one commonly has a number $m > 1$ of independent observations on X, rather than just the one value x, and obviously our theory has to allow for such multiple observations. We may also have to deal with problems involving more than one parameter. But to begin we shall avoid the notational, computational and other complications that arise with these generalizations.

In the pure Bayesian approach to the decision problem the parameter value itself is regarded as a realization of a random variable Λ with distribution function $G(\lambda)$. The distribution of Λ

is called the prior distribution. The probabilities defined by $G(\lambda)$ are not necessarily interpretable in terms of relative frequencies. A fundamental problem in the pure Bayes approach is the specification of G. The Bayes solution of a decision problem generally depends on G, and the Bayes decision function is denoted by $\delta_G(x)$ to show this dependence. An outline of the pure Bayes approach is given in section 1.2, with applications to examples which are used repeatedly later on.

Decision functions derived without appeal to the notion of a prior distribution will be described as conventional, non-Bayes or classical. A great classical literature exists. The various criteria for obtaining non-Bayes decision rules, and related special techniques, are well documented and will be assumed known. It will be seen that rules derived by the likelihood principle are prominent among the non-Bayes rules.

In the empirical Bayes approach the existence of a prior distribution is postulated, but it is taken to be susceptible to a frequency interpretation. Further, the availability of previous data, suitable for estimation of the prior distribution G is assumed. The mathematical derivations associated with the Bayes method are used to obtain a decision function $\delta_G(x)$, generally dependent on G, but then $\delta_G(x)$ is replaced by an estimate based on the previous data. Such an estimated $\delta_G(x)$ is called an empirical Bayes decision rule.

1.2 An introduction to Bayes techniques

Since the EB approach uses the techniques and results of the Bayes approach some of the standard results are reviewed in this and following sections. Applications of the EB methods described in this monograph are envisaged as occurring in repetitive experimentation with parameters varying from experiment to experiment. The notion of **expected loss** therefore seems rather natural in this context, hence our introduction to Bayes methods is based on the notion of a loss function.

Let a **loss**, $L(\delta(x), \lambda) \geqslant 0$, be incurred when the parameter value is λ and a decision $\delta(x)$ is made. For example, if $\delta(x)$ is a point estimate of λ it is common to put $L(\delta(x), \lambda) = (\delta(x) - \lambda)^2$. Or, if $\delta(x)$ is an interval estimate one may put $L(\delta(x), \lambda) = 0$ or 1 according as the interval does or does not contain λ.

The expected loss for fixed λ is the **risk**

$$R_\delta(\lambda)=\int L\{\delta(x),\lambda\}f(x|\lambda)\,dx, \tag{1.2.1}$$

where $f(x|\lambda)$ is the probability density function (p.d.f.) of X. Modification of (1.2.1) for discrete X is obvious. The selection of a decision function now becomes a matter of choosing a $\delta(x)$ whose $R_\delta(\lambda)$ has acceptable properties; see, for example, Ferguson (1967, section 1.6). Clearly, the smaller $R_\delta(\lambda)$ for any λ, the better, but it is trivially true that there is generally no $\delta^*(x)$ such that $R_{\delta^*}(\lambda)\leqslant R_\delta(\lambda)$ for all λ and every δ. Thus there is no uniformly best δ, and an additional criterion for selecting a δ has to be invoked. One of these is provided in the Bayes approach, in which the goodness of a δ is judged by the overall expected loss, or the average risk, with respect to the prior distribution $G(\lambda)$. It is given by

$$W(\delta)=\int\int L\{\delta(x),\lambda\}f(x|\lambda)\,dx\,dG(\lambda). \tag{1.2.2}$$

Now δ is chosen so as to minimize W. The δ which does minimize W will depend on G, and it is denoted by δ_G to indicate the dependence. We shall call $W(\delta_G)$ the **Bayes risk**, but different terminology is also in use. Some authors refer to $W(\delta_G)$ as the **Bayes envelope functional**.

The actual determination of δ_G can proceed in principle by noting in an obvious abbreviated notation that $W=E(L)=EE(L|x)$ so that we choose δ for every x so as to minimize

$$E(L|x)=\int L(\delta,\lambda)f(x|\lambda)\,dG(\lambda)\Big/\int f(x|\lambda)dG(\lambda). \tag{1.2.3}$$

The details of such calculations depend on L. Most of this book is devoted to problems of point estimation and decision between two hypotheses. In the former case we usually take $L=(\delta-\lambda)^2$, in the latter $L=0$ or 1 according as the right or wrong decision is made. The relevant calculations for these two cases are taken up in somewhat more detail in following sections.

1.3 Bayes point estimation: one parameter

Let $\delta(x)$ be any point estimate of λ. If δ_G is the Bayes point estimate we have

$$W(\delta)\geqslant W(\delta_G) \tag{1.3.1}$$

by definition. Now, with $L(\delta, \lambda) = (\delta - \lambda)^2$,

$$\begin{aligned} W(\delta) &= \iint \{\delta(x) - \lambda\}^2 f(x|\lambda) dG(\lambda) dx \\ &= W(\delta_G) + \iint \{\delta(x) - \delta_G(x)\}^2 f(x|\lambda)\, dG(\lambda)\, dx \qquad (1.3.2) \\ &\quad + 2\iint \{\delta(x) - \delta_G(x)\}\{\delta_G(x) - \lambda\} f(x|\lambda)\, dG(\lambda)\, dx. \end{aligned}$$

Condition (1.3.1) will be satisfied if the third term in (1.3.2) is zero, which can be arranged by putting

$$\int \{\delta_G(x) - \lambda\} f(x|\lambda)\, dG(\lambda) = 0$$

for every x. This gives

$$\delta_G(x) = \frac{\int \lambda f(x|\lambda)\, dG(\lambda)}{\int f(x|\lambda)\, dG(\lambda)}. \qquad (1.3.3)$$

Thus $\delta_G(x)$ is the mean of the posterior distribution of Λ for given $X = x$. The same result is readily obtained from (1.2.3) by differentiation with respect to δ.

The following are some notes arising from the derivation of $\delta_G(x)$:

1. In the denominator of the right-hand side of (1.3.3) we have the marginal p.d.f. of X,

$$f_G(x) = \int f(x|\lambda)\, dG(\lambda). \qquad (1.3.4)$$

   The corresponding marginal distribution function is $F_G(x)$ and sometimes it will be convenient to refer to the marginal random variable (r.v.) whose cumulative distribution function (c.d.f.) is $F_G(x)$ as X_G.
2. In the joint distribution of X_G and Λ, $\delta_G(x)$ is the regression of Λ on X_G.
3. The marginal distribution of X_G is also called a **compound** or a **mixed** distribution.
4. With $\delta_G(x)$ given by (1.3.3) relation (1.3.2) becomes

$$W(\delta) = W(\delta_G) + \int \{\delta(x) - \delta_G(x)\}^2 f_G(x)\, dx,$$

   which is often useful for calculating $W(\delta)$.

Example 1.3.1 Let $F(x|\lambda)$ be the $N(\lambda, \sigma^2)$ c.d.f. and $G(\lambda)$ the $N(\mu_G, \sigma_G^2)$ c.d.f. Then, according to standard theory for the bivariate normal distribution, the distribution of X_G is $N(\mu_G, \sigma^2 + \sigma_G^2)$, the joint distribution of Λ, and X_G is bivariate normal with correlation coefficient ρ such that $\rho^2 = \sigma_G^2/(\sigma^2 + \sigma_G^2)$ and

$$\delta_G(x) = (x/\sigma^2 + \mu_G/\sigma_G^2)/(1/\sigma^2 + 1/\sigma_G^2); \tag{1.3.5}$$

see, for example Lindley (1965, p. 2). Also,

$$W(\delta_G) = 1/(1/\sigma^2 + 1/\sigma_G^2). \tag{1.3.6}$$

Example 1.3.2 Let $p(x|\lambda)$ be the Poisson probability distribution

$$p(x|\lambda) = e^{-\lambda}\lambda^x/x!, \qquad x = 0, 1, 2, \ldots$$

and $G(\lambda)$ a gamma c.d.f. with p.d.f.

$$g(\lambda) = \{1/\Gamma(\beta)\}\alpha^\beta\lambda^{\beta-1}e^{-\alpha\lambda}, \qquad \alpha, \beta > 0.$$

Then

$$\delta_G(x) = (\beta + x)/(\alpha + 1) \tag{1.3.7}$$

and

$$W(\delta_G) = \beta/\{\alpha(\alpha + 1)\}. \tag{1.3.8}$$

In Examples 1.3.1 and 1.3.2 the type of prior distribution is given. Such knowledge will rarely be available in applications of EB methods, hence the following examples, in which the form of $G(\lambda)$ is not specified, are of particular interest. They have played an important role in the literature on EB methods.

Example 1.3.3 The Poisson case as in Example 1.3.2:

$$\begin{aligned}\delta_G(x) &= (1/x!)\int \lambda^{x+1}e^{-\lambda}dG(\lambda)\Big/\left\{(1/x!)\int \lambda^x e^{-\lambda}dG(\lambda)\right\}\\ &= (x+1)p_G(x+1)/p_G(x),\end{aligned} \tag{1.3.9}$$

where $p_G(x)$ is a mixed Poisson probability distribution.

Example 1.3.4 The geometric distribution:

$$p(x|\lambda) = (1-\lambda)\lambda^x, \qquad x = 0, 1, 2, \ldots; \quad 0 < \lambda < 1.$$

$$\begin{aligned}\delta_G(x) &= \int (1-\lambda)\lambda^{x+1}dG(\lambda)\Big/\int (1-\lambda)\lambda^x\,dG(\lambda)\\ &= p_G(x+1)/p_G(x).\end{aligned}$$

Example 1.3.5 The negative binomial distribution:

$$p(x|\lambda) = (1-\lambda)^p\{\Gamma(p+x)/\Gamma(p)\}\lambda^x/x!, \qquad x = 0, 1, 2, \ldots; \qquad \lambda, p > 0$$

with p known. By steps like those in the previous examples

$$\delta_G(x) = \left(\frac{x+1}{p+x}\right)\frac{p_G(x+1)}{p_G(x)}.$$

In the preceding three examples we found that

$$\delta_G(x) = C(x)p_G(x+1)/p_G(x) \tag{1.3.10}$$

where $C(x)$ is a known function of x. The fact that $\delta_G(x)$ is expressed in terms of marginal probabilities of the r.v. X_G turns out to be useful in EB estimation, as was pointed out by Robbins (1955). In fact, a result like (1.3.10) holds for the members of the exponential family of discrete probability distributions which can be put in the form

$$p(x|\lambda) = \lambda^x \exp\{C(\lambda) + V(x)\} \tag{1.3.11}$$

for which

$$\delta_G(x) = \exp\{V(x) - V(x+1)\}p_G(x+1)/p_G(x). \tag{1.3.12}$$

A similar result is obtainable for a continuous r.v. Y with distribution in the exponential family of p.d.f.s

$$f(y|\mu) = \exp\{A(\mu) + B(\mu)W(y) + U(y)\}. \tag{1.3.13}$$

Making the transformations $X = W(Y)$ and $\lambda = \{\exp(B(\mu)\}$ the p.d.f. of X is

$$f(x|\lambda) = \lambda^x \exp\{C(\lambda) + V(x)\}; \tag{1.3.14}$$

note also (1.3.11). Hence $\delta_G(x)$ can be written as in (1.3.12) with p replaced by f. Although this produces an expression for $\delta_G(x)$ in terms of marginal probabilities the parametrization may be somewhat unnatural, as the following example shows.

Example 1.3.6 If the distribution of X is $N(\mu, \sigma^2)$, σ^2 known, the Bayes estimator of $\lambda = e^{\mu/\sigma^2}$ is

$$\delta_G(x) = e^{(x+1/2)/\sigma^2} f_G(x+1)/f_G(x),$$

a result of the type of (1.3.12). However, in applications of the normal distribution interest will usually be centred on μ rather than $\exp(\mu/\sigma^2)$.

In the case of continuous distributions it is more natural to use differentiation rather than the differencing process leading to (1.3.12) to obtain a similar result. An illustration is given in Example 1.3.7.

Example 1.3.7 The normal distribution as in Example 1.3.6:

Since $\ln f(x|\mu) = -\ln(\sigma\sqrt{2\pi}) - \frac{1}{2\sigma^2}(x-\mu)^2$

$$\frac{1}{f(x|\mu)}\frac{\partial f(x|\mu)}{\partial x} = -\frac{(x-\mu)}{\sigma^2}$$

or

$$\mu = x + \sigma^2 \frac{1}{f(x|\mu)}\frac{\partial f(x|\mu)}{\partial x}.$$

Appropriate substitutions in (1.3.3) give the Bayes point estimate of μ as

$$\delta_G(x) = x + \sigma^2 f'_G(x)/f_G(x); \qquad (1.3.15)$$

see Miyasawa (1961).

In Example 1.3.6 the Bayes estimate of $\exp(\mu/\sigma^2)$ is expressed in terms of marginal probability densities of X_G, and in Example 1.3.7 the Bayes estimate of μ is expressed in terms of the marginal p.d.f. and its derivative. As in the case of discrete X this type of expression of Bayes estimates is useful in connection with EB estimation, and will be taken up in more detail in section 3.3.

1.4 Bayes decisions between k simple hypotheses

We consider first the case of two simple hypotheses H_1: $\lambda = \lambda_1$ and H_2: $\lambda = \lambda_2$, $\lambda_1 < \lambda_2$. The prior probabilities are $P(\Lambda = \lambda_j) = \theta_j$, $j = 1, 2$ with $\theta_1 + \theta_2 = 1$. Thus $G(\lambda)$ is a step function with jumps at λ_1 and λ_2 of sizes θ_1 and θ_2 respectively. The decision function $\delta(x)$ is defined in terms of a partition of the sample space into two regions A_1 and A_2 such that H_j is accepted when $x \in A_j$, $j = 1, 2$. The most common definition of the loss function, $L\{\delta(x), \lambda\}$, in this context is to let $L = 0$ when the correct decision is made, and 1 otherwise. Then

$$W(\delta) = \theta_1 \int_{A_2} f(x|\lambda_1)\,dx + \theta_2 \int_{A_1} f(x|\lambda_2)\,dx,$$

with the obvious modifications for discrete X. We see that $W(\delta)$ is the overall expected proportion of wrong decisions.

To minimize $W(\delta)$ we define A_1 and A_2 such that $x \in A_1$ when $\theta_2 f(x|\lambda_2) < \theta_1 f(x|\lambda_1)$. Hence the Bayes decision rule can be stated as follows: choose H_1 if the posterior probability of H_1 exceeds 1/2. Equivalently choose whichever of H_1 and H_2 has the greater posterior probability.

When $f(x|\lambda)$ is such that $f(x|\lambda_1)/f(x|\lambda_2)$ is monotonic in x it follows immediately that A_1 comprises all values of $x < \xi_G$, where $x = \xi_G$ is the solution of

$$\theta_2 f(x|\lambda_2) = \theta_1 f(x|\lambda_1), \tag{1.4.1}$$

a suitable convention being adopted when X is discrete.

Example 1.4.1 Suppose that the distribution of X is $N(\lambda, 1)$. Then (1.4.1) becomes

$$\frac{\theta_2}{\theta_1} = \exp\{-(\lambda_2 - \lambda_1)(2x - \lambda_1 - \lambda_2)/2\},$$

the right-hand side is monotonic in x, and

$$\xi_G = \frac{(\lambda_1 + \lambda_2)}{2} - \frac{1}{(\lambda_2 - \lambda_1)} \ln(\theta_2/\theta_1). \tag{1.4.2}$$

In principle the case of $2 < k < \infty$ simple hypotheses is not different from that of two simple hypotheses. The hypotheses are H_j: $\lambda = \lambda_j$; $j = 1, 2, \ldots, k$, $\lambda_1 < \lambda_2 \cdots < \lambda_k$, and the respective prior probabilities are θ_j, $j = 1, 2, \ldots, k$, with $\theta_1 + \theta_2 + \cdots + \theta_k = 1$. Defining δ such that A_j is the region of acceptance of H_j, $j = 1, 2, \ldots, k$,

$$W(\delta) = \sum_{i \neq j = 1}^{k} \theta_j \int_{A_i} f(x|\lambda_j)\, dx. \tag{1.4.3}$$

Following the argument used for deriving the Bayes rule in the case $k = 2$, $W(\delta)$ is minimized by choosing for every x the H which has the largest posterior probability.

1.5 Bayes decisions between two composite hypotheses

Let B_1 and B_2 represent a partition of the parameter space, so that we choose between H_1; $\lambda \in B_1$ and H_2; $\lambda \in B_2$. The sample space is partitioned into A_1 and A_2 so that H_j is selected when $x \in A_j$, $j = 1, 2$. Then, with the 0–1 loss structure as in section 1.4,

$$W(\delta) = \int_{x \in A_1} \int_{\lambda \in B_2} f(x|\lambda)\, dG(\lambda)\, dx + \int_{x \in A_2} \int_{\lambda \in B_1} f(x|\lambda)\, dG(\lambda). \quad (1.5.1)$$

Arguing as in section 1.4, $W(\delta)$ is minimized by assigning a point x in the sample space to A_1 if

$$\int_{\lambda \in B_2} f(x|\lambda)\, dG(\lambda) < \int_{\lambda \in B_1} f(x|\lambda)\, dG(\lambda). \quad (1.5.2)$$

If equality can occur in (1.5.2) with non-zero probability, as with discrete X, a suitable convention is adopted.

A special case of some interest is when λ is a location parameter and we let H_1: $\lambda < \lambda_0$, H_2: $\lambda \geqslant \lambda_0$. Then (1.5.2) becomes

$$\int_{\lambda_0}^{\infty} f(x|\lambda) dG(\lambda) < \int_{-\infty}^{\lambda_0} f(x|\lambda) dG(\lambda).$$

An interpretation of this result is: choose H_1 if the posterior median of Λ is smaller than λ_0, otherwise choose H_2.

An alternative loss structure for the special case H_1: $\lambda < \lambda_0$, H_2: $\lambda \geqslant \lambda_0$ has been considered by Samuel (1963) and Robbins (1964). Let

$$\text{Loss} = \begin{cases} 0 & \text{for } \lambda < \lambda_0,\ x \in A_1 \\ 0 & \text{for } \lambda \geqslant \lambda_0,\ x \in A_2 \\ \lambda_0 - \lambda & \text{for } \lambda < \lambda_0,\ x \in A_2 \\ \lambda - \lambda_0 & \text{for } \lambda \geqslant \lambda_0,\ x \in A_1. \end{cases}$$

Then

$$W(\delta) = \int_{A_1} \int_{\lambda_0}^{\infty} (\lambda - \lambda_0) f(x|\lambda)\, dG(\lambda)\, dx + \int_{A_2} \int_{-\infty}^{\lambda_0} (\lambda_0 - \lambda) f(x|\lambda)\, dG(\lambda)\, dx,$$

and it is minimized by assigning x to A_1 if

$$\frac{\int \lambda f(x|\lambda)\, dG(\lambda)}{\int f(x|\lambda)\, dG(\lambda)} < \lambda_0,$$

i.e. if the posterior mean is $< \lambda_0$.

The arguments leading to (1.5.2) are readily extended to $k > 2$ hypotheses; no further detail will be given here.

1.6 Bayes estimation of vector parameters

Much of what has gone before can be generalized quite easily when X is replaced by the vector r.v. $\mathbf{X}=(X_1\cdots X_p)^{\mathrm{T}}$ and Λ by $\boldsymbol{\Lambda}=(\Lambda_1,\Lambda_2,\ldots,\Lambda_k)^{\mathrm{T}}$. Standard examples are the multivariate normal and the multinominal distributions. Letting $L\{\boldsymbol{\delta}(\mathbf{x}),\boldsymbol{\lambda}\}$ be the loss in making decision $\boldsymbol{\delta}(\mathbf{x})$ about $\boldsymbol{\lambda}$, the discussion of section 1.2 carries over with hardly any change, and formally the Bayes decision rule $\boldsymbol{\delta}_G$ can be obtained by using (1.2.3) on replacing x by $\mathbf{x}$, etc. Of course, G is now a k-variate distribution.

For some problems of point estimation a natural generalization of squared error loss is

$$L(\boldsymbol{\delta},\boldsymbol{\lambda})=(\boldsymbol{\delta}-\boldsymbol{\lambda})^{\mathrm{T}}\mathbf{A}(\boldsymbol{\delta}-\boldsymbol{\lambda})$$

where $\mathbf{A}$ is a positive definite matrix. The Bayes point estimate $\boldsymbol{\delta}_G(\mathbf{x})$ can be obtained from the vector version of (1.3.2),

$$\begin{aligned} W(\boldsymbol{\delta})=W(\boldsymbol{\delta}_G)+\int\!\!\int &\{\boldsymbol{\delta}(\mathbf{x})-\boldsymbol{\delta}_G(\mathbf{x})\}^{\mathrm{T}}\mathbf{A}\{\boldsymbol{\delta}(\mathbf{x})\\ &-\boldsymbol{\delta}_G(\mathbf{x})\}f(\mathbf{x}|\boldsymbol{\lambda})\,dG(\boldsymbol{\lambda})\,d\mathbf{x}\\ +2\int\!\!\int &\{\boldsymbol{\delta}(\mathbf{x})-\boldsymbol{\delta}_G(\mathbf{x})\}^{\mathrm{T}}\mathbf{A}\{\boldsymbol{\delta}_G(\mathbf{x})-\boldsymbol{\lambda}\}f(\mathbf{x}|\boldsymbol{\lambda})\,dG(\boldsymbol{\lambda})\,d\mathbf{x}. \end{aligned}$$

The third term in the right-hand side of the above expression can be made equal to zero, thus ensuring $W(\boldsymbol{\delta})>W(\boldsymbol{\delta}_G)$, by letting the ith element, $\delta_{Gi}(\mathbf{x})$ of $\boldsymbol{\delta}_G(\mathbf{x})$ be

$$\delta_{Gi}(\mathbf{x})=\frac{\int\lambda_i f(\mathbf{x}|\boldsymbol{\lambda})\,dG(\boldsymbol{\lambda})}{\int f(\mathbf{x}|\boldsymbol{\lambda})\,dG(\boldsymbol{\lambda})},$$

the posterior mean of Λ_i. Remarkably this Bayes point estimate is not dependent on $\mathbf{A}$. Of course, the value of $W(\boldsymbol{\lambda})$ for any $\boldsymbol{\delta}$, including $\boldsymbol{\delta}_G$, will depend on $\mathbf{A}$.

If a single function $w(\boldsymbol{\lambda})$ of the parameters is to be estimated subject to quadratic loss its Bayes point estimate is readily seen to be the posterior mean of $w(\boldsymbol{\Lambda})$.

Example 1.6.1 Suppose that the joint distribution of $\mathbf{X}|\boldsymbol{\lambda}$ is k-variate normal with known covariance matrix $\boldsymbol{\Sigma}$ and that the prior distribution of $\boldsymbol{\Lambda}$ is k-variate normal with mean $\boldsymbol{\mu}_G$ and covariance matrix $\boldsymbol{\Sigma}_G$. Then the posterior distribution of $\boldsymbol{\Lambda}$ is k-variate normal with mean vector

$$\boldsymbol{\mu}\{\boldsymbol{\Sigma}^{-1}+\boldsymbol{\Sigma}_G^{-1}\}^{-1}\{\boldsymbol{\Sigma}^{-1}\mathbf{x}+\boldsymbol{\Sigma}_G^{-1}\boldsymbol{\mu}_G\}$$

and covariance matrix $\{\boldsymbol{\Sigma}^{-1}+\boldsymbol{\Sigma}_G^{-1}\}^{-1}$.

The marginal distribution of $\mathbf{X}_G$ is k-variate normal with mean vector $\boldsymbol{\mu}_G$ and covariance matrix $\boldsymbol{\Sigma}+\boldsymbol{\Sigma}_G$.

1.7 Bayes decision and multiple independent observations

Throughout, the discussion so far has been in terms of a single observation x or $\mathbf{x}$ being made on X or $\mathbf{X}$. Generalization to the case of m independent observations, on x or $\mathbf{x}$, is straightforward in principle. Concentrating for now on the univariate single parameter case, suppose that m independent observations are made on X. Then $f(x|\lambda)$ in (1.2.3) is replaced by the likelihood $\prod_{i=1}^m f(x_i|\lambda)$, the method otherwise remaining unchanged. If a one-dimensional sufficient statistic $t(x_1,\ldots,x_m)$ exists we have $\prod_{i=1}^m f(x_i|\lambda)=g(x_1\cdots x_m)h(t|\lambda)$. Therefore on substitution in (1.2.3) the problem is essentially reduced to the one-sample case. Most of the important distributions taken as special cases in this monograph admit one- or low-dimensional sufficient statistics.

Example 1.7.1 Let X be a Poisson r.v. with mean λ. Then

$$\prod_{i=1}^{m} p(x_i|\lambda)=e^{-m\lambda}\lambda^{\Sigma x_i}\prod_{i=1}^{m}(1|x_i!)$$

and (1.2.3) assumes the form given by a single observation $y=\sum x_i$ on a Poisson r.v. with mean $m\lambda$.

Similar simplifications can be made for univariate and multivariate normal distributions which will play an important role in the sequel.

Where low-dimensional sufficient statistics exist the form of the factor $h(t|\lambda)$ in the factorization of the likelihood is exploited to generate the class of **natural conjugate** prior distributions; see for example de Groot (1970, p. 159).

If low-dimensional sufficient statistics do not exist it is possible to reduce data by calculating maximum likelihood (ML) estimates, and then, using exact or approximate sampling distributions of these estimates to apply the Bayes techniques in a straightforward manner.

Strictly, this does not produce a Bayes rule, but the asymptotic sufficiency of MLEs, for which the reader may refer to Cox and Hinkley (1974, p. 307), provides some justification for such a procedure.

1.8 Empirical Bayes methods

Empirical Bayes methods rely on the existence of a prior distribution $G(\lambda)$ which can be given a frequency interpretation, and which can be estimated using suitable observations. Thus the EB approach can be essentially non-Bayesian in the sense of not involving subjective probabilities. In the simplest case the EB sampling scheme is as follows: a current observation x is made when the parameter value is λ, a realization of Λ, and x is to be used in a decision about λ. At the time of making the current observation there are available past observations $x_1, x_2, \ldots, x_n$ obtained with independent past realizations $\lambda_1, \lambda_2, \ldots, \lambda_n$ of Λ. In this scheme every x_i is a realization of X_i, and the X_i's are mutually independent. It is useful to represent the EB sampling scheme as in (1.8.1).

EB sampling scheme

	Previous stages	Current stage	
unknowns:	$\lambda_1 \cdots \lambda_n$	λ	
observables:	$x_1 \cdots x_n$	x	. (1.8.1)

The words 'current' and 'past' are not necessarily to be taken in a strictly temporal sense. Usually it is assumed that the actual values $\lambda_1, \lambda_2, \ldots, \lambda_n$ never become known.

The possibility of obtaining an estimate of G arises through the fact that $x_1, x_2, \ldots, x_n$ may be regarded as an independent sequence of observations on X_G whose distribution function F_G is given in (1.3.4) with f replaced by F. The empirical c.d.f. of these x-values, $F_n(x)$, is an estimate of $F_G(x)$ such that $F_n(x) \to F_G(x)$ in probability (P), as $n \to \infty$ for every x. This suggests that it might be possible to find a c.d.f. $G(\lambda)$ such that

$$F_n(x) \simeq \int F(x|\lambda)\, d\hat{G}(\lambda)$$

with the property that $\hat{G}(\lambda) \to G(\lambda)$, ($P$), for all λ as $n \to \infty$.

If such an empirical $G(\lambda)$ can be found, substituting it for $G(\lambda)$ in the derivation of a Bayes decision rule will yield an **empirical Bayes** rule, $\delta_n(x_1, x_2, \ldots, x_n; x)$. This notation emphasizes that the EB rule will generally depend on all past x-values as well as the current x. We may regard it in a broad sense as an estimate of the Bayes rule. In the case of point estimation, $\delta_n(x_1, x_2, \ldots, x_n; x)$ is a point estimate of $\delta_G(x)$. EB rules need not necessarily be obtained by directly exploiting the relation $F_G(x) = \int F(x|\lambda)\, dG(\lambda)$ to obtain an estimate of G. In section 1.9 we discuss an example which illustrates this and also motivates several questions that may be asked about EB methods.

To conclude this introduction to EB methods we draw attention to a somewhat broader interpretation of the term 'empirical Bayes' than is implied by the typical EB sampling scheme described above. Suppose that a Bayes decision rule involves a parameter ω of a prior distribution. If ω is replaced by any estimate derived from observed data we may refer to the resulting rule as an empirical Bayes rule. An example of this sort occurs in the EB analysis of contingency tables proposed by Laird (1978). In this example the parameter σ^2 of a prior distribution is estimated, but the sampling scheme is not exactly like the classical EB scheme.

1.9 An example: EB estimation in the Poisson case

In Example 1.3.3, (1.3.9) gives an expression for the Bayes point estimate of λ in terms of the marginal probabilities $p_G(x+1)$ and $p_G(x)$. Now suppose that among the past observations there are $f_n(x)$ having the value x, $x = 0, 1, 2, \ldots$. Since $x_1, x_2, \ldots, x_n$ are independent realizations of X_G with probability distribution (p.d.) $p_G(x)$ we can estimate $p_G(x)$ by $f_n(x)/n$. Including the current x we have $[1 + f_n(x)]$ observations with the value x, out of a total of $n+1$ observations, and $f_n(x+1)$ with value $x+1$. Therefore we have an estimate of the Bayes estimate given by

$$\delta_n(x_1 \cdots x_n; x) = (x+1) f_n(x+1)/[1 + f_n(x)]. \qquad (1.9.1)$$

Several comments on this estimate are in order:

1. Explicit estimation of G is not needed to obtain δ_n. Only estimates of the marginal X distribution are used. Other such EB estimates will be studied in more detail in section 3.4; they are called simple EB estimates.

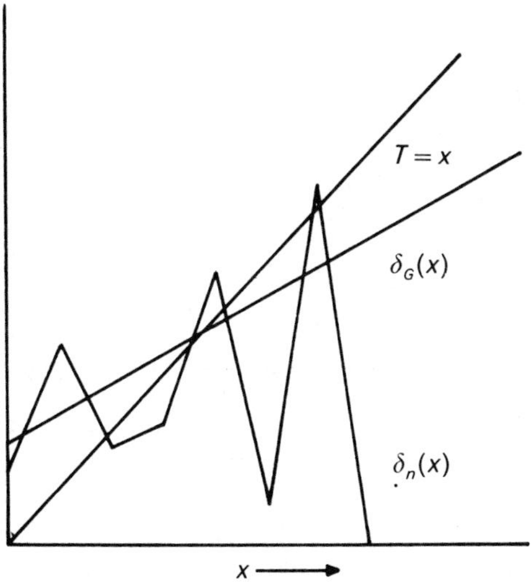

Fig. 1.1

2. It is not a smooth function of x. In a finite sample $f_n(x)$ will clearly be 0 for x large enough, and irregularities in the observed values $f_0(x)$, $f_1(x), \ldots$ will typically produce a graph of δ_n as shown in Fig. 1.1.
3. Figure 1.1 also shows graphs of the maximum likelihood estimate $T = x$, and the Bayes estimate $\delta_G(x)$ for a typical example. It seems clear that a smoothed version of δ_n will be closer to δ_G, and therefore better in some reasonable sense. Of course the Bayes estimate will not necessarily be a straight line function of x as shown in Fig. 1.2, but in the Poisson case, and many others, it is always monotonic in x. This follows by noting that

$$\delta_G(x+1) - \delta_G(x)$$
$$= \frac{\{\int \lambda^{x+2} e^{-\lambda} dG(\lambda)\}\{\int \lambda^x e^{-\lambda} dG(\lambda)\} - \{\int \lambda^{x+1} e^{-\lambda} dG(\lambda)\}^2}{\{\int \lambda^{x+1} e^{-\lambda} dG(\lambda)\}\{\int \lambda^x e^{-\lambda} dG(\lambda)\}}$$
$$= M(x)[E(\Lambda_x^2) - \{E(\Lambda_x)\}^2]$$

where Λ_x is a r.v. with p.d.f. $\propto \lambda^x e^{-\lambda} dG(\lambda)$ and $M(x) > 0$.

Ways of obtaining smooth EB estimators will be given in Chapter 3.

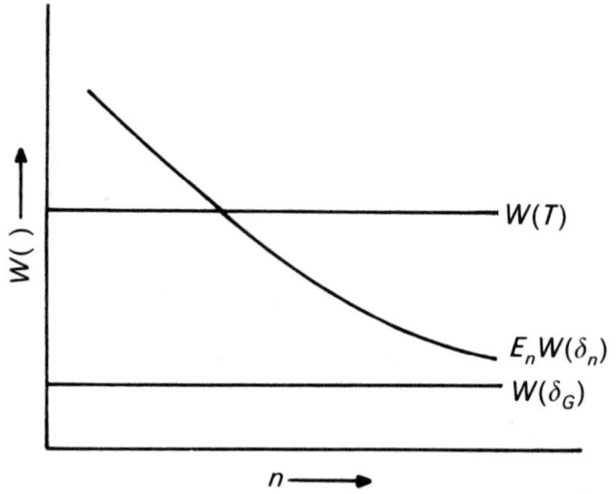

Fig. 1.2

4. As $n \to \infty$, $f_n(x)/n \to p_G(x)$ in probability, for every x. Consequently $\delta_n(x_1 \cdots x_n; x) \to \delta_G(x)$, (P), for every x. Thus δ_n may be said to be **asymptotically optimal** in this sense. There are other ways of viewing the question of asymptotic optimality. For example, the 'goodness' of a particular δ_n could be measured by $W(\delta_n)$, but with respect to past observations $W(\delta_n)$ is a random variable. Hence the goodness of the method of EB estimation represented by δ_n could be measured by $E_n W(\delta_n)$, where E_n indicates expectation with respect to past samples of size n. We could say that δ_n is asymptotically optimal if $E_n W(\delta_n) \to W(\delta_G)$ as $n \to \infty$. Asymptotic optimality will be discussed further in section 1.10 and elsewhere.

1.10 The goodness of EB procedures

The Bayes decision rule, $\delta_G(x)$, is defined as that $\delta(x)$ which minimizes $W(\delta)$ so that $W(\delta_G) \leqslant W(\delta)$ for all δ. Now, for any δ the value of $W(\delta)$ is a measure of its goodness and in the Bayes sense δ_G is best, or optimal. If δ_n is an EB rule derived from a particular set of past observations, $W(\delta_n)$ is a measure of its goodness. With respect to the past observations $W(\delta_n)$ is, of course, a random variable. Therefore an assessment of the overall goodness of an EB method should pay attention to the distribution of $W(\delta_n)$.

A natural measure of performance of an EB method, in the light of the preceding discussion, is the expectation of $W(\delta_n)$ with respect to previous samples of size n, i.e. $E_n W(\delta_n)$. We could then say that δ_n is asymptotically optimal (a.o.) if $E_n W(\delta_n) \to W(\delta_G)$ as $n \to \infty$ (Robbins, 1964). Even if δ_n is a.o., $E_n W(\delta_n)$ may be considerably greater than $W(\delta_G)$ for finite n values, and it may be greater than $W(T)$ where T is a non-Bayes rule. For example, in point estimation T may be a maximum likelihood estimator, and δ_n would not necessarily be preferred to it unless $E_n W(\delta_n) < W(T)$. Typically the relation between $E_n W(\delta_n)$, $W(T)$, $W(\delta_G)$ can be depicted as in Fig. 1.2. Usually there will be a value of n, say n_T, such that $W(T) < E_n W(\delta_n)$ for $n < n_T$. The asymptotic optimality of EB procedures has been studied in considerable detail by several authors, and more will be said about it in section 3.2 and elsewhere.

A point of notation: it will often be convenient to refer to the W values of the Bayes, EB and other rules as W(Bayes), W(EB), etc.

In choosing whether to use an EB method in preference to a non-Bayes method, criteria other than $E_n W(\delta_n)$ could be used. For example, if one were concerned that the realized δ_n should be better than T one may focus attention on $P_n\{W(\delta_n) < W(T)\}$, where P_n indicates a probability calculated with reference to previous samples of size n. Another definition of asymptotic optimality is a.o.(P): $W(\delta_n) \to W(\delta_G)$, ($P$), as $n \to \infty$. The property $E_n W(\delta_n) \to W(\delta_G)$ as $n \to \infty$ can be called a.o.(E). With some restrictions a.o.(P) will imply a.o.(E). Also a.o.(P) will imply $P_n\{W(\delta_n) < W(T)\} > 1 - \varepsilon$ for n large enough. In practice the choice between δ_n and T has to be made on the basis of known results for $E_n W(\delta_n)$ or $P_n\{W(\delta_n) < W(T)\}$ for cases similar to the problem in hand, or by trying to estimate these quantities.

1.11 Smooth EB estimates

The monotonicity of the Bayes estimator in the Poisson case has been noted in remark (3), section 1.9. Many other Bayes estimators have this property, indicating that EB estimators should also have a minimal smoothness. Smooth EB estimates can be obtained in at least two ways: (a) by smoothing a simple EB estimator obtained as indicated in the example of section 1.9; (b) by exploiting the monotonicity of the Bayes estimator and replacing G by an estimated G. Regarding (a), the smoothing can be by fitting a straight line or some other suitable curve through the observed non-smooth graph of a simple δ_n. In some instances fitting a particular functional form, like a straight line, is tantamount to assuming a particular parametric form for G. Direct smoothing of δ_n can also be done by some monotonic regression technique. Smoothing according to (b) requires estimation of G. Here we have many possibilities, the simplest being the assumption of a parametric form of distribution for G, in which case standard techniques of estimation, such as maximum likelihood, can be used.

If the form of G is not known a smooth estimator can still be obtained by taking G to belong to a certain class of distributions. For example, G may be taken to be a finite step function. The EB estimator obtained in this way will not generally be a.o. unless the assumed class contains the actual G. However, it may be that a member of the class can be a good approximation to G, in which case the derived smooth estimator, although it may be described as an approximate EB estimator, may still have satisfactory performance.

The use of an approximation to the true G is discussed further in section 1.12 and later.

1.12 Approximate Bayes and empirical Bayes methods

1.12.1 Linear Bayes estimators

Most of the discussion of this section will be in terms of point estimation although some of the ideas do carry over to other types of decisions. Any point estimator derived using the prior distribution to find a best estimator within a certain class can be called an approximate Bayes estimator if it is not the actual Bayes estimator. We consider first **linear Bayes estimators** (Hartigan (1969), Griffin and Krutchkoff (1971)).

The simplest case is when there is just one observation x on X when the parameter value is λ. We consider estimates of the form

$$\delta(\omega_0, \omega_1; x) = \delta(\boldsymbol{\omega}; x) = \omega_0 + \omega_1 x$$

where ω_0 and ω_1 are chosen to minimize $W\{\delta(\boldsymbol{\omega}; x)\}$. The terminology 'linear Bayes' is explained by the form of $\delta(\boldsymbol{\omega}; x)$ and the fact that $G(\lambda)$ plays a role in the determination of ω_0 and ω_1. Now

$$W\{\delta(\boldsymbol{\omega}; x)\} = \iint (\omega_0 + \omega_1 x - \lambda)^2 f(x|\lambda)\, dG(\lambda) \qquad (1.12.1)$$

and it is easily minimized by differentiation w.r.t. ω_0 and ω_1, the Bayes values, $\boldsymbol{\omega}_G$ being obtained as the solutions of the following equations in ω_0, ω_1:

$$\begin{bmatrix} 1 & \int E(X|\lambda)\,dG(\lambda) \\ \int E(X|\lambda)\,dG(\lambda) & \int E(X^2|\lambda)\,dG(\lambda) \end{bmatrix} \begin{bmatrix} \omega_0 \\ \omega_1 \end{bmatrix} = \begin{bmatrix} \int \lambda\,dG(\lambda) \\ \int \lambda E(X|\lambda)\,dG(\lambda) \end{bmatrix}. \tag{1.12.2}$$

Example 1.12.1 Suppose that the distribution of X is Poisson with mean λ. Then $E(X|\lambda) = \lambda$ and $E(X^2|\lambda) = \lambda^2 + \lambda$. Substituting in (1.12.2) gives

$$\omega_{0G} = \{E(\Lambda)\}^2/\{\mathrm{var}(\Lambda) + E(\Lambda)\}, \qquad \omega_{1G} = \mathrm{var}(\Lambda)/\{\mathrm{var}(\Lambda) + E(\Lambda)\}. \tag{1.12.3}$$

If the prior distribution is a Γ distribution as in Example 1.3.2, $E(\Lambda) = \beta/\alpha^2$ and substitution in (1.12.3) gives the result

$$\delta(\boldsymbol{\omega}_G; x) = \omega_{0G} + \omega_{1G}x = (\beta + x)/(\alpha + 1) = \delta_G(x)$$

as in (1.3.7). In general, of course, $\delta(\omega_G; x) \neq \delta_G(x)$.

One of the main reasons for introducing linear Bayes estimators here is that they are easily adapted to empirical Bayes estimation. Recall that for the EB estimation we require in general an estimate of $G(\lambda)$. In formula (1.12.2),

$$\int E(X^r|\lambda)\,dG(\lambda) = E(X_G^r), \qquad r = 1, 2,$$

which are obviously estimated quite readily using the observed marginal distribution of past observations. If $E(X|\lambda) = \lambda$ as in Example 1.12.1 only the first two moments of the distribution of Λ have to be estimated.

When $m > 1$ observations on X are made, linear Bayes estimation can be extended to letting the estimator be a linear function of order statistics (Lwin, 1976). For more details see section 3.8.

1.12.2 Approximations to the prior distribution

Suppose that G^* is an approximation to G. Then δ_{G^*} is an approximation to δ_G. For the purpose of empirical Bayes inference the sense in which G^* might be an approximation to G is that G^* is that member of a certain class of distributions for which the distance $D\{F_G, F_{G^*}\}$ is minimized. The distance measure D is yet to be chosen and may depend on specific applications. This definition of approximation is motivated by the fact that F_G is observable in the EB context thus making the determination of G^* feasible, at least in the sense that it can be estimated statistically.

The classes of distributions that may be considered for G^* include

1. the natural conjugate priors;
2. finite step functions.

The choice of distance measure D, determination of G^* and the goodness of δ_{G^*} as an approximation to δ_G are topics to be taken further in Chapters 2 and 3 and elsewhere, but a simple example follows.

Example 1.12.2 Suppose that $X \stackrel{\mathrm{d}}{=} N(\lambda, 1)$ (where $X \stackrel{\mathrm{d}}{=} F$: the distribution of X is F) with $\Lambda \stackrel{\mathrm{d}}{=} N(0, 1.0)$ and that $G^*(\lambda)$ is a step function with jumps at λ_1, $\lambda_2 = 0$, λ_3 each of size 1/3 and that $\lambda_1 = -\lambda_3$. One way of determining λ_1 is by equating the variances of G^* and G. This gives $\lambda_3 = -\lambda_1 = 1.2$, $W(\delta_{G^*}) = 0.5$, to compare with $W(\delta_G) = 0.5$ and $W(T = x) = 1.0$.

1.12.3 Using non-sufficient statistics

Suppose that m independent observations are made on the one-dimensional r.v. X whose distribution depends on the single parameter λ. If a one-dimensional sufficient statistic $t(x_1 \cdots x_m)$ exists the Bayes decision rule reduces to a function $\delta_G(t)$ of t. When a sufficient t does not exist calculations involving likelihoods $\prod_{i=1}^{m} f(x_i|\lambda)$ can become complicated, especially in the EB framework, and it may be contemplated to effect a reduction by basing the decision on an estimate $\hat{x}$ of λ.

To obtain a decision rule based on $\hat{x}$ the p.d.f. $f(x|\lambda)$ is replaced by the p.d.f. $h(\hat{x}|\lambda)$ of $\hat{x}$ in the formulae for obtaining Bayes estimates. Even if the exact distribution of $\hat{x}$ is used the resulting decision rule is not the actual Bayes rule; sometimes it may be possible only to obtain an approximation for $h(\hat{x}|\lambda)$. The advantage of this approach, especially in EB decisions, is that estimation of an approximation to G can be considerably simplified.

1.13 Concomitant variables

In many practical cases there will be concomitant information about the parameter values. Specifically, recall the EB sampling scheme where we have observations $(x_1, x_2, \ldots, x_n)$ when the parameter values are $(\lambda_1, \lambda_2, \ldots, \lambda_n)$. Every x_i is usually thought of as an estimate of the corresponding λ_i. Now it may happen that we also have associated with every x_i an observation c_i on a concomitant variable C. Every c_i is not necessarily an estimate of λ_i, but C and Λ may not be independent, so that taking account of the observed c should improve the estimate of λ. However, the emphasis is still on estimating individual λ values, and not on exploring the relationship between Λ and C, for instance, through the regression of Λ on C. More details of EB analysis in this case are given in section 4.7.

1.14 Competitors of EB methods

A brief discussion of developments in EB methods has been given in section 1.8. The general idea of these developments is that the EB technique is an attractive compromise between the conventional non-Bayes approach and the fully specified Bayesian approach for the analysis of historical data arising in a sampling scheme which can be represented as in (1.8.1).

There are other ways of utilizing all the available information given by an EB sampling in an 'optimal' way. They produce 'competitors' to EB methods. Among such competitors are: (1) the compound decision (CD) approach initiated by Robbins (1951) in the hypothesis testing framework and by Stein (1955) for estimation; (2) the full Bayesian (FB) multiparameter approach initiated by Lindley (1962, 1971); and (3) a modified likelihood approach first employed by Henderson *et al.* (1959). All of these approaches treat the problem of the EB scheme as one of simultaneous decisions about all unknown λ's in all stages. Brief introductions to these approaches now follow.

1.14.1 Compound decision theory

Consider the problem of estimating the unknown λ of a $N(\lambda, \sigma^2)$ distribution discussed in Example 1.3.1. Suppose that we consider the problem of estimating all the elements of $\boldsymbol{\lambda} = (\lambda_1, \ldots, \lambda_{n+1})^{\mathrm{T}}$

simultaneously using the data given by $\mathbf{x}=(x_1,\ldots,x_{n+1})^{\mathrm{T}}$ where λ_{n+1} and x_{n+1} respectively stand for λ and x of the current stage in EB scheme (1.8.1). For notational simplicity we let $k=n+1$ and consider the k-parameter problem of estimating the mean vector $\boldsymbol{\lambda}$ using the data $\mathbf{x}$ from a k-variate normal distribution with covariance matrix $\sigma^2 I_{k\times k}$ where $I_{k\times k}$ is a $k\times k$ identity matrix.

Let $\hat{\boldsymbol{\lambda}}$ be an estimate to be sought such that the total quadratic loss

$$L(\hat{\boldsymbol{\lambda}},\boldsymbol{\lambda})=\sum_{i=1}^{k}(\hat{\lambda}_i-\lambda_i)^2 \tag{1.14.1}$$

is optimized in some sense. For example, one might minimize the expected value of L. Suppose also that the estimates of $\boldsymbol{\lambda}$ are restricted to the class defined by

$$\hat{\lambda}_i=a_0+a_1x_i, \qquad i=1,\ldots,k.$$

Then the optimal estimate $\hat{\boldsymbol{\lambda}}^*$ in this class, which minimizes the risk

$$W(\hat{\boldsymbol{\lambda}},\boldsymbol{\lambda})=\int\cdots\int\left\{\sum_{i=1}^{k}(a_0+a_1x_i-\lambda_i)^2\right\}\prod_{i=1}^{k}f(x_i|\lambda_i)\,dx_1\cdots dx_k, \tag{1.14.2}$$

is given by

$$a_0=\bar{\lambda}-a_1\bar{\lambda}$$
$$a_1=\sigma_\lambda^2/(\sigma^2+\sigma_\lambda^2)$$

where

$$\bar{\lambda}=\sum_{i=1}^{k}\lambda_i/k$$
$$\sigma_\lambda^2=\sum_{i=1}^{k}(\lambda_i-\bar{\lambda})^2/k.$$

Hence the optimal linear estimate of $\boldsymbol{\lambda}$ is $\hat{\boldsymbol{\lambda}}^*$, elements given by

$$\tilde{\lambda}_i^*=\bar{\lambda}+(1-c_\lambda)(x_i-\bar{\lambda}), \qquad i=1,2,\ldots,k, \tag{1.14.3}$$

where

$$c_\lambda=1-a_1=\sigma^2/(\sigma^2+\sigma_\lambda^2).$$

The optimal linear estimator $\hat{\boldsymbol{\lambda}}^*$ depends on the unknown quantities $\bar{\lambda}$ and σ_λ^2. However, they can be estimated from the data by noting that

$$\left.\begin{aligned} E(\bar{X}) &= \bar{\lambda} \\ E(S_{xx}) &= (k-1)\sigma^2 + k\sigma_\lambda^2 \end{aligned}\right\} \tag{1.14.4}$$

where

$$\bar{X} = \sum_{i=1}^{k} X_i/k$$

$$S_{xx} = \sum_{i=1}^{k} (X_i - \bar{X})^2.$$

The relationships (1.14.4) provide unbiased estimates of $\bar{\lambda}$ and σ_λ^2 as

$$\left.\begin{aligned} \hat{\bar{\lambda}} &= \bar{x} \\ \hat{\sigma}_\lambda^2 &= s_{xx}/k - (k-1)\sigma^2/k \end{aligned}\right\}. \tag{1.14.5}$$

Alternatively, one can estimate c_λ directly using the class of estimates

$$\hat{c}_\lambda(v) = v\sigma^2 s_{xx}^{-1}. \tag{1.14.6}$$

Under the assumption of normality of the X_i's, v can be chosen to obtain an unbiased estimate of c_λ, i.e. $v = (k-2)$. Use of either (1.14.5) or (1.14.6) provides an estimator of c_λ in (1.14.2) and hence an estimate of λ_i based on the optimal linear estimate can be constructed as

$$\hat{\lambda}_i^+ = \bar{x} + (1 - \hat{c}_\lambda)(x_i - \bar{x}). \tag{1.14.7}$$

The type of estimate (1.14.7) is very similar to the EB estimate derived from (1.12.3). In the literature of compound decision theory, such estimates are called compound estimates as contrasted with 'simple estimates' which use only the data of ith stage to estimate unknowns at the ith stage. A more detailed discussion of compound decision theory is given in Chapter 7. The main aim of the above development is to demonstrate that it is not necessary to assume the existence of a prior distribution of λ's to obtain estimates of the EB type. Such estimates can be constructed in a purely non-Bayes setting so long as the loss structure reflects the fact that decisions at all stages are considered simultaneously. Using the total loss as in (1.14.1) is one way of doing this. Estimates of the form (1.14.7) were first given by James and Stein (1960).

1.14.2 The full Bayesian approach to the EB scheme

Originally, the introduction of James–Stein type estimators aroused suspicion and confusion, but it stimulated further work aimed at clarifying the 'intriguing aspects' of estimators of the form (1.14.7). Lindley (1962), in the discussion of Stein's (1962) paper, formulated a full Bayesian (FB) approach to the problem of simultaneous estimation of $\boldsymbol{\lambda}$ in the FB scheme. This approach employs the assumptions and data of the EB approach according to (1.14.1). In addition, the FB approach assumes that the unknown λ's are exchangeable so that their joint prior distribution is a mixture of a distribution G by a hyper prior distribution. Hence the joint density function of $\boldsymbol{\lambda}$ is given by

$$g(\boldsymbol{\lambda}\,|\,P) = \int \prod_{i=1}^{k} g(\lambda_i\,|\,\phi)\,dP(\phi)$$

where P is the hyper prior distribution function of the parameter ϕ of the distribution function G. Thus the joint distribution of $(\mathbf{x}, \boldsymbol{\lambda})$ has a density function given by

$$a(\mathbf{x}, \boldsymbol{\lambda}) = \int \left\{ \prod_{i=1}^{k} f(x_i\,|\,\lambda_i)\,g(\lambda_i\,|\,\phi) \right\} dP(\phi).$$

The FB approach then proceeds to obtain the posterior distribution of $\boldsymbol{\lambda}$ given $\mathbf{x}$. Its density function is given by

$$b(\boldsymbol{\lambda}\,|\,\mathbf{x}) = a(\mathbf{x}, \boldsymbol{\lambda}) \bigg/ \int a(\mathbf{x}, \boldsymbol{\lambda})\,d\boldsymbol{\lambda}. \qquad (1.14.8)$$

Inference on $\boldsymbol{\lambda}$ then proceeds by looking at various aspects of (1.14.8). In particular the posterior mean of $\boldsymbol{\lambda}$ is of the interest. But the posterior mode was advocated for two reasons. Firstly, this provides a treatment parallel to that of maximizing the likelihood function in the non-Bayes approach. Secondly, for more realistic cases, the mode seems to be more tractable than the mean.

We demonstrate below an application of the FB approach to the problem considered in Example 1.9.2. Here $f(x\,|\,\lambda)$ is $N(\lambda, \sigma^2)$ and it is assumed that $g(\lambda\,|\,\boldsymbol{\phi})$ is the $N(\mu_G, \sigma_G^2)$ density so that the hyper-parameter vector is $\boldsymbol{\phi} = (\mu_G, \sigma_G^2)$. By assuming exchangeability of the λ_i's, the prior distribution of $\boldsymbol{\lambda}$ is obtained for a specific choice of the hyper-parameter density function $p(\mu_G, \sigma_G^2)$. For simplicity, we assume further that σ_G^2 and σ^2 are known and μ_G is taken to have a uniform prior distribution on the real line. Thus the prior distribution of $\boldsymbol{\lambda}$ becomes

$$g(\boldsymbol{\lambda}|P) = \int \left\{ \prod_{i=1}^{k} g(\lambda_i|\mu_G, \sigma_G^2) \right\} d\mu_G.$$

The posterior distribution of $\boldsymbol{\lambda}$ can be shown to be (Lindley, 1971) a multivariate normal distribution with mean vector $\hat{\boldsymbol{\lambda}}^*$ and covariance matrix $\mathbf{v}^*$. The ith element of $\hat{\boldsymbol{\lambda}}^*$ is given by

$$\tilde{\lambda}_i^* = \bar{x} + (1 - c(\sigma_G^2))(x_i - \bar{x}) \tag{1.14.9}$$

where

$$c(\sigma_G^2) = \sigma^2/(\sigma^2 + \sigma_G^2).$$

The elements of $\mathbf{v}^*$ depend only on σ_G^2, σ^2 and k. Thus the FB approach in this simple case gives an estimate $\hat{\lambda}_i^*$ for λ_i which is similar to the EB estimate.

In Chapter 7, the FB approach will be discussed in more detail. It will be seen that in general the EB and FB approaches do not lead to identical results although they both lead to non-simple estimates or decision rules. The FB approach has been unified, and extended to a number of situations, in Lindley and Smith (1972), Smith (1973) and Deely and Lindley (1981) thus providing formidable rival procedures for the EB methods. However, the FB approach seems to be heavily dependent on the parametric assumptions in general and normality in particular. It seems too early to decide which of the EB and FB approaches is more practically attractive.

1.14.3 A modified likelihood approach

The EB sampling scheme was introduced and the EB type methods were proposed with a general assertion that the standard non-Bayes methods cannot deal with information in the form of previous data; in particular the usual likelihood approach was regarded as lacking such a mechanism. However, there has been a development which may be termed a modification of the usual likelihood for handling data from an EB scheme. This appeared in connection with the random effects model in the analysis of variance (Henderson *et al.*, 1959; Nelder, 1972; Finney, 1974). Briefly, the approach defines the likelihood function of the unknown $\boldsymbol{\lambda}$ as the joint probability of

$(\mathbf{x}, \boldsymbol{\lambda})$ in the EB scheme, i.e. the likelihood function is

$$L(\mathbf{x}, \boldsymbol{\lambda}) = \left\{ \prod_{i=1}^{k} f(x_i|\lambda_i) g(\lambda_i|\boldsymbol{\phi}) \right\}.$$

This likelihood function is then maximized with respect to $\boldsymbol{\lambda}$ and $\boldsymbol{\phi}$.

In the special case when $f(x_i|\lambda_i)$ is the $N(\lambda_i, \sigma^2)$ density with known σ^2 and $g(\lambda_i|\boldsymbol{\phi})$ is the $N(\mu_G, \sigma_G^2)$ density with known σ_G^2, the 'maximum likelihood' estimate is given by

$$\hat{\mu}_G(L) = \bar{x}$$

$$\hat{\lambda}_i(L) = \hat{\mu}_G + (1 - c(\sigma_G^2))(x_i - \hat{\mu}_G)$$

where $c(\sigma_G^2)$ is given by (1.14.9).

The above approach poses new questions. Is it justified to use $L(\mathbf{x}, \boldsymbol{\lambda})$ in constructing estimates of $\boldsymbol{\lambda}$ and/or $\boldsymbol{\phi}$? What sort of properties does this 'likelihood function' have? Does it possess a local maximum? Note that it is not a likelihood function in the usual sense of the word since $\boldsymbol{\lambda}$ is an unobservable random variable. More research seems to be necessary to answer these questions satisfactorily. A possible answer to the first question is suggested in a more detailed discussion in Chapter 7.

CHAPTER 2

Estimation of the prior distribution

2.1 Introduction

In this chapter we shall deal with one of the two basic technical tools required in implementing the EB approach. As mentioned in Chapter 1, EB decision rules can be constructed via two main approaches. The first is based on an explicit estimation of the unknown prior distribution. The second is based on a method of expressing the Bayes estimate or decision rule in terms of functionals of G and estimating the Bayes rule itself directly. We shall also see that, in general, smooth EB rules obtained by the former approach can be 'better' than those of the latter. The feasibility of estimating a prior distribution G depends on the possibility of finding a distribution function (d.f.) G satisfying the relationship

$$H(x) = \int F(x|\lambda)\, dG(\lambda) \tag{2.1.1}$$

where $H(x)$ and $F(x|\lambda)$ are given d.f.s. The d.f. $H(x)$ is often called a 'mixture' of $F(x|\lambda)$ type d.f.s, while $F(x|\lambda)$ and $G(\lambda)$ are referred to as the **kernel** and **mixing** distributions, respectively. The general mathematical problem of finding G, given H and F connected by (2.1.1), is of interest in its own right, and has received attention from many authors. For example, see Medgyessy (1961), Teicher (1961) for early studies and Tallis and Chesson (1982) for a more recent study. When $F(x|\lambda)$ is a continuous d.f., (2.1.1) is a Fredholm integral equation of the first kind. The study of existence and determination of a solution of (2.1.1) leads to the concept of identifiability which is characterized by the existence of a unique solution of (2.1.1).

In the empirical Bayes context $H(x)$ becomes $F_G(x)$, but an additional complication arises since $F_G(x)$ is not known exactly. The form of $F(x|\lambda)$ will usually be assumed known. The assumption of identifiability is at the heart of the estimation of the mixing distribution G. Its practical importance becomes obvious when it is seen that estimation procedures for G are not likely to be well defined without identifiability. Direct information on $F_G(x)$ is supplied only by n observations on the r.v. X_G. They can be used to construct an empirical d.f. $F_n(x)$ which is an estimate of $F_G(x)$. Thus the problem here is twofold. First, it is necessary to examine the question of determining G, exactly or approximately, with $H(x)$ known exactly; this is the study of identifiability of various specified forms of $F(x|\lambda)$. Second, it is necessary to examine the estimation of G, again exactly or approximately, with $H(x)$ replaced by $F_n(x)$.

In the discrete case when the kernel is $p(x|\lambda)$ the corresponding analogue of (2.1.1) is obtained by replacing $F(x|\lambda)$ by $p(x|\lambda)$ and $H(x)$ by $p_G(x)$, i.e.

$$p_G(x) = \int p(x|\lambda)\, dG(\lambda). \tag{2.1.2}$$

2.2 Identifiability

An immediate question, assuming that a solution of (2.1.1) exists is whether it is unique. The concept of identifiability has been defined as follows: G is said to be **identifiable** in the mixture H, if a unique solution, G, of (2.1.1) can be found. Lack of identifiability is not uncommon, as is shown by considering the binomial distribution,

$$p(x|\lambda) = \binom{n}{x} \lambda^x (1-\lambda)^{n-x}, \qquad x = 0, 1, \ldots, n. \tag{2.2.1}$$

We see that, for every x, $p_G(x)$ is a linear function of the first n moments, $\mu'_r = \int \lambda^r\, dG(\lambda)$, $r = 1, \ldots, n$, of $G(\lambda)$. Consequently any other $G^*(\lambda)$ with the same first n moments will yield the same mixed distribution $p_G(x)$.

In general, identifiability depends on both the kernel distribution and the family of distributions to which G is assumed to belong. Restrictions on this family can render G identifiable, and as we shall see, for certain $F(x|\lambda)$ or $p(x|\lambda)$, G is always identifiable. We shall accordingly examine identifiability of G under two broad headings: identifying restrictions on G, and special families of kernel distributions.

In more extensive studies of identifiability by Teicher (1963), Tallis (1969) and others, the single parameter λ has been replaced by a vector parameter. Except to report isolated results on multiparameter kernel distributions, we shall confine attention to the one-parameter case. Our main concern is with distributions which have been studied in the EB field.

2.3 Parametric G families

Perhaps the simplest, and also the most severe restriction on the family of G distributions, is that G belongs to a certain parametric family of distributions. By such a family we mean distributions $G(\lambda; \boldsymbol{\alpha})$ of known form, depending on a finite-dimensional vector parameter $\boldsymbol{\alpha}$. Common examples are the normal and gamma families. In the EB context such a restriction is perhaps somewhat unrealistic, but it deserves consideration because the experimenter may have good reason for faith in a certain type of prior distribution. For example, the 'naturally conjugate' prior distribution may be considered appropriate (cf. Raiffa and Schlaifer, 1961). Technically the parametric G families enable one to check identifiability more readily as is seen below (cf. Tallis, 1969).

Let $k(x, \lambda; \gamma)$ be the joint density function of the pair (X, Λ); γ is a finite-dimensional parameter. Then we can write

$$k(x, \lambda; \gamma) = f(x|\lambda; \boldsymbol{\theta})g(\lambda; \boldsymbol{\alpha}) = h(x; \boldsymbol{\delta})b(\lambda|x; \boldsymbol{\beta})$$

where $\boldsymbol{\alpha}$, $\boldsymbol{\beta}$, $\boldsymbol{\delta}$, $\boldsymbol{\theta}$ are finite-dimensional parameter vectors which index the respective density functions, $h(x; \boldsymbol{\delta})$ is the density function of the mixture $H(x)$ and $b(\lambda|x; \boldsymbol{\beta})$ is the density function of the posterior distribution of Λ. If $\boldsymbol{\beta}$ can be expressed as a function of $\boldsymbol{\delta}$, $\boldsymbol{\theta}$, say,

$$\boldsymbol{\beta} = \psi(\boldsymbol{\delta}, \boldsymbol{\theta}) \tag{2.3.1}$$

then knowledge of $\boldsymbol{\delta}$ and $\boldsymbol{\theta}$ gives complete determination of g. Since we have

$$b(\lambda|x; \boldsymbol{\beta}) = g(\lambda; \boldsymbol{\alpha})f(x|\lambda; \boldsymbol{\theta})/h(x, \boldsymbol{\delta}),$$

this means that $\boldsymbol{\alpha}$ must necessarily be some unique function of $\boldsymbol{\theta}$ and $\boldsymbol{\delta}$ if the condition (2.3.1) holds, i.e.

$$\boldsymbol{\alpha} = \phi(\boldsymbol{\theta}, \boldsymbol{\delta}). \tag{2.3.2}$$

But the form of the density $g(\lambda; \boldsymbol{\alpha})$ is known. Thus if the condition

(2.3.1) holds, knowledge of $f(x|\lambda)$ and $h(x;\boldsymbol{\delta})$ uniquely determines the prior density $g(\lambda;\boldsymbol{\alpha})$. The condition (2.3.1) readily extends to the case when X and Λ themselves are vector random variables. Of course (2.3.1) need not be checked if a unique function ϕ can be found in (2.3.2).

We give some simple examples below using conjugate prior distributions. They indicate that condition (2.3.1) can be readily checked so that the question of identifiability can usually be settled easily.

Example 2.3.1 Let $p(x|\lambda)$ be the geometric kernel

$$p(x|\lambda)=(1-\lambda)\lambda^x, \qquad x=0,1,\ldots,\infty.$$

The conjugate prior density is of beta type:

$$g(\lambda|p,q)=\lambda^{p-1}(1-\lambda)^{q-1}\,d\lambda/B(p,q), \qquad p,q,>0.$$

Then

$$p_G(x)=B(p+x,q+1)/B(p,q).$$

Further the posterior density is also of beta type:

$$b(\lambda|x)=\lambda^{(p+x)-1}(1-\lambda)^{(q+1)-1}\,d\lambda/B(p+x,q+1).$$

Thus the parameters of the posterior density are $(p+x,q+1)$, which can be expressed as a simple function of the prior distribution parameters p, q. Since there are no other parameters in $p(x|\lambda)$ except λ, specifying p and q in $p_G(x)$ leads to a complete specification of the parameters of $dB(\lambda|x)$. Thus G is identifiable if it does belong to the specified class.

Example 2.3.2 Let $p(x|\lambda)$ be the binomial kernel given by equation (2.2.1) and $g(\lambda;\boldsymbol{\alpha})$ the beta density of the preceding example. Then, the posterior density of λ given x is also of beta type with parameters $p+x$, $q+(n-x)$. In this example using the notation of (2.3.1), $\boldsymbol{\beta}=\{p,q,n\}$, $\boldsymbol{\delta}=\{p,q\}$, $\boldsymbol{\theta}=\{n\}$. Thus knowledge of $\boldsymbol{\theta}$ and $\boldsymbol{\delta}$ completely determines $\boldsymbol{\beta}$. Thus G is identifiable when it belongs to the beta family.

Although identifiability has been established in this case, it is worth noting that in section 2.4 we treat the case where the parametric G family is of a particular discrete type, and in that case G is not always identifiable.

Example 2.3.3 Let $f(x|\lambda)$ be the $N(\lambda,\sigma^2)$ density and $g(\lambda;\boldsymbol{\alpha})$ the

$N(\mu, \tau^2)$ density. Then X_G has a $N(\mu, \sigma^2 + \tau^2)$ density. Further, the posterior distribution of Λ given x is $N(\mu^*, \sigma^{*2})$ where

$$\mu^* = (x/\sigma^2 + \mu/\tau^2)/(1/\sigma^2 + 1/\tau^2)$$

$$\sigma^{*2} = (1/\sigma^2 + 1/\tau^2)^{-1}.$$

In the notation of (2.3.1), we have $\boldsymbol{\delta} = \{\mu, \sigma^2 + \tau^2\}$, $\boldsymbol{\theta} = \{\sigma^2\}$, $\boldsymbol{\beta} = \{\sigma^2, \tau^2, \mu\}$. Thus G is readily identified.

2.4 Finite mixtures

Let $F(x|\lambda_j)$, $j = 1, \ldots, k$ be a family of d.f.s. Then the mixed d.f.

$$F_G(x) = \sum_{j=1}^{k} \theta_j F(x|\lambda_j), \tag{2.4.1}$$

where $\theta_i > 0$ for all i, $\theta_1 + \cdots + \theta_k = 1$, is an example of a finite mixture of d.f.s. This finite mixture can also be regarded as a type of parametrization of G. For example, when $\lambda_1, \ldots, \lambda_k$ are known, the form of G is known and it depends on the finite number of parameters $\theta_1, \ldots, \theta_k$. But it is more versatile than the strict parametric approach mentioned in section 2.3 where a functional form of G needs to be assumed. More generally, we may have a mixture of a family of d.f.s, $F_j(x)$, $j = 1, 2, \ldots, k$, which are not necessarily of the same form. Since these do not normally play any part in EB problems, we shall consider only families in (2.4.1). We shall also introduce further simplifications as follows: given that $F_G(x)$ is a finite mixture of the type (2.4.1), both θ_j and λ_j, $j = 1, \ldots, k$ may be unspecified. However, we shall concentrate on the two simpler cases where either the θ_j's or the λ_j's are given.

Finite mixtures arise in problems of deciding between a finite number of alternative hypotheses. They are also important as probability models to describe some heterogeneous populations which can be regarded as being composed of a finite number of more homogeneous subpopulations (see e.g. Titterington, Smith and Makov, 1986). In the context of EB estimation, they are important because the mixing distribution G, being a step-function G_k, with jumps of size θ_j at the points λ_j, can be used as an approximation to any $G(\lambda)$. This is a standard mathematical procedure, whose exploitation in EB work has been motivated by the work of Teicher (1963), and others, on the identifiability of finite mixtures.

2.4.1 *Finite mixtures with* $\lambda_1, \ldots, \lambda_k$ given

In this case the mixture (2.4.1) is a special case of the general mixture

$$H(x) = \sum_{j=1}^{k} \theta_j F_j(x) \tag{2.4.2}$$

where $F_j(x)$ is the jth component distribution function; $F_1, \ldots, F_k$ are also assumed to be distinct members of a family $\mathscr{F}$ of known distribution functions. In this context, a general sufficient condition for the identifiability of H is that the F_j's are linearly independent (Yakowitz and Spragins, 1968; Tallis, 1969). A sufficient condition for the linear independence of $F_1, \ldots, F_k$ is that (Teicher, 1963) at least k distinct values $x_1, \ldots, x_k$ exist such that the determinant,

$$\begin{vmatrix} F_1(x_1) & F_2(x_1) & \cdots & F_k(x_1) \\ F_1(x_2) & F_2(x_2) & \cdots & F_k(x_2) \\ \vdots & \vdots & & \vdots \\ F_1(x_k) & F_2(x_k) & \cdots & F_k(x_k) \end{vmatrix} \neq 0. \tag{2.4.3}$$

The case with known λ_i's is the special case with $F_j(x_i) = F(x_i | \lambda_j)$ in (2.4.3); $F(x|\lambda)$ can also be replaced by the p.d.f. $f(x|\lambda)$ or the discrete p.d. $p(x|\lambda)$.

An equivalent condition, useful in certain cases considered below, is that for a certain value of an auxiliary variable t in an interval $(-\delta, \delta)$, with finite $\delta > 0$, the relation,

$$\sum_{j=1}^{k} \lambda_j \psi_j(t) = 0, \tag{2.4.4}$$

implies that $\lambda_j \equiv 0$ for all j. Here $\psi_j(t)$ is the characteristic function of F_j.

Example 2.4.1 The binomial kernel distribution. We have noted the non-identifiability in general of G in the case of binomial mixtures. However, when G is the finite step-function under discussion, condition (2.4.3) is satisfied when $k \leqslant n$. The determinant in (2.4.3) can be reduced to the form

$$\begin{vmatrix} (1-\lambda_1)^{k-1} & \cdots & (1-\lambda_k)^{k-1} \\ \lambda_1(1-\lambda_1)^{k-2} & \cdots & \lambda_k(1-\lambda_k)^{k-2} \\ \vdots & & \vdots \\ \lambda_1^{k-1} & \cdots & \lambda_k^{k-1} \end{vmatrix} \prod_{s=0}^{k-1} \lambda_s^n (1-\lambda_s)^{n-k+1}, \tag{2.4.5}$$

when we put $x_1 = 0$, $x_2 = 1$, etc. If the largest λ_j is $\neq 1$, the determinant is $\neq 0$, because the determinant, D, in (2.4.5) is $\neq 0$ if all λ_j are distinct. This follows because, fixing $\lambda_2, \lambda_3, \ldots, \lambda_k$, means that putting $D = 0$ yields a polynomial equation of degree $k-1$ in λ_1. It has at most $k-1$ solutions which we know to be $\lambda_2, \lambda_3, \ldots, \lambda_k$. Hence $D = 0$ only if λ_1 is equal to one of the other λ's, which possibility is excluded by hypothesis. Further, when say $\lambda_k = 1$, we need consider only D with the last row and column deleted, and a similar argument applies; we then also use

$$\sum_{j=1}^{k} \theta_j = 1.$$

Example 2.4.2 The geometric distribution. Application of (2.4.3) with $x = 0$, $x = 1$, etc., leads to the determinant

$$(1-\lambda_1)(1-\lambda_2)\cdots(1-\lambda_k)\begin{vmatrix} 1 & 1 & \cdots & 1 \\ \lambda_1 & \lambda_2 & \cdots & \lambda_k \\ \vdots & \vdots & & \vdots \\ \lambda_1^{k-1} & \lambda_2^{k-1} & \cdots & \lambda_k^{k-1} \end{vmatrix}$$

which is $\neq 0$ for distinct $\lambda_1, \lambda_2, \ldots, \lambda_k$, by the argument above.

Example 2.4.3 The normal distribution, $N(\lambda, 1)$. Application of (2.4.3) to this case is awkward, but identifiability of G can be established in various ways. For example, putting

$$\mu_r' = \int x^r f_G(x)\,dx,$$

we have

$$\int \lambda\, dG(\lambda) = \mu_1'$$

$$\int \lambda^2\, dG(\lambda) = \mu_2' - 1$$

$$\int \lambda^3\, dG(\lambda) = \mu_3' - 3\mu_1'$$

etc.

Thus we have linear equations in $\theta_1, \theta_2, \ldots, \theta_k$ of the form

$$\sum_{j=1}^{k} \theta_j \lambda_j^r = \alpha_r, \qquad r = 1, 2, \ldots, k,$$

the coefficient matrix of which has a determinant of the same form as in Examples 2.4.1 and 2.4.2.

Example 2.4.4 Translation parameter mixtures. Let $F(x|\lambda)$ be a member of a translation parameter family indexed by a translation parameter λ; i.e. $F(x|\lambda) = F(x-\lambda)$ where $F(\cdot)$ is of a known form. Application of (2.4.4) gives the result that H is identifiable if

$$\sum_{j=0}^{k} \alpha_j \psi(t, \lambda_j) = 0$$

implies $\alpha_j = 0$, $j = 1, \ldots, k$ for some t in the interval $(-\delta, \delta)$ with $\delta > 0$. Now $\psi(t, \lambda)$ satisfies the relation

$$\psi(t, \lambda) = e^{it\lambda} \psi(t, 0).$$

Since $\psi(t, 0)$ is continuous and $\psi(0, 0) = 1$, there exists a region $(-\delta, \delta)$ with $\delta > 0$ such that $\psi(t, 0) > 0$ for $t \in (-\delta, \delta)$. Hence H is identifiable if for $t \in (-\delta, \delta)$,

$$\sum_{j=1}^{k} \alpha_j \exp(it\lambda_j) = 0$$

implies $\alpha_j = 0$ $(j = 1, \ldots, k)$. This has been shown to be the case by Yakowitz and Spragins (1968). Thus the finite mixtures of translation parameter families are identifiable.

Example 2.4.5 Scale parameter mixtures. Let $F(x|\lambda)$ be a member of the scale parameter family indexed by a scale parameter λ, i.e. $F(x|\lambda) = F(x/\lambda)$, where $F(\cdot)$ is of a known form. Let $\lambda = \exp(\theta)$, $X = \exp(Y)$. Then the distribution of Y is

$$\Pr\{Y \leqslant y\} = \Pr\{X \leqslant e^y\} = F\{\exp(y - \theta)\}$$

which is a location parameter distribution. Hence finite mixtures of the distribution of Y are identifiable. The relationship between X and Y requires that finite mixtures of the distribution of X are identifiable provided $F(\cdot)$ possesses an rth moment for some $r > 0$ (Behboodian, 1975).

2.4.2 *Finite mixtures with $\theta_1, \ldots, \theta_k$ given*

Our main application of mixtures of this type is to problems where an unknown G is approximated by a step-function G_k. In this case assuming $\theta_j = 1/k$ is convenient and represents no great loss in generality. A useful result is obtained by considering kernel distributions such that

$$\int x^r \, dF(x|\lambda) = \sum_{j=0}^{r} A_j \lambda^j, \tag{2.4.6}$$

a polynomial of degree r in λ. Let

$$\mu_r' = \int x^r \, dF_G(x) \, dx.$$

Now the equation

$$\sum \theta_j F(x|\lambda_j) = F_G(x)$$

gives

$$\sum_j \theta_j \left\{ \sum_{i=0}^{r} A_i \lambda_j^i \right\} = \mu_r'$$

$$\sum_{i=0}^{r} A_i \sum_j^k \theta_j \lambda_j^i = \mu_r', \qquad r = 1, 2, \ldots.$$

Thus we have in general

$$\sum \theta_j \lambda_j^i = \alpha_i$$

where the α_i's depend on A_j and μ_i'. Since $\theta_j = 1/k$ we obtain the equations for λ_j as

$$\begin{aligned} \lambda_1 + \lambda_2 + \cdots + \lambda_k &= \alpha_1 \\ \lambda_1^2 + \lambda_2^2 + \cdots + \lambda_k^2 &= \alpha_2 \\ &\vdots \\ \lambda_1^k + \lambda_2^k + \cdots + \lambda_k^k &= \alpha_k \end{aligned} \tag{2.4.7}$$

which have $k!$ solutions. But, owing to the symmetry of the equations w.r.t. $\lambda_1, \lambda_2, \ldots, \lambda_k$, the solution is unique if we impose the restriction $\lambda_1 < \lambda_2 < \cdots < \lambda_k$.

Example 2.4.6 The Poisson kernel, $p(x|\lambda) = \lambda^x \exp(-\lambda)/x!$. The rth factorial moment of X is

$$\sum_x x(x-1)(x-2)\cdots(x-r+1)p(x|\lambda)=\lambda^r,$$

so that the rth moment is a polynomial of degree r in λ.

Example 2.4.7 The normal kernel, $N(\lambda, 1)$. The central moments do not depend on λ, hence the moments about the origin are polynomials in λ of the required order.

Example 2.4.8 $F(x|\lambda)=F\{(x-\lambda)/\sigma\}$. When the scale parameter, σ, is known, the argument of Example 2.4.7 applies here.

2.4.3 *Finite mixtures, continuous G*

As we remarked above, finite mixing distributions occur naturally in problems involving a finite number of simple hypotheses. In problems of testing composite hypotheses, continuous approximating d.f.s may be regarded as more suitable than discrete ones. A continuous finite approximation which has been used in such problems (Maritz, 1968), is depicted in Fig. 2.1, along with a corresponding step-function approximation.

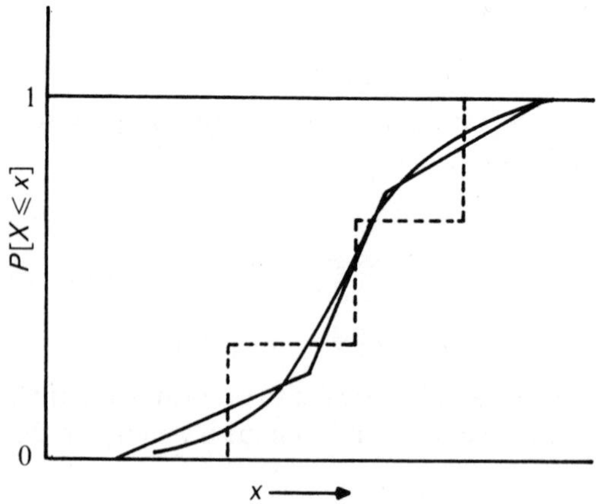

Fig. 2.1

This finite d.f. has the p.d.f.

$$dG_k^* = \frac{\theta_j d\lambda}{(\lambda_{j+1} - \lambda_j)}, \qquad \text{for } \lambda_j < \lambda \leqslant \lambda_{j+1}, \qquad j = 1, 2, \ldots, k-1. \tag{2.4.8}$$

As before, we shall assume that either the θ's or the λ's are given.

When the λ's are fixed, a condition for identifiability, similar to (2.4.3), can be written down. Putting

$$Q_j(x) = \int_{\lambda_j}^{\lambda_{j+1}} f(x|\lambda) d\lambda, \qquad j = 1, 2, \ldots, k-1, \tag{2.4.9}$$

the condition is that $k-1$ distinct values, $x_1, x_2, \ldots, x_{k-1}$, of x should exist such that the determinant $|Q_j(x_i)| \neq 0$.

This condition is rather cumbersome to apply in general, and again we consider the restricted class of distributions of equation (2.4.6). For such distributions we have

$$\mu_r' = \int x^r dF_G(x) = \int (A_0 + A_1\lambda + \cdots + A_r\lambda^r) dG_k^*(\lambda)$$

$$= \sum_{j=1}^{k-1} \frac{\theta_j}{(\lambda_{j+1} - \lambda_j)} \sum_{s=0}^{r} \frac{A_s}{(s+1)} [\lambda_{j+1}^{s+1} - \lambda_j^{s+1}], \tag{2.4.10}$$

leading to conditions like those in Examples 2.4.1 and 2.4.2.

2.5 General mixtures: identifiability

The general mixture of a distribution function $F(x|\lambda)$ or a p.d. $p(x|\lambda)$ is given by (2.1.1) or (2.1.2), where $G(\lambda)$ is a distribution function whose support is an interval or a countably infinite set of values. Necessary and sufficient conditions analogous to (2.4.3) exist in these more general cases (see Tallis, 1969; Tallis and Chesson, 1982) under certain conditions such as continuity and square integrability of $\partial F(x|\lambda)/\partial\lambda$ in the case of continuous G. Application of such conditions is not easy in general. For some special classes of general mixtures, more special but easier techniques are available as shown below.

2.5.1 *Scale and location parameter families*

Sections 2.3 and 2.4 have dealt mainly with identifiability when the family of mixing distributions is restricted in some way, although equation (2.4.6) defined a limited class of distributions $F(x|\lambda)$. We shall now give some attention to special types of kernel distributions. In this section they are of the form $K((x-\lambda)/\sigma)$, where λ and σ are, respectively, the location and scale parameters. Since we are mainly concerned with mixtures on one parameter, we shall assume that σ is a known constant; mixtures with σ variable and λ constant are also of general interest, but in EB work they are less common.

Example 2.5.1 The normal distribution, $N(\lambda, \beta^2)$. Let $\beta = 1$ without loss of generality, and consider the mixture

$$f_G(x) = \int f(x-\lambda)dG(\lambda),$$

where $f(x-\lambda) = (1/\sqrt{2\pi})\exp\{-\frac{1}{2}(x-\lambda)\}$. The characteristic function of the mixture is

$$\int e^{ixt} f_G(x)dx = \int\int e^{ixt} f(x-\lambda)dG(\lambda)$$

$$= e^{-t^2/2} \int e^{i\lambda t} dG(\lambda),$$

or

$$e^{t^2/2} \int e^{ixt} f_G(x)dx = \int e^{i\lambda t} dG(\lambda). \tag{2.5.1}$$

Equation (2.5.1) shows that the l.h.s. function, which is uniquely determined by $f_G(x)$, is the characteristic function of $G(\lambda)$. Hence, by the inversion theorem for characteristic functions (c.f.s), $G(\lambda)$ is determined. Thus G is identifiable in all mixtures of normal distributions when the scale parameter is held constant. The identifiability of finite mixtures of normal distributions, noted before, is included in this result.

More generally, let us consider mixtures

$$H(x) = \int K(x-\lambda)dG(\lambda), \tag{2.5.2}$$

where we have put $\sigma = \text{constant} = 1$. A d.f. H defined by the r.h.s. of (2.5.2), is called the **convolution** of F and G, and is written $H = K * G$. It is well known that the c.f., $\psi_H(t)$, of H is

$$\psi_H(t) = \psi_K(t)\psi_G(t), \tag{2.5.3}$$

where ψ_K and ψ_G are the c.f.s of $K(x)$ and $G(\lambda)$ respectively (see e.g. Lukacs, 1960).

Now suppose that two distributions G_1 and G_2 generate the same mixture H. Then

$$\psi_K(t)\psi_{G1}(t) = \psi_K(t)\psi_{G2}(t), \tag{2.5.4}$$

so that, unless $\psi_K(t) = 0$ over a finite interval,

$$\psi_{G1}(t) = \psi_{G2}(t),$$

thus implying the identifiability of G. This general treatment covers the cases of the following example with $\sigma = 1$.

Example 2.5.2 The following distributions are well-known examples of the location-scale type:

(i) the normal distribution:

$$f(x|\lambda,\sigma) = C\exp\left\{-\left(\frac{x-\lambda}{\sigma}\right)^2\right\},$$

(ii) the 'extreme-value' distribution:

$$F(x|\lambda,\sigma) = \exp\{-\exp[-(x-\lambda)/\sigma]\},$$

(iii) the Pearson Type VII distribution:

$$f(x|\lambda,\sigma) = \frac{1}{\sigma B(\frac{1}{2}, m-\frac{1}{2})}\left[1+\frac{(x-\lambda)^2}{\sigma^2}\right]^{-m}, \qquad \text{for fixed } m.$$

2.5.2 *Identifiability: additively closed families*

The d.f. $F(x|\lambda)$ is said to belong to the additively closed family of distributions if

$$F(x|\lambda_1) * F(x|\lambda_2) = F(x|\lambda_1+\lambda_2). \tag{2.5.5}$$

In other words, if the independent r.v.s X_1 and X_2 have d.f.s $F(x|\lambda_1)$ and $F(x|\lambda_2)$, then the d.f. of $X_1 + X_2$ is $F(x|\lambda_1+\lambda_2)$.

Example 2.5.3 A well known and important example is the Poisson distribution, which we consider first. The factorial moment generating function (m.g.f.) of the mixed Poisson distribution is

$$\sum_x \int (1+t)^x e^{-\lambda}\frac{\lambda^x}{x!}dG(\lambda) = \int e^{\lambda t}dG(\lambda).$$

Hence $G(\lambda)$ is identifiable if the factorial m.g.f. of the mixture exists.

For a more general result we follow Teicher (1961) and again make use of characteristic functions. Denoting the c.f.s of $F(x|\lambda_1), F(x|\lambda_2)$ by $\psi(t, \lambda_1)$ and $\psi(t, \lambda_2)$ condition (2.5.5) leads to

$$\psi(t, \lambda_1)\psi(t, \lambda_2) = \psi(t, \lambda_1 + \lambda_2),$$

and the only solution of this functional equation has the form

$$\psi(t, \lambda) = e^{\lambda C(t)}.$$

Thus

$$\psi(t, 1) = e^{C(t)},$$

or

$$\psi(t, \lambda) = [\psi(t, 1)]^{\lambda}. \tag{2.5.6}$$

In this argument $\lambda > 0$; negative values are excluded, for with negative values a suitable choice of λ_1 and λ_2 would yield

$$F(x|\lambda_1) * F(x|\lambda_2) = \text{a degenerate distribution,}$$

which is impossible unless $F(x|\lambda)$ is itself degenerate.

Using (2.5.6) we see that the c.f. of the mixture $H(x)$ can be expressed as

$$\psi_H(t) = \int_{\lambda \geqslant 0} [\psi(t, 1)]^{\lambda} dG(\lambda) \tag{2.5.7}$$

where

$$\int_{\lambda > 0} dG(\lambda) = 1.$$

Now the transform

$$\psi(z; G) = \int_{\lambda > 0} z^{\lambda} dG(\lambda)$$

is analytic, at least in the annulus $0 < |z| < 1$. If the d.f.s G_1 and G_2 yielded the same mixture H, then $\psi(z; G_1)$ and $\psi(z; G_2)$ would coincide for $z = \psi(t, 1)$, and consequently throughout $0 < |z| < 1$. This would entail

$$\psi(\rho e^{it}; G_1) = \psi(\rho e^{it}; G_2),$$

for $\rho < 1$, and hence, by the dominated convergence theorem, for

$\rho = 1$. Thus

$$\int_{\lambda \geq 0} e^{it\lambda} dG_1(\lambda) = \int_{\lambda \geq 0} e^{it\lambda} dG_2(\lambda) \tag{2.5.8}$$

implying $G_1(\lambda) = G_2(\lambda)$, and the identifiability of G.

Example 2.5.4 Poisson kernel. Identifiability also follows from the preceding argument.

Example 2.5.5 Mixtures of Γ-distributions. Let

$$dF(x|\lambda, \gamma) = \gamma^{\lambda} \frac{x^{\lambda-1}}{\Gamma(\lambda)} e^{-\gamma x} dx, \tag{2.5.9}$$

The c.f. of this distribution is

$$\psi(t; \lambda, \gamma) = [1 - (it/\gamma)]^{-\lambda}.$$

Thus

$$\psi(t; \lambda_1, \gamma)\psi(t; \lambda_2, \gamma) = \psi(t; \lambda_1 + \lambda_2, \gamma),$$

and the family of distributions (2.5.9), with γ fixed, is additively closed.

2.6 Identifiability of multiparameter mixtures

In a general formulation of the problem of finite mixtures Teicher (1963) allows the distribution of X to depend on more than one parameter. The case of mixtures of one-dimensional d.f.s $F(x|\boldsymbol{\lambda})$, where $\boldsymbol{\lambda}$ is p-vector, was considered. For a fixed set of points $\boldsymbol{\lambda}_1, \boldsymbol{\lambda}_2, \ldots, \boldsymbol{\lambda}_k$ with which are associated probabilities $\theta_1, \theta_2, \ldots, \theta_k$, the criterion (2.4.3) with $\boldsymbol{\lambda}_i$ in place of λ still holds for identifiability of the discrete p-dimensional mixing distribution. The case when x is replaced by a p-vector has been studied by Yakowitz and Spragins (1968). Tallis (1969) and Tallis and Chesson (1982) also considered the vector variate–vector parameter case in a more general framework covering countably infinite mixtures and continuous mixing distributions G. The necessary and sufficient conditions for these more general cases are given; however with greater generality they become more difficult to apply.

In EB applications we often have two-dimensional distributions of the r.v.s X_1 and X_2 depending on two parameters α and β, usually such that $E(X_1|\alpha, \beta) = \beta$, and $E(X_2|\alpha, \beta) = \alpha$. Typical cases are X_1,

X_2 being estimates of the slope and intercept in a linear regression, and X_1, X_2 being the sample mean and variance of observations from a normal population. Some examples of identifiable mixtures follow.

Example 2.6.1 The joint distribution of X_1 and X_2 is normal with mean vector (α, β) and known covariance matrix Σ. Then the joint mixed m.g.f. of X_{1G} and X_{2G} is

$$\int e^{it'x} dF_G(x) = \iint e^{it'x} dF(\mathbf{x}|\alpha, \beta) dG(\alpha, \beta)$$

$$= e^{-t'\Sigma t/2} \int e^{i(t_1\alpha + t_2\beta)} dG(\alpha, \beta) \qquad (2.6.1)$$

where $G(\alpha, \beta)$ is the two-dimensional prior d.f. By the arguments of section 2.5, G is identifiable.

Example 2.6.2 Suppose that X_1 and X_2 are independent for given (α, β) such that

$$F(x_1, x_2|\alpha, \beta) = F_1(x_1|\beta) F_2(x_2|\alpha, \beta). \qquad (2.6.2)$$

This form holds, for example, when X_2 is the mean and X_1 is the variance in sampling from a normal population. Then the marginal mixed X_1-distribution has the p.d.f.

$$f_G(x_1) = \int_{x_2} f_G(x_1, x_2) dx_2$$

$$= \int_\alpha \int_\beta \int_{x_2} f_1(x_1|\beta) f_2(x_2|\alpha, \beta) dx_2 dG(\alpha, \beta)$$

$$= \int_\alpha \int_\beta f_1(x_1|\beta) dG(\alpha, \beta) = \int_\beta f_1(x_1|\beta) dG_1(\beta).$$

Now if $G(\alpha, \beta)$ is a finite distribution with masses θ_j concentrated at the known points (α_j, β_j), $j = 1, 2, \ldots, k$, then the θ_j are determined by the marginal X_1-distribution if the one-dimensional $G_1(\beta)$ is identifiable. If no two α's have the same associated β, $G(\alpha, \beta)$ is thus identified. When the β's for different α's are not necessarily distinct, it is still possible to establish identifiability. For example, if we have four masses θ_{ij}, $i, j = 1, 2$ concentrated at the points (α_i, β_j), then

$\theta_{\cdot 1} = \theta_{11} + \theta_{21}$ and $\theta_{\cdot 2} = \theta_{12} + \theta_{22}$ are determined by the marginal X_1-distribution.

Now,

$$\begin{aligned}\mu'_{01} &= \int_{x_1}\int_{x_2} x_2 dF_G(x_1, x_2) \\ &= \int_{x_1}\int_{x_2}\int_{\alpha}\int_{\beta} x_2 dF_1(x_1|\beta)dF_2(x_2|\alpha, \beta)dG(\alpha, \beta) \\ &= (\alpha_1 - \alpha_2)\theta_{11} + (\alpha_1 - \alpha_2)\theta_{12} + \alpha_2(\theta_{\cdot 1} + \theta_{\cdot 2}),\end{aligned}$$

and similarly,

$$\mu'_{11} = (\alpha_1 - \alpha_2)\beta_1\theta_{11} + (\alpha_1 - \alpha_2)\beta_2\theta_{12} + \alpha_2\beta_1\theta_{\cdot 1} + \alpha_2\beta_2\theta_{\cdot 2}.$$

Hence we have two linear equations

$$\begin{aligned}\theta_{11} + \theta_{12} &= [\mu'_{01} + \alpha_2(\theta_{\cdot 1} + \theta_{\cdot 2})]/(\alpha_1 - \alpha_2) \\ \beta_1\theta_{11} + \beta_2\theta_{12} &= [\mu'_{11} + \alpha_2\beta_1\theta_{\cdot 1} + \alpha_2\beta_2\theta_{\cdot 2}]/(\alpha_1 - \alpha_2),\end{aligned}$$

from which we can find θ_{11} and θ_{12} uniquely.

Example 2.6.3 Let the prior distribution have four equal masses at the points

$$\begin{matrix}(\beta_1, \alpha_{11}), & (\beta_1, \alpha_{12}) \\ (\beta_2, \alpha_{21}), & (\beta_2, \alpha_{22})\end{matrix}$$

and let $F(x_1, x_2|\alpha, \beta)$ be of the form (2.6.2). Then, by the same reasoning as above, β_1 and β_2 are determined by the marginal X_1-distribution. Now suppose that $F_2(x_2|\alpha, \beta)$ is such that

$$\int x_2^r dF_2(x_2|\alpha, \beta) = B_{0r}(\beta) + B_{1r}(\beta)\alpha + \cdots + B_{rr}(\beta)\alpha^r,$$

i.e. a polynomial of degree r in α. Then putting $\alpha_{1\cdot} = \alpha_{11} + \alpha_{12}$, $\alpha_{2\cdot} = \alpha_{21} + \alpha_{22}$, we have

$$\begin{aligned}B_{11}(\beta_1)\alpha_{1\cdot} + B_{11}(\beta_2)\alpha_{2\cdot} &= 4\mu'_{01} - 2[B_{01}(\beta_1) + B_{01}(\beta_2)] \\ \beta_1 B_{11}(\beta_1)\alpha_{1\cdot} + \beta_2 B_{11}(\beta_2)\alpha_{2\cdot} &= 4\mu'_{11} - 2[\beta_1 B_{01}(\beta_1) + \beta_2 B_{01}(\beta_2)],\end{aligned}$$

and these two equations yield solutions for $\alpha_{1\cdot}$ and $\alpha_{2\cdot}$ when $\beta_1 \neq \beta_2$. Similarly, we can solve for $(\alpha_{11}^2 + \alpha_{12}^2)$ and $(\alpha_{21}^2 + \alpha_{22}^2)$ from

$$\begin{aligned}B_{22}(\beta_1)(\alpha_{11}^2 + \alpha_{12}^2) + B_{22}(\beta_2)(\alpha_{21}^2 + \alpha_{22}^2) &= D_1 \\ \beta_1 B_{22}(\beta_1)(\alpha_{11}^2 + \alpha_{12}^2) + \beta_2 B_{22}(\beta_2)(\alpha_{21}^2 + \alpha_{22}^2) &= D_2,\end{aligned}$$

where

$$D_1 = 4\mu'_{02} - 2[B_{02}(\beta_1) + B_{02}(\beta_2)] - B_{12}(\beta_1)\alpha_{1\cdot} - B_{12}(\beta_2)\alpha_{2\cdot}$$
$$D_2 = 4\mu'_{12} - 2[\beta_1 B_{02}(\beta_1) + \beta_2 B_{02}(\beta_2)]$$
$$- \beta_1 B_{12}(\beta_1)\alpha_{1\cdot} - \beta_2 B_{12}(\beta_2)\alpha_{2\cdot}.$$

Thus we have two sets of equations

$$\alpha_{11} + \alpha_{12} = C_{11} \qquad \alpha_{21} + \alpha_{22} = C_{21}$$
$$\alpha_{11}^2 + \alpha_{12}^2 = C_{12} \qquad \alpha_{21}^2 + \alpha_{22}^2 = C_{22}$$

which, owing to their symmetry, give unique solutions for the α's.

Examples 2.6.2 and 2.6.3 can be generalized in an obvious way to $k = r \times s$ mass points. Another useful extension is obtained if we relax the requirement that X_1 and X_2 are independent for given α and β, as expressed by equation (2.6.2). If the marginal distribution of X_1, for given α and β, depends only on β and not also on α, then we again have

$$f_G(x_1) = \int f_1(x_1|\beta)dG_1(\beta).$$

Consequently, in a finite mixture of the type considered in Example 2.6.3, the β's are determined by the marginal X_1-distribution if $G_1(\beta)$ is identifiable. Now if

$$\iint x_1^s x_2^r dF(x_1, x_2|\alpha, \beta) = B_{0sr}(\beta) + B_{1sr}(\beta)\alpha + \cdots + B_{rsr}(\beta)a^r,$$

then the α's are determined by steps similar to those of Example 2.6.3.

2.7 Determination of $G(\beta)$

Our survey of results relating to the question of identifiability of G has shown that, in theory, G can be determined when either G or F is known to belong to certain classes of distributions. We observe that these classes contain most of the distributions occurring commonly in applications. However, the actual determination of $G(\beta)$ remains a non-trivial task, and the difficulties are increased when H is known approximately.

Let us examine the class of translation parameter distributions considered in section 2.4. We found that

$$\psi_H(t) = \psi_K(t)\psi_G(t), \tag{2.7.1}$$

every ψ being a characteristic function. Thus

$$\psi_G(t) = \psi_H(t)/\psi_K(t)$$

is a c.f., and by implementing the inversion theorem for c.f.s G can be determined. In EB problems such a direct procedure is not possible. The d.f. H is approximated by the empirical d.f. $F_n(x)$, a step-function, and it may happen that $\psi_{F_n}(t)/\psi_K(t)$ is not a c.f. Indeed, it is obvious that in those cases where $F(x|\lambda)$ is continuous at all x for every λ, any mixture of $F(x|\lambda)$ is also continuous, so that no G exists such that

$$F_n(x) = \int F(x|\lambda)dG(\lambda) \qquad (2.7.2)$$

is satisfied exactly. We shall return to this question in section (2.7.2) and consider a method of constructing an approximate solution to the basic equation (2.1.1) which can also be applied to (2.7.2).

When G is restricted to certain narrow classes of distributions as in section 2.3, it is usually possible to determine G rather easily. Examples 2.3.1, 2.3.2 and 2.3.3 illustrate how this can be done. In the case of finite mixtures with λ_j given, θ_j can be found by solving suitable linear equations. Alternatively, with θ_j given, solutions to equations like (2.4.7) may be sought. In either case, lack of precise knowledge of H will mean that the r.h.s. of the equations are not known exactly, and that exact solutions do not exist. We may then consider minimizing the differences between the l.h.s. and the r.h.s. by suitable choice of the unknown (λ_j or θ_j). In such a prodedure, some criterion, for example weighted least squares, would have to be used in the actual minimizing process. Of course, when the r.h.s. is known exactly, the process should yield a 'residual sum of squares' of zero. The methods which are discussed in section (2.7.2) are logically analogous to the above proposals, are motivated by the method of maximum likelihood, and may be regarded as one way of overcoming the problem of choosing the weights for the least squares procedure.

2.7.1 *A measure of the distance between two distributions*

Consider two sequences of probabilities p_j and z_j ($j = 1, \ldots, k$), such that

$$\sum_{j=1}^{k} p_j = 1 = \sum_{j=1}^{k} z_j.$$

Let the p be fixed and let

$$I(p, z) = \sum_{j=1}^{k} p_j \log(p_j/z_j) \tag{2.7.3}$$

be a function of $z_1, \ldots, z_n$. Then $I(p, z) > 0$, except when $z_j = p_j$, $j = 1, \ldots, k$, in which case $I(p, z) = 0$. To prove this we put $r_j = p_j/z_j$ so that

$$I(p, z) = \sum_{j=1}^{k} (r_j \log r_j) z_j.$$

Now the function $\phi(t) = t \log t$ has $\phi(1) = 0$, and

$$\phi'(t) = 1 + \log t$$
$$\phi''(t) = 1/t,$$

so that

$$\begin{aligned} \phi(t) &= \phi(1) + (t-1)\phi'(1) + \tfrac{1}{2}(t-1)^2 \phi''(u) \\ &= (t-1) + \tfrac{1}{2}(t-1)^2 \phi''(u), \end{aligned}$$

where u lies between t and 1. Hence

$$\begin{aligned} I(p, z) &= \sum_{j=1}^{k} (r_j - 1) z_j + \tfrac{1}{2} \sum_{j=1}^{k} (r_j - 1)^2 \phi''(u_j) z_j \\ &= \tfrac{1}{2} \sum_{j=1}^{k} (r_j - 1)^2 \phi''(u_j) z_j, \end{aligned} \tag{2.7.4}$$

where u_j lies between r_j and 1, so that $u_j > 0$. We note that $\phi''(t) = 1/t > 0$ for $t > 0$, and therefore $I(p, z) \geqslant 0$.

The non-negative $I(p, z)$ may be regarded as a measure of 'distance' between the two probability distributions p_j and z_j, $j = 1, \ldots, k$ (Kullback, 1959). Its definition can be extended to the more general discrete case as $k \to \infty$, and to continuous distributions. In the latter case, let $h(x)$ and $w(x)$ denote two p.d.f.s, then

$$\begin{aligned} I(h, w) &= \int \log \left\{ \frac{h(x)}{w(x)} \right\} dH(x) \\ &= \int \log \{h(x)\} dH(x) - \int \log \{w(x)\} dH(x) \\ &\geqslant 0. \end{aligned} \tag{2.7.5}$$

We observe that minimizing $I(h, w)$ for fixed $h(x)$, by suitable choice of

$w(x)$, is equivalent to maximizing

$$L(h, w) = \int \log\{w(x)\}dH(x).$$

It is of interest, and relevant, to note a connection with the method of maximum likelihood. Suppose a sample $x_1, \ldots, x_n$ is drawn from a population with d.f. H and that the empirical d.f. H_n is constructed. If we believe that H is actually a member of a certain parametric family of d.f.s, say the normal family with p.d.f.

$$w(x; \mu, \sigma^2) = (1/\sigma\sqrt{2\pi})\exp\{-(x-\mu)^2/2\sigma^2\},$$

we can estimate μ and σ^2 by selecting them in such a way as to maximize $L(h, w)$, with H replaced by H_n. This leads to maximization of

$$\frac{1}{n}\sum_{j=1}^{k} \log w(x_j; \mu, \sigma^2)$$

by varying μ and σ^2, and corresponds to the usual maximum likelihood estimation.

2.7.2 *Determination of mixtures by minimizing $I(\cdot, \cdot)$*

Let $H(x)$ be the mixed distribution defined by (2.1.1) and put

$$F_{G^*}(x) = \int F(x|\lambda)dG^*(\lambda). \tag{2.7.6}$$

Then $H = F_G$. Now suppose we seek $G(\lambda)$ such that $I(H, F_{G^*})$ is minimized. We know that $I(H, F_G) = 0$, thus the minimization will yield $G^*_{\min}$ such that

$$\int F(x|\lambda)dG^*_{\min}(\lambda) = H(x),$$

and if G is identifiable, $G^*_{\min} = G$.

The above procedure may seem rather roundabout, but it has the obvious advantage that it can be applied when $H(x)$ is not known exactly, but is estimated by an empirical d.f., as happens in EB applications. The result is that a maximum likelihood estimate of G is obtained. Another useful aspect of this approach is that it affords a way of constructing a solution to (2.1.1) by successive approximation. Motivation of this approach may be found in the work of Teicher (1963) and Robbins (1964).

First we observe that a sequence $\{G_k^{**}\}$ of step-functions with jumps of size $1/k$ at $\lambda_1^{**} \leqslant \cdots \leqslant \lambda_k^{**}$, can easily be constructed such that $G_k^{**} \to G$ (weakly) as $k \to \infty$. For example, let $G(\lambda_j^{**}) = (j - \frac{1}{2})/k$, $j = 1, \ldots, k$. This is adequate when G is continuous, but suitable conventions for defining the λ_j^{**} in other cases are easily defined. For example, if $G(\lambda' + 0) = G(\lambda'' - 0) = (r - \frac{1}{2})/k$, put $\lambda_r^{**} = (\lambda' + \lambda'')$. Now, since $G_k^{**} \to G$, it follows from the Helly–Bray theorem that

$$H_k^{**}(x) = \int F(x|\lambda) dG_k^{**}(\lambda) \to H(x)$$

for all x, and $I(H, H_k^{**}) \to 0$. Certain restrictions on $dF_{G_k^{**}}$, dF_G and G are required for the truth of these statements (Maritz, 1967).

We also define the sequence $\{G_k\}$ of d.f.s with jumps of size $1/k$ at $\lambda_1 \leqslant \cdots \leqslant \lambda_k$, such that, for every k, $I(H, H_k)$ is minimized. If G is identifiable, $G_k \to G$ (weakly). For, suppose that $G_k \not\to G$, then $I(H, H_k^{**}) \not\to 0$. But we know by definition that

$$0 \leqslant I(H, H_k) \leqslant I(H, H_k^{**}) \to 0,$$

thus contradicting $I(H, H_k) \not\to 0$.

An advantage of constructing a solution by this method of successive approximation is that it can be used when H is not known exactly, but is estimated by an empirical d.f., H_n. Other measures of 'distance' between two distributions can, of course be used, but the principle of successive approximation remains essentially the same (cf. Deely and Kruse, 1968). We note, as in section 2.7.1, that, in practice we would maximize $L(\cdot, \cdot)$.

Another aspect of this process becomes important when we consider the most realistic situation of nothing being known about G; in particular, its identifiability may be in doubt. We can then adopt a slightly more conservative approach, observing that $I(H, H_k) \to 0$ implies $|dH(x)/dH_k(x)| \to 1$, except possibly over a set B_k such that $\int_{B_k} dH(x) \to 0$. This means that if a reasonably small finite k yields a small $I(H, H_k)$, we shall have $|dH(x)/dH_k(x)|$ small, except possibly in the 'tails' of the H distribution. Let us now suppose that, over a λ-range, R, such that

$$\int_R dG(\lambda) > 1 - \varepsilon,$$

with ε small, we can approximate $f(x|\lambda)$ (or $p(x|\lambda)$) by a polynomial,

$$f(x|\lambda) = A_0(x) + A_1(x)\lambda + \cdots + A_s(x)\lambda^s.$$

Then we have

$$\sum_{j=0}^{s} A_j(x)\lambda^j dG_k(\lambda) \approx \sum_{j=0}^{s} A_j(x)\lambda^j dG(\lambda). \tag{2.7.7}$$

Thus, if a number $\geqslant (s+1)$ of values of x exist such that the determinant

$$|A_j(x_r)| \neq 0, \tag{2.7.8}$$

then we have,

$$\int \lambda^j dG_k(\lambda) = \int \lambda^j dG(\lambda), \qquad j = 1, \ldots, s. \tag{2.7.9}$$

Identifiability of G only affects the determination of G indirectly. Any restrictions on G_k to render it determinable are clearly permissible, but the fulfilment of (2.7.8) will dictate the accuracy of the approximation of G by G_k which can be achieved. For example, if $p(x|\lambda)$ is the binomial kernel $B(x; n, \lambda)$ then there are only $(n+1)$ distinct values of x, and hence s can at most be n. This clearly places a limitation on the accuracy of the approximation, because, as (2.7.8) shows, we can approximate no more than the first n moments of $G(\lambda)$. We shall see, in section 3.7, that in the case of EB point estimation, the question of identifiability of G becomes unimportant, especially when var (Λ) is small.

While we have been concerned mainly with the properties of G_k as $k \to \infty$, the arguments clearly suggest that it may be possible to obtain a satisfactory approximation to G by a G_k with reasonably small k. It may be remarked that such a process is analogous to the approximation of a curve by a polynomial of increasing order.

2.8 Estimation of G: parametric G families

In EB problems we shall mostly be concerned with cases where $H(x)$ is not known exactly, but is estimated by an empirical d.f., H_n, obtained from a sample of n observations on the r.v. whose d.f. is H. Thus we have instead of the exact equation

$$H(x) = \int F(x|\lambda) dG(\lambda), \tag{2.8.1}$$

the approximate relationship

$$H_n(x) \simeq \int F(x|\lambda)dG(\lambda), \tag{2.8.2}$$

Following the idea of minimizing $I(\cdot,\cdot)$, that is, maximizing $L(\cdot,\cdot)$, presented in section 2.7, we now obtain an estimate $\hat{G}$ of G by maximizing $L(H_n, F_G)$. The search for G would be an impractical task unless some restriction were imposed on the function G. In particular, it must be a d.f. As we have noted in section 2.7, the problem is largely overcome if it can be assumed that G belongs to a given parametric family $G(\lambda; \alpha, \beta, \ldots)$ of distributions, where $\alpha, \beta, \ldots$ are unknown parameters to be estimated. Maximizing $L(H_n, F_G)$ then means maximizing

$$\sum_{i=1}^{n} \log f(x_i; \alpha, \beta, \ldots) \tag{2.8.3}$$

where

$$f_G(x) = f(x; \alpha, \beta, \ldots) = \int f(x|\lambda)dG(\lambda; \alpha, \beta, \ldots).$$

In other words, since $F_G(x)$ now belongs to a certain parametric family, we can use the method of maximum likelihood to estimate $\alpha, \beta, \ldots$. In these circumstances, other 'conventional' methods of estimating $\alpha, \beta, \ldots$, are, of course, feasible, and may be preferred. The preference may be merely for computational reasons, if many estimates are to be calculated as a matter of routine.

2.8.1 *The method of moments*

One may use the observations $x_1, \ldots, x_n$ to estimate the parameters $\xi = (\alpha, \beta, \ldots)$ of the marginal p.d.f. $f_G(x)$ by the usual method of moments. We note that if

$$\mu_r'(\lambda) = \int x^r dF(x|\lambda), \qquad r = 1, 2, \ldots,$$

the rth-order moment of $F(x|\lambda)$, and if

$$\mu_r'(\xi) = \int x^r df_G(x) = \int x^r h(x|\xi)dx,$$

then

$$\mu_r'(\xi) = E_G \mu_r'(\Lambda), \qquad r = 1, 2, \ldots. \tag{2.8.4}$$

The left-hand side of (2.8.4) can be estimated directly by the sample moment

$$m'_r = \sum_{i=1}^{n} x_i^r/n.$$

Thus if the functional dependence of $\mu'_r(\xi)$ on ξ can be determined explicitly, a sufficient number of estimating equations can be written down for determining elements of ξ.

2.8.2 *The method of maximum likelihood*

For many parametric d.f.s $G(\lambda|\xi)$ satisfying some regularity conditions, it is possible to develop a general iterative technique to estimate the unknown parameter ξ by the method of maximum likelihood. Consider the likelihood function (2.8.3). The likelihood equations for ξ_j can be written as

$$0 = \frac{\partial \ln L}{\partial \xi_j} = \sum_{i=1}^{n} \frac{1}{h(x_i|\xi)} \int f(x_i|\lambda) \frac{\partial \ln g(\lambda|\xi)}{\partial \xi_j} g(\lambda|\xi) d\lambda \tag{2.8.5}$$

provided the interchange of differentiation and integration can be effected in the expression given by

$$\frac{\partial}{\partial \xi_j} \int f(x|\lambda) g(x|\xi) d\lambda,$$

where $g(\lambda|\xi)$ is the p.d.f. of $G(\lambda|\xi)$. The equations (2.8.5) can be rewritten as

$$0 = \sum_{i=1}^{n} E\left(\frac{\partial \ln g(\Lambda|\xi)}{\partial \xi_j} \middle| x_i \right)$$

where the expectation is with respect to the posterior d.f. of Λ given $X = x_i$. We can expand the quantity

$$\frac{\partial \ln g(\lambda|\xi)}{\partial \xi_j}$$

at $\xi = \xi^{(0)}$, a known initial value of ξ, to obtain

$$\frac{\partial \ln g(\lambda|\xi)}{\partial \xi_j} = \frac{\partial \ln g(\lambda|\xi^{(0)})}{\partial \xi_j} + \sum_{u=1}^{q} (\xi_u - \xi_u^{(0)}) \left[\frac{\partial^2 \ln g(\lambda|\xi)}{\partial \xi \partial \xi_j} \right]_{\xi = \xi^{(0)}}.$$

Thus we have a system of equations for ξ_u's in terms of initial values $\hat{\xi}_u^{(0)}$'s as

$$0=\sum_{i=1}^{n} E\left[\frac{\ln g(\Lambda|\xi)}{\partial \xi_j}\middle| X=x_i\right]_{\xi=\xi^{(0)}}$$

$$+\sum_{i=1}^{n}\sum_{u=1}^{q}(\xi_u-\hat{\xi}_u^{(0)})E\left[\frac{\partial^2 \ln g(\Lambda|\xi)}{\partial \xi_u \partial \xi_j}\middle| X=x_i\right]_{\xi=\xi^{(0)}}$$

$$j=1,\ldots. \tag{2.8.6}$$

Hence a system of iterative equations for updating an initial estimate $\boldsymbol{\xi}^{(0)}$ of $\boldsymbol{\xi}$ is obtained by rewriting (2.8.6) in a matrix form:

$$\boldsymbol{\xi}^{(i+1)}=\boldsymbol{\xi}^{(i)}+\{K(\boldsymbol{\xi}^{(i)},\mathbf{x})\}^{-1}\sqcup(\boldsymbol{\xi}^{(i)},\mathbf{x}), \qquad i=0,1,\ldots, \tag{2.8.7}$$

where $\sqcup(\boldsymbol{\xi},\mathbf{x})$ is a $q\times 1$ vector whose jth element is

$$\sum_{i=1}^{n} E\left(\frac{\partial \ln g(\Lambda|\xi)}{\partial \xi_j}\middle| X=x_i\right)$$

and $K(\boldsymbol{\xi},\mathbf{x})$ is a $q\times q$ matrix whose (j,t)th element is

$$\sum_{i=1}^{n} E\left(\frac{\partial^2 \ln g(\Lambda|\xi)}{\partial \xi_j \partial \xi_t}\middle| X=x_i\right).$$

The equation (2.8.7) can be used to obtain an iterative process by setting $i=1,2,\ldots$. The above iterative procedure is a result of an application of the EM algorithm (cf. Dempster, Laird and Rubin, 1977). The details are given in a more general setting in section 2.12.

Example 2.8.1 Let

$$p(x|\lambda)=e^{-\lambda}\lambda^x/x!, \qquad x=0,1,2,\ldots,$$

$$dG(\lambda;\alpha,\beta)=\frac{\alpha^\beta}{\Gamma(\beta)}\lambda^{\beta-1}e^{-\alpha\lambda}d\lambda, \qquad \alpha,\beta>0,$$

then

$$p_G(x)=\left(\frac{\alpha}{\alpha+1}\right)^\beta\frac{\Gamma(\beta+x)}{\Gamma(\beta)x!}\left(\frac{1}{\alpha+1}\right)^x, \qquad x=0,1,2,\ldots,$$

the negative binomial distribution. Its mean and variance are

$$\beta/\alpha \quad\text{and}\quad \beta/\alpha+\beta/\alpha^2$$

respectively; the distribution is often reparametrized by putting $\beta/\alpha = m$, $\beta = p$, so that

$$\left.\begin{aligned} \text{mean} &= m \\ \text{variance} &= m + m^2/p \end{aligned}\right\}. \tag{2.8.8}$$

The moment relations (2.8.8) can be used to estimate m and p. If $\bar{x}$ and s^2 are the sample mean and variance of $x_1, \ldots, x_n$, we have estimates $\check{m}$ and $\check{p}$ of m and p as

$$\left.\begin{aligned} \check{m} &= \bar{x} \\ \check{p} &= \begin{cases} \bar{x}^2/(s^2 - \bar{x}), & s^2 > \bar{x} \\ +\infty, & \text{otherwise.} \end{cases} \end{aligned}\right\} \tag{2.8.9}$$

The maximum likelihood estimates of m and p are $\hat{m}$ and $\hat{p}$, where

$$\hat{m} = \bar{x},$$

and $\hat{p}$ is that positive value of p which maximizes

$$\left(\frac{p}{p + \bar{x}}\right)^p \left\{\prod_{i=1}^{n} \frac{\Gamma(p + x_i)}{\Gamma(p) x_i!}\right\} \left(\frac{\bar{x}}{p + \bar{x}}\right)^{n\bar{x}};$$

$p = +\infty$ is a permissible solution.

Example 2.8.2 Let

$$\begin{aligned} f(x \mid \lambda) &= N(\lambda, 1), \\ dG(\lambda) &= N(\mu, \sigma^2), \end{aligned}$$

then $f_G(x) = N(\mu, 1 + \sigma^2)$, and the 'natural' moment estimates of μ and σ^2 are provided by

$$\begin{aligned} \check{\mu} &= \bar{x} \\ 1 + \check{\sigma}^2 &= s^2, \end{aligned}$$

where $\bar{x}$ and s^2 are the sample mean and variance of the past observations. Since $\sigma^2 \geqslant 0$, our estimate of σ^2 is $\max(0, s^2 - 1)$. The maximum likelihood estimates of μ and σ^2 are

$$\begin{aligned} \hat{\mu} &= \bar{x} \\ \hat{\sigma}^2 &= \max\left\{0, \left(\frac{n-1}{n}\right)s^2 - 1\right\}. \end{aligned}$$

2.9 Estimation of G: finite approximation of G

Realistically we may expect to have reliable knowledge of the functional form of G only in exceptional cases. More commonly, our knowledge of G will be rather vague, consisting perhaps of the information that λ is limited to a certain finite interval, for example, as in the case of the binomial $p(x|\lambda)$. The discussion of section 2.7.2 suggests a possible solution to the problem, namely, to replace the G of the unknown functional form by a finite G_k. By making k large enough, the approximation can be made arbitrarily good, according to some reasonable criterion. In fact, as we shall see in Chapter 3, for certain purposes k can be quite small.

Implementation of this suggestion requires that we treat the observations $x_1, x_2, \ldots, x_k$ as if they are the results of random sampling from a population with d.f.

$$F_{G_k}(x) = \sum_{j=1}^{k} \theta_j F(x|\lambda_j) \tag{2.9.1}$$

(Maritz, 1967). The mixture likelihood can be approximated by

$$L_{G_k} = \sum_{j=1}^{n} \ln\{f_{G_k}(x_i)\}. \tag{2.9.2}$$

Estimation of the λ's (or θ's) in this approximate model by the method of maximum likelihood corresponds, again, to an application of the method of section 2.7.2, with the l.h.s. of (2.7.1) replaced by the empirical d.f. of the observations $x_1, \ldots, x_n$.

2.9.1 *$\lambda_1, \ldots, \lambda_k$ given; $\theta_1, \ldots, \theta_k$ unknown*

The EM algorithm described by Dempster, Laird and Rubin (1977) is particularly useful for finding the maximum likelihood estimates of the θ's in this case. Starting with initial trial values $\theta_r^{(0)}$, $r = 1, 2, \ldots, k$, new values $\theta_r^{(1)}$ are obtained as follows:

$$\theta_{ri}^{(1)} = \theta_r^{(0)} f(x_i|\lambda_r) \Big/ \sum_{j=1}^{k} \theta_j^{(0)} f(x_i|\lambda_j)$$

$$\theta_r^{(1)} = \sum_{i=1}^{n} \theta_{ri}^{(1)}/n. \tag{2.9.3}$$

Similar iterative sequences are obtained by the methods of Day (1969) or Behboodian (1975). For starting values one can take $\theta_r^{(0)} = 1/k$ $(r = 1, \ldots, k)$.

Lindsay (1983a) linked this procedure with the general approach of the vertex direction method explained in more detail in section 2.10. In particular it was pointed out that the EM algorithm terminates at restricted maxima and hence a check for global maximality should be carried out. In particular a simple first-order check was given as a verification of the condition

$$D''(\lambda; \hat{G}_k) \leqslant 0 \quad \text{at} \quad \lambda = \lambda_1, \ldots, \lambda_k \tag{2.9.4}$$

where

$$D''(\lambda; \hat{G}_k) = \sum_{j=1}^{k} n_j \left[\frac{\partial^2}{\partial \lambda^2} f(x_j^* \mid \lambda) / f_{G_k}(x_j^*) \right]. \tag{2.9.5}$$

The quantities n_j and x_j^* are as defined for $D(\lambda, G)$ mentioned in relation to the vertex direction method.

2.9.2 $\theta_1, \ldots, \theta_k$ *given;* $\lambda_1, \ldots, \lambda_k$ *unknown*

For convenience we shall take $\theta_r = 1/k$ for $r = 1, 2, \ldots, k$, and for the sake of identifiability we impose the restriction $\lambda_1 \leqslant \lambda_2 \leqslant \cdots \leqslant \lambda_k$. We now have to maximize L_{G_k} in (2.9.2) with respect to $\lambda_1, \ldots, \lambda_k$. As in section 2.9.1, the EM algorithm is useful here and particularly easy to apply for certain forms of 'regular' $f(x \mid \lambda)$. Starting with trial values $\lambda_r^{(0)}$ we first compute

$$z_{ri}^{(0)} = f(x_i \mid \lambda_r^{(0)}) \Big/ \sum_{j=1}^{k} f(x_i \mid \lambda_j^{(0)}),$$

for $r = 1, 2, \ldots, k$ and $i = 1, 2, \ldots, n$. Then new values $\lambda_r^{(1)}$ are obtained as the solutions of the equations, in λ_r,

$$\sum_{i=1}^{n} z_{ri}^{(0)} \partial \ln f(x_i \mid \lambda_r) / \partial \lambda_r = 0.$$

Example 2.9.1 If $f(x \mid \lambda)$ is the $N(\lambda, 1)$ density straightforward substitution gives

$$\lambda_r^{(1)} = \sum_{i=1}^{n} z_{ri}^{(0)} x_i \Big/ \sum_{i=1}^{n} z_{ri}^{(0)}.$$

Applications of the 'G_k-ML' process to specific EB problems are given in later chapters.

At this juncture we observe that, having decided to treat the observations as being generated by the d.f. $F(x; \lambda_1, \lambda_2, \ldots, \lambda_k)$, we may contemplate using other 'conventional' statistical methods for estimating the λ's. In certain instances the method of moments has appeal.

Example 2.9.2 Let $f(x|\lambda)$ be as in Example 2.9.1, and suppose that $k = 3$ with $\theta_1 = \theta_2 = \theta_3 = \frac{1}{3}$. Then the low-order moments of X_G are

$$\mu_1' = [\lambda]$$
$$\mu_2 = 1 + [\lambda^2] - [\lambda]^2$$
$$\mu_3 = [\lambda^3] - 3[\lambda][\lambda^2] + 2[\lambda]^3,$$

where $[\lambda^r] = (\lambda_1^r + \lambda_2^r + \lambda_3^r)/3$. This suggests the following method of determining approximate λ's using the sample moments, denoted by m's, of $x_1, x_2, \ldots, x_n$: if $m_2 > 1$ it is always possible to find $\breve{\lambda}$'s such that $m_1' = [\breve{\lambda}] = \breve{\mu}_1'$, and $m_2 = \breve{\mu}_2$. Hence choose λ's satisfying these equations which minimize $|m_3 - \breve{\mu}_3|$. When $m_2 \leqslant 1$, put all three λ's equal to m_1'.

Note that employment of the method of moments is tantamount to using a certain measure of distance between distributions. Still other measures of distance have been considered in connection with the same problem. A measure based on the χ^2 test of goodness of fit appears in Maritz (1967); Choi (1966) uses the Wolfowitz distance; Bartlett and Macdonald (1968) use the method of least squares; while Deely and Kruse (1968) propose

$$\text{'distance'} = \| H(x) - F_{G_k}(x) \| = \sup |H(x) - F_{G_k}(x)|.$$

They remark that once the λ's are chosen, the problem of finding θ's to minimize the distance can be reduced to a linear programming problem.

2.10 Estimation of G: continuous mixtures

We now consider the estimation of a general mixing distribution without assuming any parametric form for it or approximating it by a finite distribution. The method of estimation employed is maximization of the criterion $L(h, w)$ of Section 2.7.1 as applied to obtaining the distance between F_G and the observed empirical d.f. H_n. We have

$$L(H_n, F_G) = \int \ln\{f_G(x)\}dH_n(x)$$

$$= n^{-1} \sum_{i=1}^{n} \ln\{f_G(x_i)\}.$$

Maximization of the criterion $L(H_n, F_G)$ is of course effected by maximization of the mixture likelihood

$$L_G = \sum_{i=1}^{n} \ln\{f_G(x_i)\}. \tag{2.10.1}$$

In the above $f_G(x) = \int f(x|\lambda)dG(\lambda)$ when X is continuous. When X is a discrete r.v. $f(x|\lambda)$ is replaced by $p(x|\lambda)$.

The nature of the mixture likelihood L_G has been investigated in detail by Lindsay (1983a) for the general case when the form of $f(x|\lambda)$ is not specified and by Lindsay (1983b) for the case when $f(x|\lambda)$ belongs to the exponential family. The main results of the papers are concerned with the existence of the maximum likelihood estimator under certain conditions which specify the nature of the curve

$$\Gamma = \{\mathbf{f}_\lambda : \lambda \in \Omega\} \tag{2.10.2}$$

where

$$\mathbf{f}_\lambda = (f(x_1^*|\lambda), \ldots, f(x_s^*|\lambda)) \tag{2.10.3}$$

is the vector of likelihoods of s ($\leqslant n$) distinct values $x_1^*, \ldots, x_s^*$ of the observations $x_1, \ldots, x_n$ and Ω is the parameter space of λ. As to the determination of an ML estimate $\hat{G}$ when it exists, Lindsay (1983a) provided a vertex direction method (VDM) for constructing such a $\hat{G}$ iteratively.

2.10.1 *Vertex direction method*

Let n_i be the number of times x_j^* in (2.10.3) appears in the sample. In addition to (2.10.3) define the quantities:

$$\mathbf{f}_G = (f_G(x_1^*), \ldots, f_G(x_s^*))^{\mathrm{T}}$$

$$D(\lambda; G) = \sum_{j=1}^{s} n_j\{f(x_j^*|\lambda)/f_G(x_j^*) - 1\}$$

and

$$J(\mathbf{u}) = \sum_{i=1}^{s} \ln u_i$$

for any vector $\mathbf{u} = (u_1, \ldots, u_s)^{\mathrm{T}}$.

Let $G^{(i)}$ be an estimator of G at the ith step.

Step (i): Find $\lambda^{(i)}$ to maximize $D(\lambda; G)$ at $G = G^{(i)}$.
Step (ii): Find $\varepsilon^{(i)}$ to maximize

$$J(\mathbf{f}_G(1-\varepsilon) + \varepsilon \mathbf{f}_\lambda)$$

at $G = G^{(i)}$ and $\lambda = \lambda^{(i)}$.
Step (iii): Define

$$\delta(z \mid \lambda^{(i)}) = \begin{cases} 1 & \text{if } z = \lambda^{(i)} \\ 0 & \text{otherwise.} \end{cases}$$

Set the $(i+1)$th step estimator of G to be

$$G^{(i+1)}(z) = (1 - \varepsilon^{(i)})G^{(i)}(z) + \varepsilon^{(i)}\delta(z \mid \lambda^{(i)}).$$

Step (iv): Repeat the iteration until $G^{(i)}$ converges to $\hat{G}$.

Practical implementation of the VDM algorithm requires numerical work. A closely related algorithm has been given earlier for the special case of Poisson data d.f. by Simar (1976). An algorithm for nonparametric estimation of G has been developed for a more general case by Der Simonian (1986).

2.11 Estimation of G: miscellaneous methods

Rutherford and Krutchkoff (1967) take a position intermediate between those of sections 2.8 and 2.10. They assume that G is a member of the Pearson family of distributions. The d.f. $F(x \mid \lambda)$ is taken to be such that known functions $h_k(x)$, $k = 1, 2, 3, 4$ exist giving

$$\int h_k(x) dF(x \mid \lambda) = \lambda^k.$$

Then it follows that

$$\int h_k(x) dF_G(x) = \int \lambda^k dG(\lambda),$$

and $\int \lambda^k dG(\lambda)$ can be estimated by

$$M_{k,n} = (1/n) \sum_{i=1}^{n} h_k(x_i).$$

If $\int \lambda^4 dG(\lambda) < \infty$, $M_{k,n} \to \int \lambda^k dG(\lambda)$, in probability, as n increases.

Hence, for n large enough it will be possible to identify the particular Pearson curve which obtains, and to estimate its parameters using the $M_{k,n}$. The authors give an example of the application of this procedure.

Rolph (1968) examines the special case where λ lies in the interval $[0, 1]$, and X is a discrete integer-valued r.v. such that

$$p(x|\lambda) = \sum_{j=1}^{k_x} a_{xj}\lambda^j,$$

a polynomial in λ. The mixed p.d., $p_G(x)$, is then characterized directly by the moments of G. Rolph's procedure consists in proposing a prior distribution for the moments of G, and developing Bayes estimates of the moments.

Lord (1969) studies the case where X is discrete, assuming the values $0, 1, 2, \ldots, n$, and λ lies in the interval $[a, b]$, with the p.d.f. $g(\lambda)$. Let $t_r(\lambda)$ be any function such that the integrals

$$m_{rx} = \int_a^b t_r(\lambda)p(x|\lambda)dG(\lambda), \qquad r, x = 0, 1, 2, \ldots, n,$$

and the inverse of the matrix $\|m_{rx}\|$ exist. Putting $\|m_{rx}\|^{-1} = \|m^{xr}\|$, and

$$w_r = \sum_{r=0}^{n} m^{xr} p_G(x),$$

it can be easily verified that

$$g(\lambda) = \sum_{r=0}^{n} w_r t_r(\lambda)$$

is a solution of (2.1.1). Lord demonstrates the existence of functions $t_r(\lambda)$, defines smoothness criteria for $g(\lambda)$, and develops a method for determining w_r using the calculus of variations. He extends his method to the case where $p_G(x)$ is estimated by relative frequencies, introducing as a measure of distance between the 'observed' distributions a criterion similar to the χ^2 criterion for goodness of fit.

2.12 Estimation with unequal component sample sizes

In practical EB problems there may be unequal numbers of observations at different components. Also, the parameter $\boldsymbol{\lambda}$ of the data d.f. $F(x|\boldsymbol{\lambda})$ may be a p-vector, a realization of a vector r.v. $\boldsymbol{\Lambda}$ with d.f. $G(\boldsymbol{\lambda}|\boldsymbol{\xi})$. The corresponding to realization $\boldsymbol{\lambda}_i$ $(i=1,\ldots,n)$ of Λ, there will in general be m_i $(\geqslant 1)$ observations $x_{i1},\ldots,x_{im_i}$ denoted by $\mathbf{x}_i$ on X. Techniques of estimating G need to be extended to cover these realistic situations. The method of ML estimation discussed in sections 2.8, 2.9 and 2.10 is readily extended to deal with EB schemes of unequal component sample sizes.

2.12.1 *Parametric G families*

Suppose that G belongs to a parametric family $G(\boldsymbol{\lambda}|\boldsymbol{\xi})$ indexed by an unknown parameter vector $\boldsymbol{\xi}=(\xi_1,\ldots,\xi_q)$. The likelihood function of $\boldsymbol{\xi}$ can be obtained by using the marginal p.d.f. of $\mathbf{x}_i$ $(i=1,\ldots,n)$ given by

$$h(\mathbf{x}_i;\boldsymbol{\xi},m_i)=\int s(\mathbf{x}_i|\boldsymbol{\lambda},m_i)dG(\boldsymbol{\lambda}|\boldsymbol{\xi}) \tag{2.12.1}$$

where

$$s(\mathbf{x}_i|\boldsymbol{\lambda},m_i)=\prod_{j=1}^{m_i} f(x_{ij}|\boldsymbol{\lambda}). \tag{2.12.2}$$

Thus the log-likelihood function of $\boldsymbol{\xi}$ based on the previous data $\mathbf{x}_1,\ldots,\mathbf{x}_n$ is

$$\ln L=\sum_{i=1}^{n}\ln h(\mathbf{x}_i;\boldsymbol{\xi},m_i). \tag{2.12.3}$$

We can maximize (2.12.3) with respect to $\boldsymbol{\xi}$, by using a direct optimization algorithm. Alternatively, we can apply an EM algorithm (see Dempster, Laird and Rubin, 1977), to obtain an ML estimate $\hat{\boldsymbol{\xi}}_m$ of $\boldsymbol{\xi}$ based on (2.12.3).

The joint p.d.f. of $(\boldsymbol{\lambda}_i,\mathbf{x}_i)$ $(i=1,\ldots,n)$ from an EB scheme can be written as

$$g(\boldsymbol{\lambda}_i|\boldsymbol{\xi})s(\mathbf{x}_i|\boldsymbol{\lambda}_i,m_i). \tag{2.12.4}$$

In the terminology of the EM approach, the 'complete data' is $\{(\boldsymbol{\lambda}_1,\mathbf{x}_1),\ldots,(\boldsymbol{\lambda}_n,\mathbf{x}_n)\}$ of which the 'unobservables' are $\boldsymbol{\lambda}_1,\ldots,\boldsymbol{\lambda}_m$. The log-likelihood function of the complete data is

$$\ln L(\boldsymbol{\lambda}_1,\ldots,\boldsymbol{\lambda}_n;\mathbf{x}_1,\ldots,\mathbf{x}_n;\boldsymbol{\xi})=\sum_{i=1}^{m}\ln g(\boldsymbol{\lambda}_i|\boldsymbol{\xi})+\sum_{i=1}^{m}\ln s(\mathbf{x}_i|\boldsymbol{\lambda}_i,m_i). \tag{2.12.5}$$

Now the posterior d.f. of $\mathbf{\Lambda}$ given $(\mathbf{x}_i, m_i)$ is

$$dB(\boldsymbol{\lambda}|\mathbf{x}_i, G, m_i) = \{h(\mathbf{x}_i, G, m_i)\}^{-1}\{s(\mathbf{x}_i|\boldsymbol{\lambda}, m_i)\}dG(\boldsymbol{\lambda}|\boldsymbol{\xi}). \quad (2.12.6)$$

The E-step of the EM algorithm is carried out by taking the conditional expectation of the quantity (2.12.5) given the data $\mathbf{x}_1, \ldots, \mathbf{x}_n$ and the prior d.f. $G(\boldsymbol{\lambda}|\boldsymbol{\xi}^+)$ where $\boldsymbol{\xi}^+$ is a given initial estimate of $\boldsymbol{\xi}$. We have

$$E\{\ln L(\boldsymbol{\lambda}_1, \ldots, \boldsymbol{\lambda}_n; \mathbf{x}_1, \ldots, \mathbf{x}_n)|\mathbf{x}_1, \ldots, \mathbf{x}_n, \boldsymbol{\xi}^+\}$$

$$= \sum_{i=1}^{n} E\{\ln g(\mathbf{\Lambda}|\boldsymbol{\xi})|\mathbf{x}_i, \boldsymbol{\xi}^+\} + \sum_{i=1}^{n} E\{\ln s(\mathbf{x}_i|\mathbf{\Lambda}, n_i)|\mathbf{x}_i, \boldsymbol{\xi}^+\}$$

$$= \sum_{i=1}^{n} E\{\ln g(\mathbf{\Lambda}|\boldsymbol{\xi})|\mathbf{x}_i, \boldsymbol{\xi}^+\} + Z(\mathbf{x}_1, \ldots, \mathbf{x}_n, \boldsymbol{\xi}^+), \quad (2.12.7)$$

where the term $Z(\mathbf{x}_1, \ldots, \mathbf{x}_n, \boldsymbol{\xi}^+)$ is a function not depending on $\boldsymbol{\xi}$ and the expectations are taken with respect to d.f. of $\mathbf{\Lambda}$ given $(\mathbf{x}_i, m_i)$ in (2.12.6).

The M-step of the EM algorithm is accomplished by maximizing the quantity (2.12.7) with respect to $\boldsymbol{\xi}$ for fixed $\mathbf{x}_i$'s and $\boldsymbol{\xi}^+$. This leads to a set of equations

$$\sum_{i=1}^{n} \frac{\partial}{\partial \xi_j} E\{\ln g(\mathbf{\Lambda}|\boldsymbol{\xi})|\mathbf{x}_i, \boldsymbol{\xi}^+\} = 0, \qquad j = 1, \ldots, q.$$

Under the regularity conditions which allow for the interchange of differentiation and integration in the above set of equations, we obtain

$$\sum_{i=1}^{n} E\left\{\frac{\partial}{\partial \xi_j} \ln g(\mathbf{\Lambda}|\boldsymbol{\xi})|\mathbf{x}_i, \boldsymbol{\xi}^+\right\} = 0, \qquad j = 1, \ldots, q. \quad (2.12.8)$$

We can expand the quantity under the expectation sign in (2.12.8) in a Taylor series of ξ_i's. We then have

$$\sum_{i=1}^{n} E\left[\left\{\frac{\partial}{\partial \xi_j} \ln g(\mathbf{\Lambda}|\boldsymbol{\xi})\right\}_{\boldsymbol{\xi}=\boldsymbol{\xi}^+} |\mathbf{x}_i, \boldsymbol{\xi}^+\right]$$

$$+ \sum_{i=1}^{n} \sum_{u=1}^{q} E\left[(\xi_u - \xi_u^+)\left\{\frac{\partial^2 \ln g(\mathbf{\Lambda}|\boldsymbol{\xi})}{\partial \xi_j \partial \xi_u}\right\}_{\boldsymbol{\xi}=\boldsymbol{\xi}^+} |\mathbf{x}_i, \boldsymbol{\xi}^+\right] + \cdots$$

$$= 0 \qquad (j = 1, \ldots, q).$$

Using the Newton–Raphson approach, we obtain an iterative

sequence for updating ξ^+ as

$$\xi^{(i+1)} = \xi^{(i)} - \mathbf{J}^{-1}\sqcup, \qquad i = 1, 2, \ldots \tag{2.12.9}$$

where $\sqcup$ is a $q \times 1$ vector whose tth element is

$$\sqcup_t(\mathbf{x}_1, \ldots, \mathbf{x}_n; \xi^{(i)}) = \sum_{u=1}^{n} E\left[\left\{\frac{\partial \ln g(\Lambda|\xi)}{\partial \xi_t}\right\}_{\xi=\xi^{(i)}} |\mathbf{x}_u, \xi^{(i)}\right]$$

and $\mathbf{J}$ is a $q \times q$ matrix whose (j, t)th element is

$$\mathbf{J}_{jt}(\mathbf{x}_1, \ldots, \mathbf{x}_n; \xi^{(i)}) = \sum_{i=1}^{n} E\left[\left\{\frac{\partial \ln g(\Lambda|\xi)}{\partial \xi_j \partial \xi_t}\right\}|\mathbf{x}_u, \xi^{(i)}\right].$$

In the iteration process, (2.12.9), $\xi^{(1)}$ is taken as the initial estimate ξ^+. When the iteration converges, the resulting quantity $\hat{\xi}_m$ is taken as the ML estimator of ξ.

In the following chapters, the bias vector and covariance matrix of $\hat{\xi}_m$ will be denoted respectively by

$$\psi(\xi) = E(\hat{\xi}_m) - \xi, \tag{2.12.10}$$

and

$$V(\xi) = E(\hat{\xi}_m - E\hat{\xi}_m)(\hat{\xi}_m - E\hat{\xi}_m)^{\mathrm{T}}. \tag{2.12.11}$$

The mean square error matrix is then given by

$$\Gamma(\xi) = V(\xi) + \psi(\xi)\psi(\xi)^{\mathrm{T}} = E(\hat{\xi}_m - \xi)(\hat{\xi}_m - \xi). \tag{2.12.12}$$

Exact evaluation of the bias and mean squared error of $\hat{\xi}_m$ is seldom possible. However, approximate quantities to terms of $O(n^{-1})$ can be obtained by using standard likelihood methods (see e.g. Bowman and Shenton, 1973) as applied to (2.12.3).

2.12.2 *Finite step-function approximation to G*

Suppose next that G can be approximated by a finite distribution $G_k(\lambda)$ having probability masses $\theta_1, \ldots, \theta_k$ at the points $\boldsymbol{\alpha}_1, \ldots, \boldsymbol{\alpha}_k$ of the p-dimensional parameter space Ω of λ. The masses satisfy the constraints $(\theta_1 + \cdots + \theta_k = 1, 0 < \theta_i < 1)$. We consider here only the case where $\boldsymbol{\alpha}_1, \ldots, \boldsymbol{\alpha}_k$ are assumed to be known and only $\theta_1, \ldots, \theta_k$ are to be estimated. This is the case where identifiability of G_k is most easily established. Related cases such as when $\theta_1, \ldots, \theta_k$ are known and $\boldsymbol{\alpha}_1, \ldots, \boldsymbol{\alpha}_k$ are unknown or all θ_i's and $\boldsymbol{\alpha}_i$'s are unknown can also be treated in principle, but it is quite difficult to establish conditions for identifiability. The approximate likelihood function approach demonstrated for the single-parameter case in section 2.9 is readily extended to the more general case of p-dimensional $\boldsymbol{\lambda}$ and also of unequal sample sizes at various stages of EB scheme. The analysis of the single-parameter case can be repeated and an iterative sequence for θ_u's can be obtained as follows:

$$\theta_u^{(i+1)} = \sum_{t=1}^{n} \theta_{ut}^{(i+1)}/n$$

with

$$\theta_{ut}^{(i+1)} = \theta_u^{(i)} s(\mathbf{x}_i|\boldsymbol{\alpha}_u, m_i) \Big/ \sum_{j=1}^{k} \theta_j^{(i)} s(\mathbf{x}_i|\boldsymbol{\alpha}_u, m_i)$$

where $s(\mathbf{x}_i|\boldsymbol{\alpha}_u, m_i)$ is computed from (2.12.2) with $\boldsymbol{\alpha}_u$ in place of $\boldsymbol{\lambda}$.

CHAPTER 3

Empirical Bayes point estimation

3.1 Introduction

An outline of Bayes point estimation has been given in sections 1.3 and 1.6 and we recall that the Bayes estimate of λ under the quadratic loss structure is

$$\delta_G(x) = \int \lambda dF(x|\lambda)dG(\lambda) \Big/ \int dF(x|\lambda)dG(\lambda); \tag{3.1.1}$$

see also (1.3.3). The expected loss for any estimator $\delta(x)$, $W(\delta)$, is related to $W(\delta_G)$ by

$$\begin{aligned} W(\delta) &= W(\delta_G) + \int \{\delta(x) - \delta_G(x)\}^2 dF_G(x) \\ &= W(\delta_G) + K(\delta, \delta_G). \end{aligned} \tag{3.1.2}$$

Any estimate of $\delta_G(x)$ based on past data can be thought of as an empirical Bayes estimator. Most detailed studies of empirical Bayes estimators (EBEs) have been in the framework of the sampling scheme described in section 1.8: independent past observations $x_1, x_2, \ldots, x_n$, obtained with independent realizations $\lambda_1, \lambda_2, \ldots, \lambda_n$ of Λ comprise the past data. For the most part this scheme will be adhered to in this chapter. The overall expected loss of an EBE $\delta_n(x)$ is defined as $E_n W(\delta_n)$, as in section 1.10.

In this chapter details of certain methods of obtaining EB point estimates are given, and they are applied to some of the standard distributions. The behaviour of $W(\delta_n)$ is clearly a topic of importance in assessing the performance of EBEs. Reasonably general statements can be made about asymptotic optimality of δ_n, but studies of individual cases seem to be needed when n is not large. Some such case studies are reported, involving distributions like the Poisson, normal and several others.

Much of the writing on pure Bayes estimation is occupied with considerations of **diffuse** or **non-informative** prior distributions. The philosophical issue is reconciliation of the use of Bayes's theorem for inference, while not claiming sharp prior knowledge of parameter values. In the EB approach, and EB point estimation in particular, our attitude is different. The introductory discussion of the performance of EBEs, section 1.10, suggests that an EBE would only be considered a serious competitor for a more conventional estimator, T, if $W(\delta_G)$ is considerably smaller than $W(T)$. Take the example of $X \overset{d}{=} N(\lambda, \sigma^2)$, $\Lambda \overset{d}{=} N(\mu_G, \sigma_G^2)$, $T = X$, where $W(T) = \sigma^2$, $W(\delta_G) = 1/\{1/\sigma^2 + 1/\sigma_G^2\}$. If $\sigma_G^2 = \sigma^2$ we have $W(\delta_G) = (1/2)W(T)$ and it would seem that some worthwhile gain may be possible with an EBE. But if $\sigma_G^2 = 10\sigma^2$ we have $W(\delta_G) = (10/11)W(T)$, and EB is unlikely to be much better than conventional estimation. Thus, the flavour of discussions in this chapter is that the dispersion of the distribution of Λ is small in some sense that is relevant to point estimation for the particular data distribution $F(x|\lambda)$.

3.2 Asymptotic optimality

3.2.1 Consistent estimation of $\delta_G(x)$

Suppose that $\delta_n(x)$ is an EBE, that is, a consistent estimate of $\delta_G(x)$ in the sense that $\delta_n(x) \to \delta_G(x)$, (P), for every x. Without being practically unrealistic some of the mathematical arguments to do with asymptotic optimality can be simplified by truncating $\delta_n(x)$ at finite lower and upper limits L and U respectively. This will also ensure that $E_n\{\delta_n(x) - \delta_G(x)\}^2 \to 0$ for all x such that $\delta_G(x) \in (L, U)$.

Suppose also that

$$\int \{\delta_G(x)\}^2 dF_G(x) < \infty. \tag{3.2.1}$$

This condition is not very restrictive because we can write

$$W(\delta_G) = \int \lambda^2 dG(\lambda) - \int \delta_G^2(x) dF_G(x),$$

so that finiteness of $\int \lambda^2 dG(\lambda)$ will ensure the validity of (3.2.1). In the light of the discussion in section 3.1 restricting attention to prior distributions with finite variances when considering EBEs given by (3.1.1) seems entirely reasonable.

Consider the set of x-values such that $\delta_G(x) \in (L, U)$, and, if necessary, a subset contained in an x-interval of finite length. Suppose such an interval to be $I(L, U)$. Then, since $\delta_n(x) \to \delta_G(x), (P)$, for all $x \in I(L, U)$ we can state that

$$\int_{x \in I(L,U)} \{\delta_n(x) - \delta_G(x)\}^2 dF_G(x) \to 0, (P).$$

Also

$$\int_{x \notin I(L,U)} \{\delta_n(x) - \delta_G(x)\}^2 dF_G(x)$$

$$\leqslant \int_{x \notin I(L,U)} \{\delta_G(x)\}^2 dF_G(x).$$

Therefore, noting condition (3.2.1) we can choose $I(L, U)$ so that

$$W(\delta_n) \to W(\delta_G) + \varepsilon, (P),$$

where ε is arbitrarily small.

Arguing similarly, using $E_n\{\delta_n(x) - \delta_G(x)\}^2 \to 0$ for x such that $\delta_G(x) \in (L, U)$, we can show that

$$E_n W(\delta_n) \to W(\delta_G) + \varepsilon',$$

again with ε' arbitrarily small.

Example 3.2.1 If the distribution of X is Poisson (λ) we have

$$\delta_G(x) = (x+1) p_G(x+1)/p_G(x)$$

and it is readily seen that

$$\delta_n(x) = (x+1) f_n(x+1)/\{1 + f_n(x)\}$$

given by (1.9.1) is a consistent estimate of $\delta_G(x)$ for every x.

3.2.2 *Consistent estimation of an approximation to $\delta_G(x)$*

Approximations to the Bayes estimate have been discussed in sections 1.12 and 2.7. Let $\delta_G^*(x)$ be an approximation of $\delta_G(x)$ having the property that $W(\delta_G^*) - W(\delta_G) = \Delta$, where Δ is usually small. Suppose that $\hat{\delta}_n^*$ is an empirical version of δ_G^*, i.e. an estimate of δ_G^* such that $\hat{\delta}_n^*(x) \to \delta_G^*(x), (P)$ for every x. Then by arguing as in section 3.2.1, in particular by truncation of $\hat{\delta}_n^*$ if necessary, we can show that

$$E_n W(\delta_n^*) \to W(\delta_G) + \Delta + \varepsilon$$

and refer to this property as Δ-asymptotic optimality. See also Rutherford and Krutchkoff (1969).

Example 3.2.2 Suppose that $X \overset{d}{=} \text{Poisson}(\lambda)$ and $\Lambda \overset{d}{=} U(0, A)$. Then $\delta_G(x) = (x+1)I_{x+1}/I_x$ where

$$I_x = 1 - e^{-A}(1 + A + A^2/2! + \cdots + A^x/x!).$$

Let $\delta_G^*(x)$ be the linear Bayes estimator as given in Example 1.12.1. Substituting $E(\Lambda) = A/2$, $\text{var}(\Lambda) = A^2/12$,

$$\delta_G^*(x) = (3 + x)/(1 + 6/A).$$

Suppose that $\hat{A} = 2\bar{X}$ where $\bar{X}$ is the mean of the past x-values. Then $\hat{A} \to A, (P)$ and $\hat{\delta}_n^*(x) = (3+x)/(1 + 6/\hat{A}) \to \delta_G^*(x), (P)$, for every x. Therefore $\hat{\delta}_n^*$ is Δ-asymptotic optimal, the actual value of Δ depending on A. For example, if $A = 3$, $\delta_G^*(x) = 1 + x/3$, $W(\delta_G) = 0{\cdot}4725$, $W(\delta_G^*) = 0.5$, $\Delta = 0.0275$.

3.2.3 The rate of convergence of $E_n W(\delta_n)$

If δ_n is a.o. (E) the rate of convergence of $E_n W(\delta_n)$ will clearly depend on the rate of convergence of $\delta_n(x)$ to $\delta_G(x)$ or $E\{\delta_n(x) - \delta_G(x)\}^2$ to zero. A very common situation corresponds to well established practice in point estimation where $E\{\delta_n(x) - \delta_G(x)\}^2 = O(1/n)$. The majority of regular estimation problems allow this sort of statement, but we shall indicate some exceptions relevant to EB estimation.

Now, if it is true that $\delta_n(x)$ is a consistent estimator of $\delta_G(x)$ with the property that for each x,

$$E\{\delta_n(x) - \delta_G(x)\}^2 = O(1/n), \tag{3.2.2}$$

then $E_n W(\delta_n) = W(\delta_G) + O(1/n)$. This can be established by arguing as in section 3.2.1, and, if necessary, truncating $\delta_n(x)$ as before.

Example 3.2.3 Consider the Poisson case again, as in Example 3.2.1. The joint distribution of $f_n(x)$ and $f_n(x+1)$ is trinomial $\{n, p_G(x), p_G(x+1)\}$ and by using Taylor type expansions it is readily established that (3.2.2) holds.

The rate of approach to Δ-asymptotic optimality can be dealt with in similar fashion. Suppose that $\delta_n^*(x)$ is a consistent estimator of $\delta_G^*(x)$ satisfying a relation like (3.2.2). Then $E_n W(\delta_n^*) = W(\delta_G) + \Delta + O(1/n)$.

Example 3.2.4 Refer to Example 3.2.2 and note that $\hat{A}$ is an unbiased estimator of A with variance $A^2/3n$, $\delta_n^*(x) \cdot (3 + x)\hat{A}/(\hat{A} + 6)$, $\delta_G^*(x) = (3 + x)A/(A + 6)$, and straightforward calculations show that (3.2.2) holds. The distribution of $\hat{A}$ is asymptotically $N(A, A^2/3n)$ and the distribution of $\delta_n^*(x)$ is asymptotically $N(\delta_G^*(x); 12A^2/n(A + 6)^2)$ (Serfling, 1980, p. 118).

Examples of EBEs for which (3.2.2) does not necessarily hold occur when the distribution of X is continuous and the EBE is expressed in terms of estimates of $f_G(x)$ and $f'_G(x)$. Typically kernel estimates of $f_G(x)$ and $f'_G(x)$ may be used. They are discussed briefly in section 3.4.6. The rates of convergence of these estimates depend on factors such as the choice of window width.

3.3 Robustness with respect to the prior distribution

The question we address is: suppose the prior distribution is G and belonging to a class $\mathcal{G}$, but in attempting to estimate G it is taken to belong to a class $\mathcal{G}^*$. Suppose that the estimating procedure is such that a $\hat{G}^*$ is obtained as an estimate of G^* which is, in some defined sense, the closest member of $\mathcal{G}^*$ to G. Then $\delta_{\hat{G}^*}$ is an estimate of δ_{G^*}, the latter being an approximation to δ_G. Can something be said about the magnitude of $\Delta = W(\delta_{G^*}) - W(\delta_G)$?

Before attempting to answer the question we note that the notion of Δ-asymptotic optimality was used in section 3.2.2; obviously one would say that an EB estimation procedure is relatively robust if it guarantees Δ-optimality with Δ small. The smallness of Δ is relative to $W(\delta_G)$. The notion of Δ-asymptotic optimality is used also in connection with procedures such as linear EB estimators, based on linear Bayes estimators. In a given problem there is not necessarily a class $\mathcal{G}^*$ such that δ_{G^*} is linear in x. Thus, while the ideas of robustness and Δ-asymptotic optimality have much in common they also have to do with slightly different facets of EB estimation.

In section 3.1 we argued in favour of relatively non-diffuse priors in the context of EB estimation, and shall adopt that approach here. We begin by assuming, as we shall do throughout this chapter, that Λ has

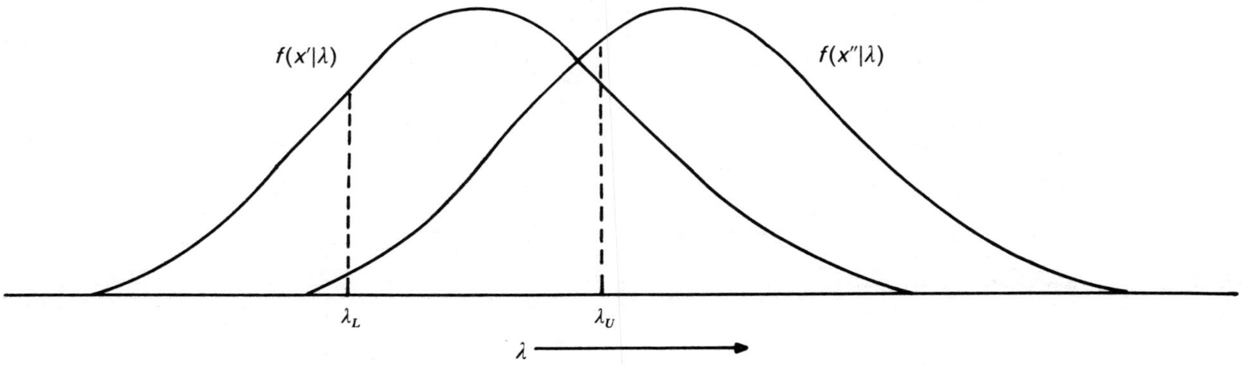

Fig. 3.1

finite variance and that we can set two limits λ_L, λ_U such that $P\{\lambda_L < \Lambda < \lambda_U\} = 1 - \varepsilon$ with ε small. Crudely we can now regard calculation of the Bayes estimate according to formula (3.1.1) as averaging $\lambda f(x|\lambda)$ and $f(x|\lambda)$ w.r.t. to $G(\lambda)$ truncated over a window whose endpoints are λ_L and λ_U.

Assume that

1. G^* and G have identical first, second and third moments; and
2. over the interval (λ_L, λ_U) it is possible to approximate $f(x|\lambda)$ by a quadratic for every x. In Fig. 3.1 $f(x|\lambda)$ is depicted for two values of x, x' and x''. For each of these the approximation of $f(x|\lambda)$ by a quadratic in λ would be quite reasonable. Thus

$$f(x|\lambda) \simeq A_{x0} + A_{x1}\lambda + A_{x2}\lambda^2 = \check{f}(x|\lambda)$$

and

$$\check{\delta}_G(x) = \frac{\int \lambda \check{f}(x|\lambda)dG(\lambda)}{\int \check{f}(x|\lambda)dG(\lambda)} \simeq \delta_G(x).$$

Now, by assumption 1 of equality of the first three moments,

$$\delta_{G^*}(x) \simeq \check{\delta}_{G^*}(x) = \check{\delta}_G(x) \simeq \delta_G(x).$$

In these conditions Δ could be quantified in terms of the maximum difference between $f(x|\lambda)$ and $\check{f}(x|\lambda)$, but the discussion shows qualitatively that a high degree of robustness w.r.t. the choice of G^* can be expected with some modest regularity requirements of $f(x|\lambda)$ as a function of λ.

Example 3.3.1 Let $X \overset{d}{=} N(\lambda, 1)$, $\Lambda \overset{d}{=} N(0, 1/9)$ and consider three classes of distributions $\mathscr{G}_r^*$, $r = 1, 2, 3$ from which an approximation for G might be chosen.

$$\mathscr{G}_1^*\text{:uniform; } \mathscr{G}_2^*\text{:triangular; } \mathscr{G}_3^*: g(\lambda) = (1/\sigma)\exp(-|\lambda - \mu|/\sigma).$$

Then, adjusting the parameters so that the first two moments are the same as those of the true G,

$$G_1^* \text{ is } U\left(-\frac{1}{\sqrt{3}}, \frac{1}{\sqrt{3}}\right)$$

$$G_2^* \text{ is triangular } (-\sqrt{2/3},\ +\sqrt{2/3})$$

$$G_3^* \text{ has p.d.f. } g_3^*(\lambda) = (3/\sqrt{2})\exp\{-|3\sqrt{2}\lambda|\}.$$

Typical results are shown in Table 3.1.

Table 3.1

x	$\delta_G(x) = x/10$	$\delta_{G_1^*}(x)$	$\delta_{G_2^*}(x)$	$\delta_{G_3^*}(x)$
0	0·0	0	0	0
1	0·1	0·1539	0·0963	0·0903
2	0·2	0·2783	0·1873	0·2024
3	0·3	0·3649	0·2683	0·3631
4	0·4	0·4219	0·3384	0·6291
$W(\ )$	0·1000	0·1011	0·1000	0·1001

These results, especially the values of $W(\cdot)$ for the different approximations, support the expectations of reasonable robustness w.r.t. G.

Similar robustness studies have been reported by Rubin (1977). Some more details of approximations of Bayes estimators by estimators of type δ_{G*} will be given in later sections dealing with particular data distributions. For further literature on robustness in this context the reader is referred to Berger (1986).

3.4 Simple EB estimates

3.4.1 *Introduction*

Conceptually the most direct way of obtaining an EB estimate is by constructing an estimate, $\hat{G}$, of the prior G and then replacing G by $\hat{G}$ in the formula for δ_G. In special cases it turns out that δ_G can be estimated without explicitly estimating G. The Poisson example of section 1.9 is a case in point. An apparent advantage of adopting such a procedure, where possible, is that it is distribution free w.r.t. G. No assumption is needed about the class of distributions to which G might belong, or in which a good approximation to G might be found. A disadvantage of these estimators is that they are typically not smooth.

We shall refer to all EB estimates which are derived without explicit estimation of G as **simple** EBEs. Simple EBEs are obtainable whenever the Bayes estimate can be expressed in terms of the probabilities or density of the marginal distribution F_G, or transforms of these.

3.4.2 *The discrete exponential family*

Let

$$p(x|\lambda) = \lambda^x B(x) \exp\{A(\lambda)\}, \qquad x = 0, 1, 2, \ldots \tag{3.4.1}$$

This is a somewhat special form of the exponential family of discrete distributions. The Poisson distribution can be written as in (3.4.1) with $B(x) = 1/x!$, $A(\lambda) = -\lambda$, but in its natural parametrization the binomial distribution does not take the form (3.4.1).

Substituting $p(x|\lambda)$ in (3.1.1) gives

$$\delta_G(x) = \{B(x)/B(x+1)\} p_G(x+1)/p_G(x) \tag{3.4.2}$$

as the Bayes point estimate of λ. Now, if $f_n(x)$ of the past observations $x_1, x_2, \ldots, x_n$ have the value x we can estimate $p_G(x)$ and $p_G(x+1)$ by $(1 + f_n(x))/(n+1)$ and $f_n(x+1)/(n+1)$ giving the EBE

$$\delta_n(x) = \{B(x)/B(x+1)\} f_n(x+1)/\{1 + f_n(x)\}. \tag{3.4.3}$$

It is a simple EBE because $\delta_G(x)$ depends only on the marginal probabilities $p_G(x)$ and $p_G(x+1)$ which can be estimated as indicated above.

The joint distribution of $f_n(x)$ and $f_n(x+1)$ is trinomial, as noted in Example 3.2.3, and $\delta_n(x)$ is clearly a consistent estimate of $\delta_G(x)$. Therefore $\delta_n(x)$ is a.o. (E) in the slightly restricted sense of the argument in section 3.2. Moreover, calculation of $E_n W(\delta_n)$ is in principle straightforward, although it might be tedious in special cases. Calculations can be simplified somewhat by using

$$E_n\left\{\frac{f_n(x+1)}{1+f_n(x)}\right\} = \frac{q_{x+1}}{q_x}\{1 - (1-q_x)^n\}$$

$$E_n\left\{\frac{f_n(x+1)}{1+f_n(x)}\right\}^2 = n(n-1)q_{x+1}^2 \sum_{r=0}^{n-2} \frac{1}{(r+1)^2} B(n-2, q_x, r) + nq_{x+1} \sum_{r=0}^{n-1} \frac{1}{r+1} B(n-1, q_x, r)$$

where $q_x = p_G(x)$, $B(n, \theta, r) = \binom{n}{r} \theta^r (1-\theta)^{n-r}$, $r = 0, 1, 2, \ldots, n$.

For large n one can use the approximation

$$E_n\left\{\frac{f_n(x+1)}{1+f_n(x+1)} - \frac{p_G(x+1)}{p_G(x)}\right\}^2 \simeq \{p_G(x+1)p_G(x) + p_G^2(x+1)\}/\{np_G^3(x)\}. \tag{3.4.4}$$

Some standard examples of distributions of the type (3.4.1) are the Poisson, geometric, negative binomial, as given in section 1.3. Numerical results for $E_n W(\delta_n)$ in some special cases are given in section 3.7.

3.4.3 *The continuous exponential family*

Consider the special form of the continuous exponential family of distributions with p.d.f.

$$f(x|\lambda) = \exp\{\lambda A(x) + B(\lambda) + C(x)\}, \tag{3.4.5}$$

for which

$$\{\partial f(x|\lambda)/\partial x\}/f(x|\lambda) = \lambda A'(x) + C'(x)$$

and

$$\lambda = \{1/A'(x)\}[\{\partial f(x|\lambda)/\partial x\}/f(x|\lambda) - C'(x)].$$

Substituting this form of λ in (3.1.1), assuming that we can write $f'_G(x) = \int\{\partial f(x|\lambda)/\partial x\}dG(\lambda)$,

$$\delta_G(x) = \{1/A'(x)\}\{f'_G(x)/f_G(x) - C'(x)\}. \tag{3.4.6}$$

As an important special case recall Example 1.3.7, the normal distribution where $A(x) = x/\sigma^2$, $B(x) = -\lambda^2/2\sigma^2$, $C(x) = -x^2/2\sigma^2 + \ln(\sigma\sqrt{2\sigma})$, giving

$$\delta_G(x) = x + \sigma^2 f'_G(x)/f_G(x).$$

Now let $\hat{f}_n(x)$ and $\hat{f}'_n(x)$ be estimates of the density $f_G(x)$ and its derivative $f'_G(x)$. These estimates are obtainable as explained in section 3.4.6 as functions of the past x-observations without any reference to the possible form of G. Consequently

$$\hat{\delta}_n(x) = \{1/A'(x)\}\{\hat{f}'_n(x)/\hat{f}_n(x) - C'(x)\} \tag{3.4.7}$$

is a simple EBE of λ. The estimators $\hat{f}_n(x)$ and $\hat{f}'_n(x)$ are somewhat more complicated than the straightforward estimators of probabilities $p_G(x)$ in the discrete case. But, if they are consistent, asymptotic optimality can be established as in section 3.2. Rates of convergence of $E_n W(\hat{\delta}_n)$ have been studied by Lin (1975), and found to be generally slower than $1/n$.

3.4.4 *General construction of simple EB estimators*

Two examples are useful for motivating the approach of this section.

Example 3.4.1 $X \overset{d}{=} \mathrm{Bin}(n, \lambda)$, so that

$$p(x|\lambda) = \binom{n}{x} \lambda^x (1-\lambda)^{n-x}$$
$$= \exp\left\{ \ln\binom{n}{x} + x \ln\left(\frac{\lambda}{1-\lambda}\right) + n \ln(1-\lambda) \right\},$$

verifying that this distribution is also a member of the discrete exponential family. To put it in the form (3.4.1) we have to reparametrize setting $\theta = \lambda/(1-\lambda)$, in which case we can obtain a simple EBE of θ, but not of λ. In other words we can obtain a direct estimate of $E(\Theta|x)$, but not of $E(\Lambda|x)$.

Example 3.4.2 Let $f(x|\lambda) = (1/\lambda)e^{-x/\lambda}$, $x \geqslant 0$, and 0 otherwise. Then

$$\int_0^x f(u|\lambda)du = 1 - e^{-x/\lambda}$$

or

$$e^{-x/\lambda} = 1 - \int_0^x f(u|\lambda)du.$$

Substituting in (3.1.1) we obtain

$$\delta_G(x) = \{1 - F_G(x)\}/f_G(x). \tag{3.4.8}$$

In order to use (3.4.5) we would have to reparametrize to $\theta = 1/\lambda$ and the Bayes estimator of θ is expressed in terms of f_G and f'_G. The form (3.4.7) has two advantages, one that in some circumstances λ rather than θ may be the natural parameter to estimate. Another is that estimation of F_G is simpler than estimation of f'_G.

The two examples suggest that it may be worthwhile to contemplate more general operations on $F(x|\lambda)$ in order to construct EBEs. Also, one need not necessarily consider estimating λ directly. A function of λ, or even a function of λ and x could be estimated.

Let T_1 and T_2 be operators on distribution functions such that $T_r F(x|\lambda) = h_r(x|\lambda) f(x|\lambda)$, $r = 1, 2$. Then

$$\alpha(x) = \frac{T_1 F_G(x)}{T_2 F_G(x)} = \frac{E_{\Lambda,x}\{h_1(x, \Lambda)\}}{E_{\Lambda,x}\{h_2(x, \Lambda)\}} \tag{3.4.9}$$

where $E_{\Lambda,x}$ is expectation w.r.t. the posterior distribution of Λ given x

(see Maritz and Lwin, 1975). The method of implementation of (3.4.8) in simple EB estimation relies on the fact that $\alpha(x)$ can be estimated from past observations because both $T_1 F_G(x)$ and $T_2 F_G(x)$ are just properties of the marginal X-distribution.

There are two problems, one to do with the details of estimating $\alpha(x)$, but as we shall show by example, the greater freedom of choice of T_r can make estimation of $\alpha(x)$ relatively straightforward. The other problem can be discussed somewhat more effectively if we simplify (3.4.8) by letting T_2 be $\partial/\partial x$ so that $T_2 F(x|\lambda) = f(x|\lambda)$, i.e. $h_2(x, \lambda) = 1$. Then

$$\alpha(x) = E_{\Lambda,x}\{h_1(x, \Lambda)\}. \tag{3.4.10}$$

So, the estimate of $\alpha(x)$ provides an estimate of the posterior mean of $h_1(x, \Lambda)$ whereas we actually want the posterior mean of Λ, i.e. $E_{\Lambda,x}(\Lambda) = \delta_G(x)$.

A first approximation for $\delta_G(x)$ can be taken as the solution of

$$\alpha(x) = h_1\{x, \delta_G(x)\}$$

and an improved approximation may be obtained as the solution of

$$\alpha(x) = h_1\{x, \delta_G(x)\} + \frac{\sigma_x^2}{2}\left\{\frac{\partial^2 h_1(x, \lambda)}{\partial \lambda^2}\right\}_{\lambda = \delta_G(x)} \tag{3.4.11}$$

where σ_x^2 is the posterior variance of Λ. In EB estimation $\alpha(x)$ is, of course, replaced by its estimate $\hat{\alpha}(x)$.

Example 3.4.3 Suppose we let $T_1 B(x) = \int_{-\infty}^{x} u dB(u)$, $T_2 B(x) = B(x)$. Then $\alpha(x) = \int_{-\infty}^{x} u dF_G(u)/F_G(u)$, the mean of the X_G distribution right-truncated at x. A natural estimator of $\alpha(x)$ is obtained from order statistics $x_{(i)}$, $i = 1, 2, \ldots, n$ as

$$\hat{\alpha}(x_{(r)}) = \sum_{i=1}^{r} x_{(i)}/r.$$

This example shows that density estimation problems can be avoided, although other problems, relating to the use of (3.4.10) may be introduced.

Examples of other operators T_1 and T_2 applied to certain types of distributions are to be found in Rutherford and Krutchkoff (1969), Nichols and Tsokos (1972) and Cressie (1982).

3.4.5 *Smoothing of simple EB estimators*

Most Bayes estimators, $\delta_G(x)$, are smooth functions of x in some obvious sense. Whether X is continuous or discrete, $\delta_G(x)$ is usually monotonic in X, and if X is continuous $\delta_G(x)$ is usually also differentiable w.r.t. x. For example, in the Poisson X–gamma Λ case $\delta_G(x) = (\beta + x)/(\alpha + 1)$ as in (1.3.7), and in the normal X–normal Λ case $\delta_G(x) = (x/\sigma^2 + \mu_G/\sigma_G^2)/(1/\sigma^2 + 1/\sigma_G^2)$. But, as we have pointed out before, $\delta_n(x)$ need not be smooth. For example, in Table 3.2, values of $\delta_n(x)$ are shown for a particular set of data generated by a Poisson X–gamma Λ model. A plot of $\delta_n(x)$ against x would produce an irregular graph as in Fig. 1.1.

Since $\delta_G(x)$ is generally smooth it seems sensible, when a non-smooth $\delta_n(x)$ has been obtained, to smooth it by fitting a straight line or some other curve through observed $\delta_n(x)$ values. Methods for such direct smoothing of simple EBEs, $\delta_n(x)$, have received little attention, and we limit this discussion to just one suggestion, put forward in Maritz (1967) for discrete X in the exponential family.

Suppose that $\delta_G(x) = A + Bx$. Then fitting of a straight line $\delta_n^*(x) = A^* + B^*x$ would be indicated by the data. A method of fitting is suggested by the following:

$$\begin{aligned}\sum\{\delta_n(x) - \delta_G(x)\}^2 p_G(x) &\\ = \sum\{\delta_n(x) - \delta_n^*(x)\}^2 p_G(x) &+ \sum\{\delta_n^*(x) - \delta_G(x)\}^2 p_G(x)\\ &+ 2\sum(\delta_n(x) - A^* - B^*x)\{(A^* - A) + (B^* - B)x\}p_G(x).\end{aligned} \tag{3.4.12}$$

By setting the third term in (3.4.12) equal to 0 we can ensure that $W(\delta_n^*) \leqslant W(\delta_n)$, and this can be done by letting A^* and B^* be the solutions of

$$\begin{bmatrix} 1 & \sum x p_G(x) \\ \sum x p_G(x) & \sum x^2 p_G(x) \end{bmatrix} \begin{bmatrix} A^* \\ B^* \end{bmatrix} = \begin{bmatrix} \sum \delta_n(x) p_G(x) \\ \sum x \delta_n(x) p_G(x) \end{bmatrix}. \tag{3.4.13}$$

Clearly (3.4.13) cannot be applied directly in practice since the $p_G(x)$ values are not known. Either estimates using the observed frequencies $f_n(x)$ can be used to replace $p_G(x)$ values, or smoothed approximations can be used in an iterative scheme. Begin by fitting a straight line by eye or using estimates of $p_G(x)$ based on $f_n(x)$, $x = 0, 1, 2, \ldots$. Let this line be $y_x^{(0)} = A^{(0)} + B^{(0)}x$. Then using formula (3.4.2) in reverse, calculate numbers $p^0(x)$ proportional to the estimated $p_G(x)$

according to

$$p^0(1) = p^0(0)y_0^{(0)}, \qquad p^0(2) = \tfrac{1}{2}p^0(0)y_0^{(0)}y_1^{(0)}, \text{ etc.} \tag{3.4.14}$$

Substituting $p^0(x)$ for $p_G(x)$ in (3.4.12), new values $A^{(1)}$, $B^{(1)}$ can be calculated, etc.

Another method of fitting is indicated by (3.4.14). Dividing the trial $p^0(x)$ by a suitable factor fitted probabilities, estimates of the $p_G(x)$ are obtained. The likelihood of the past observations can then be calculated as

$$L = \prod_{i=1}^{n} p^{(0)}(x_i), \tag{3.4.15}$$

a function of $A^{(0)}$ and $B^{(0)}$. Applying the maximum likelihood principle, $A^{(0)}$ and $B^{(0)}$ are adjusted to maximize L.

Both methods described for fitting a straight line can be extended in an obvious manner to fit higher-order polynomials.

An important contribution to the idea of smoothing of simple (or any EB) estimators was made by van Houwelingen (1977) for the exponential family of discrete distributions. Starting with an arbitrary estimator $d(x)$, a monotonized version $d^*(x)$ of $d(x)$ is constructed having the property that $R_d^*(\lambda) \leqslant R_d(\lambda)$ for all λ, i.e. that d^* dominates d. Following van Houwelingen let $d(x)$ be a randomized estimator represented by a distribution function $D(a, x)$. Thus if x is observed an estimate a generated by $D(a, x)$ is made.

The construction of the estimator d^* is according to the following steps:

$$P(x|\lambda) = \sum_{y=0}^{x} p(x|\lambda); \qquad P(-1|\lambda) = 0$$

where $p(x|\lambda)$ is given by (3.4.1);

$$\mathscr{L}(a) = \sum_{x} D(a, x)p(x|a)$$

$$D^*(a, x) = \begin{cases} 0 & \text{if } \mathscr{L}(a) < P(x-1|a) \\ \dfrac{\{\mathscr{L}(a) - P(x-1|a)\}}{p(x|a)} & \text{if } P(x-1|a) \leqslant \mathscr{L}(a) \leqslant P(x|a) \\ 1 & \text{if } P(x|a) < \mathscr{L}(a) \end{cases}$$

$$d^*(x) = \int_A a\,dD^*(a, x).$$

van Houwelingen (1977) shows that $d^*(x)$ is monotonic in x and dominates $d(x)$. Particular examples of the application of this procedure to simple EBEs suggest that the improvement in performance of $d^*(x)$ relative to $d_n(x)$ can be quite dramatic. There is, of course, no guarantee that $d^*(x)$ will necessarily be better than some other smooth EBE, but the theoretical, and practical, advantage of van Houwelingen's method is that $d^*(x)$ is guaranteed to be better than the simple EBE $\delta_n(x)$. An example is given in section 3.7.

3.4.6 *Density estimation*

Let $x_1, x_2, \ldots, x_n$ be independent observations on a continuous random variable X with distribution function $F(x)$ and density function $f(x)$. The **kernel** density estimate of $f(x)$ proposed by Parzen (1962) is $f_n(x)$ given by

$$f_n(x) = \frac{1}{nh(n)} \sum_{j=1}^{n} K\left(\frac{x - x_j}{h(n)}\right),$$

where K is typically chosen to be one of the standard density functions, for example, a $N(0, 1)$ density. The divisor $h(n)$ is called the **window width**, and for consistent estimation we must have $h(n) \to 0$ and $nh(n) \to \infty$ as $n \to \infty$.

Density estimation is an important topic in several branches of statistics and much further useful information will be found in Silverman (1986).

3.5 EB estimation through estimating G

3.5.1 *Introduction*

A good deal of the literature on EB estimation has been devoted to simple EBEs, i.e. where no knowledge of G is assumed. In some cases it would seem perfectly reasonable to suppose that the type of G is known, i.e. the family $\mathscr{G}$ of distributions to which G belongs. Then, if G is indexed by a low-dimensional parameter vector, the marginal X-distribution $F_G(x)$ is similarly indexed, its form is known, and estimation of G becomes a standard problem of parameter estimation.

Even if it is agreed that knowledge of the exact parametric form of G will rarely be available there is value in examining such parametric G

EBEs. As we have remarked, they have a place of their own, but aside from that, knowledge of the form of G can perhaps be regarded as the most favourable state in which one can be. It is, then, useful to compare the performance of other EBEs with their parametric G counterparts.

In earlier discussions it has also emerged that Bayes estimators could be approximated by choosing a G^* belonging to a family $\mathscr{G}^*$; see sections 1.12, 2.7, 3.3.2. A particular family of some interest is $\mathscr{G}_k^*$ comprising step-functions with k steps. One of the advantages of obtaining approximate EBEs of this sort is that they are automatically smooth, and some of the problems of simple EBEs are avoided.

3.5.2 *G belonging to a known parametric family*

Suppose that $G(\lambda) = G(\lambda; \alpha, \beta)$ of known form. Then the mixed distribution $F_G(x) = F(x; \alpha, \beta)$. The past observations $x_1, x_2, \ldots, x_n$ now enable one to calculate estimates $\hat{\alpha}, \hat{\beta}, \ldots$ of the parameters $\alpha, \beta, \ldots$ by any one of the standard methods of estimation. The Bayes estimate can be expressed as $\delta_G(x) = \delta(x; \alpha, \beta, \ldots)$ and the EBE is $\hat{\delta}_G(x) = \delta(x; \hat{\alpha}, \hat{\beta}, \ldots)$.

In most standard conditions the usual results for estimation will apply to $\hat{\delta}_G$. If $\hat{\alpha}, \hat{\beta}$ are consistent for α, β in the sense that $\hat{\alpha}, \hat{\beta} \to \alpha, \beta(P)$ as $n \to \infty$ we have asymptotic optimality in the sense of section 3.2. We shall also have $E_n\{\hat{\delta}_G(x) - \delta_G(x)\}^2 = O(1/n)$ in most standard situations, so that $E_n W(\hat{\delta}_G) = W(\delta_G) + O(1/n)$, with possibly some moderate restrictions on $\hat{\delta}_G$ as in section 3.2.

Example 3.5.1 Let $X \overset{\text{d}}{=} U(\lambda - 1, \lambda + 1)$ and $\Lambda \overset{\text{d}}{=} U(A - 1, A + 1)$. Then $\delta_G(x) = (x + A)/2$. The marginal X-distribution has

$$f_G(x) = \begin{cases} (\tfrac{1}{4})(x - A + 2), & A - 2 \leqslant x \leqslant A \\ (\tfrac{1}{4})(A - x + 2), & A \leqslant x \leqslant A + 2 \end{cases}$$

i.e. a triangular density centred at A. An obvious estimator of A is $\bar{A} = \bar{X}$, the sample mean, which is unbiased with variance $2/3n$. So $\hat{\delta}_G(x) = (x + \bar{A})/2$, and

$$E_n \int_{A-2}^{A+2} \{\hat{\delta}_G(x) - \delta_G(x)\}^2 f_G(x) dx = E_n \tfrac{1}{4}(\bar{A} - A)^2 = 1/6n,$$

so, $E_n W(\hat{\delta}_G) = W(\delta_G) + 1/6n$.

In more realistic examples the calculation of $E_n W(\hat{\delta}_G)$ is less straightforward, and compact formulae for $E_n W(\hat{\delta}_G)$ appear to be obtainable only as approximations for large n.

Example 3.5.2 $X \stackrel{d}{=} N(\lambda, 1)$, $\Lambda \stackrel{d}{=} N(\mu_G, \sigma_G^2)$.

(i) σ_G^2 *known:* $X_G \stackrel{d}{=} N(\mu_G, \sigma_G^2 + 1)$ and we estimate μ_G by the sample mean $\bar{x}$ of past observations. Then

$$\hat{\delta}_G(x) = (x + \bar{x}/\sigma_G^2)/(1 + 1/\sigma_G^2), \tag{3.5.1}$$

and

$$E_n \int \left\{ \frac{x + \bar{x}/\sigma_G^2}{1 + 1/\sigma_G^2} - \frac{x + \mu_G/\sigma_G^2}{1 + 1/\sigma_G^2} \right\}^2 f_G(x)dx = \frac{1}{n(\sigma_G^2 + 1)}.$$

(ii) σ_G^2 *unknown*: Since $\text{var}(X_G) = \sigma_G^2 + 1$ we estimate σ_G^2 by $\hat{\sigma}_G^2 = \max(0, s^2 - 1)$, where s^2 is the sample variance of past observations. Now

$$E_n \int \left\{ \frac{x + \bar{x}/\hat{\sigma}_G^2}{1 + 1/\hat{\sigma}_G^2} - \frac{x + \mu_G/\sigma_G^2}{1 + 1/\sigma_G^2} \right\}^2 f_G(x)dx$$

$$= \frac{1}{n} E_n \left(\frac{1}{\hat{\sigma}_G^2 + 1} \right) + E_n \frac{(\hat{\sigma}_G^2 - \sigma_G^2)^2}{(1 + \sigma_G^2)(1 + \hat{\sigma}_G^2)^2}. \tag{3.5.2}$$

For small n evaluation of the r.h.s. of (3.5.2) is numerically fairly straightforward, using the fact that $(n-1)s^2/(1 + \sigma_G^2) \stackrel{d}{=} \chi_{n-1}^2$ and observing the condition $\hat{\sigma}_G^2 = \max(0, s^2 - 1)$. Some results are given in section 3.7.3. For $n \to \infty$, $P(s^2 - 1 < 0) \to 0$ and one can evaluate (3.5.2) approximately by replacing $1 + \hat{\sigma}_G^2$ by s^2 and ignoring the truncation. Then $E_n(1/s^2) = (n-1)/(n-3)(1 + \sigma_G^2)$ and $E(1/s^4) = (n-1)^2/(n-3)(n-5)(1 + \sigma_G^2)^2$, giving

$$E_n W(\hat{\delta}_G) \simeq \left\{ \frac{1}{n} \left(\frac{(n-1)^2}{(n-3)(n-5)} \right) + \frac{2n+6}{(n-3)(n-5)} \right\} \Big/ (1 + \sigma_G^2) + W(\delta_G). \tag{3.5.3}$$

In Examples 3.5.1 and 3.5.2 estimation of the parameters of $F(x; \alpha, \beta, \ldots)$ is straightforward and calculation of $E_n W(\hat{\delta}_G)$ fairly easy. Generally estimation of $\alpha, \beta, \ldots$ will usually be by the method of maximum likelihood or some other standard procedure. If the ML estimates of $\alpha, \beta, \ldots$ are $\hat{\alpha}, \hat{\beta}, \ldots$ then $\delta(x; \hat{\alpha}, \hat{\beta}, \ldots)$ is the MLE of δ_G. For large n one can express $E_n W(\hat{\delta}_G)$ approximately in terms of the elements of the information matrix of estimating $\hat{\alpha}, \hat{\beta}, \ldots$.

3.5.3 *Approximation of G by a parametric distribution*

Suppose that the form of G is unknown and that it is decided to approximate it by a member of a family $\mathscr{G}^*$ of distributions. According to the discussion of section 2.7.1 the best approximating $G^*(\lambda; \alpha, \beta, \ldots)$ is that member of $\mathscr{G}^*$ for which

$$\int \ln f_{G^*}(x; \alpha, \beta, \ldots) dF_G(x)$$

is maximized. The empirical version of G^* is $G^*(\lambda; \hat{\alpha}, \hat{\beta}, \ldots)$ where $\hat{\alpha}, \hat{\beta}, \ldots$ are obtained by maximizing

$$\sum \ln f_{G^*}(x_i; a, b, \ldots)$$

w.r.t. $a, b, \ldots$. In other words, we act as if the marginal X-distribution has density $f_{G^*}(x; \alpha, \beta, \ldots)$ and $\alpha, \beta, \ldots$ are estimated by the ML method. The approximate EBE δ_{G^*} is given by (3.1.1) with G replaced by $G^*(\lambda; \hat{\alpha}, \hat{\beta}, \ldots)$; it can be written $\delta_{G^*}(x) = \delta^*(x; \hat{\alpha}, \hat{\beta}, \ldots)$.

As $n \to \infty$, $\delta^*(x; \hat{\alpha}, \hat{\beta}, \ldots) \to \delta^*(x; \alpha, \beta, \ldots)$ and the goodness of $\delta^*(x; \hat{\alpha}, \hat{\beta}, \ldots)$ will depend on the goodness of the approximation of $\delta_G(x)$ by $\delta^*(x; \alpha, \beta, \ldots)$. The questions of Δ-asymptotic optimality and robustness discussed in sections 3.2 and 3.3 are relevant here. As we have seen in section 3.5.1, calculation of $E_n W(\hat{\delta}_G)$ can be difficult, and the same applies here. It is possible to obtain large sample formulae for $E_n W(\hat{\delta}^*)$.

Let us consider the case of G^* depending on only one parameter, i.e. $G^*(\lambda; \alpha)$. Modification for more than one parameter is fairly straightforward. The estimate $\hat{\alpha}$ of α is the solution $a = \hat{\alpha}$ of

$$S(\mathbf{x}; a) = \sum_i \frac{\partial \ln f_{G^*}(x_i; a)}{\partial a} = 0. \tag{3.5.4}$$

For n large,

$$\operatorname{var}(\hat{\alpha}) \simeq \frac{\operatorname{var}\{S(\mathbf{X}; \alpha)\}}{\left\{\dfrac{\partial ES(\mathbf{X}; a)}{\partial a}\right\}^2_{a=\alpha}}, \tag{3.5.5}$$

where α is the solution of

$$\int \frac{\partial \ln f_{G^*}(x; a)}{\partial a} f_G(x) dx = 0.$$

For a justification of (3.5.5) see, for example, Maritz (1981). Now, write $h^*(x;\alpha) = \partial \ln f_{G^*}(x;\alpha)/\partial\alpha$. Then

$$\operatorname{var}\{S(\mathbf{X},\alpha)\} = n\left[\int \{h^*(x;\alpha)\}^2 f_G(x) - \left\{\int h^*(x;\alpha) f_G(x)\right\}^2\right], \tag{3.5.6}$$

and

$$\left\{\frac{\partial ES(\mathbf{X},a)}{\partial a}\right\}_{a=\alpha} = n\left\{\int \frac{\partial h^*(x,a)}{\partial a} f_G(x)dx\right\}_{a=\alpha}. \tag{3.5.7}$$

Finally,

$$\operatorname{var}\{\delta^*(x;\hat{\alpha})\} \simeq \left\{\frac{\partial \delta^*(x,a)}{\partial a}\right\}^2_{a=\alpha} \operatorname{var}(\hat{\alpha}),$$

which can be used to compute the value of $E_n W(\hat{\delta}_{G^*})$ approximately.

From the results given above we obtain

$$E_n W(\hat{\delta}_{G^*}) = W(\delta_G) + \Delta + O(1/n).$$

3.5.4 *Step-function approximation of G*

The idea of approximating G by a step-function is discussed in section 2.7. Here we shall consider only the approximating $G_k(\lambda)$ having k steps of equal size $1/k$ at $\lambda_1, \lambda_2, \ldots, \lambda_k$. A method of determining G_k is described in section 2.9.

When using G_k in EB estimation it seems desirable to keep k fairly small especially for small n. Numerical examples have indicated that very small values like $k = 3$ can give quite adequate approximation of the Bayes estimator. This is in accord with the discussion of robustness of Bayes and EB estimators in section 3.3. In particular, if G is symmetrical it is clearly possible to have exact agreement of the first three moments of G and G_k.

The approximate Bayes estimator given by G_k is

$$\psi_k(x) = \sum_{j=1}^{k} \lambda_j f(x|\lambda_j) \Big/ \sum_{j=1}^{k} f(x|\lambda_j) \tag{3.5.8}$$

and its goodness is measured by $W(\psi_k) = W(\delta_G) + \Delta_k$. As we have seen before, we can obtain a sequence $\Delta_k \to 0$ as $k \to \infty$. Numerical values of $W(\psi_k)$ for certain special cases are given in section 3.7.

In EB applications $\lambda_1, \lambda_2, \ldots, \lambda_k$ are estimated from the past data. This can be done, for example, by the distance-minimizing method which in effect treats the x-values as having been generated by $F_{G_k}(x)$

and using the ML method. Other methods implying the use of other distance measures can be used, for example the method of moments, which is easily implemented when k is small. According to the pseudo-ML method $\hat{\lambda}_1, \hat{\lambda}_2, \ldots, \hat{\lambda}_k$ are obtained by maximizing $\sum_{i=1}^{n} \ln\{(1/k)\sum_{j=1}^{k} f(x_i|\lambda_j)\}$ w.r.t. $\lambda_1, \lambda_2, \ldots, \lambda_k$. The EBE, $\hat{\psi}_k(x)$, is given by (3.5.8) with $\hat{\lambda}_j$ replacing λ_j. We can also write

$$\sum_{i=1}^{n} \ln\left\{\frac{1}{k}\sum_{j=1}^{k} f(x_i|\lambda_j)\right\} = \int \ln\left\{\frac{1}{k}\sum_{j=1}^{k} f(x|\lambda_j) dF_n(x)\right\} \qquad (3.5.9)$$

and note that $F_n(x) \to F_G(x), (P)$, for every x as $n \to \infty$. Hence, by the arguments of section 2.7 $\hat{\lambda}_j \to \lambda_j$, (P), as $n \to \infty$. In other words, $\hat{\lambda}_j$ is a consistent estimate of λ_j, and $\hat{\psi}_k(x)$ is a consistent estimate of $\psi_k(x)$. Following the arguments in section 3.5.2 we have

$$E_n W(\hat{\psi}_k) = W(\delta_G) + \Delta_k + O(1/n).$$

3.6 Linear EB estimation

An outline of linear Bayes estimation is given in section 1.12.1 and (1.12.2) summarizes the method of calculating a linear Bayes estimator. In the l.h.s. matrix of (1.12.2) we have elements

$$E(X_G^r) = \int E(X^r|\lambda) dG(\lambda), \qquad r = 1, 2,$$

both of which can obviously be estimated simply by the first two sample moments of the observations $x_1, x_2, \ldots, x_n$.

In general estimation of the elements of the r.h.s of (1.12.2) is more difficult although it can be easy in special cases. Let us suppose we can find functions $U_1(X), U_2(X)$ such that

$$\begin{aligned} &\int U_1(x) f(x|\lambda) dx = \lambda \\ &\int U_2(x) f(x|\lambda) dx = \mathscr{L}(\lambda) \qquad \text{where } \mathscr{L}(\lambda) = \lambda E(X|\lambda). \end{aligned} \qquad (3.6.1)$$

Then the r.h.s. elements of (3.6.1) are estimated by

$$\sum_{i=1}^{n} U_r(x_i)/n, \qquad r = 1, 2.$$

Clearly, solution of the integral equations (3.6.1) is non-trivial, but certain forms of $f(x|\lambda)$ do lead to straightforward estimation. Very often the distribution $f(x|\lambda)$ can be parametrized so that $E(x|\lambda)=\lambda$, giving $\mathscr{L}(\lambda)=\lambda^2$ so that the choice of U_1 and U_2 is sometimes obvious.

Example 3.6.1 $X \stackrel{d}{=} \text{Poisson}(\lambda)$. In this case

$$\mathscr{L}(\lambda)=\lambda E(x|\lambda)=\lambda^2$$

and we can take $U_1(X)=X, U_2(X)=X^2-X$. Straightforward calculations show that the linear Bayes estimator is

$$\delta_G(x)=[\{E(X_G)\}^2+x\{\text{var}(X_G)-E(X_G)\}]/\text{var}(X_G), \quad (3.6.2)$$

with the moments of X_G estimated directly from the observations $x_1,\ldots,x_n$.

3.7 EB estimation for special univariate distributions: one current observation

In each of the examples of this section the natural conjugate distribution is used to generate numerical data, and in the parametric G EB estimation, for the most part only the family of conjugate priors is considered. This is done partly for simplicity, and partly because the conjugate families are thought to be flexible enough to provide good approximations to the actual G in many applications.

3.7.1 The Poisson distribution

We begin this section with an example, based on the Poisson distribution, illustrating some of the methods of constructing EBEs, and the calculation of $W(\cdot)$ values. Table 3.2 shows in the second column observed frequencies $f_n(x)$ generated by a Poisson distribution mixed by a gamma(α,β) prior G, with p.d.f. given as in Example 1.3.2. For these data $\bar{x}=5{\cdot}00$, $s^2=m_2=9{\cdot}12$.

(a) Parametric G, $\mathscr{G}$ known to be gamma(α,β): estimating α,β by the method of moments using $E(X_G)=\beta/\alpha$, $\text{var}(X_G)=\beta/\alpha+\beta/\alpha^2$ we obtain $\bar{\alpha}=1{\cdot}214$, $\bar{\beta}=6{\cdot}068$.

(b) Parametric G, $\mathscr{G}$ known: ML estimation of α,β:

$$\hat{\alpha}=1{\cdot}310, \qquad \hat{\beta}=6{\cdot}548$$
$$\hat{\delta}_G(x)=2{\cdot}835+0{\cdot}433x.$$

Table 3.2 *Observed frequencies $f_n(x)$ generated by the Poisson-gamma model when G is gamma(2, 10), Columns headed (c), (d), (f), (g) are EBE's derived by the methods given in the text*

x	$f_n(x)$	$p_G(x)$	(c) Moments	(d)	(f)	(g)	$\delta_G(x)$
0	—	0·017	3·46	3·00	1·34	3·42	3·33
1	3	0·058	3·51	4·00	2·20	3·42	3·67
2	8	0·106	3·60	3·33	2·80	3·42	4·00
3	10	0·141	3·78	0·73	3·33	3·42	4·33
4	2	0·153	4·12	18·33	3·66	5·00	4·67
5	11	0·143	4·66	2·00	4·75	5·00	5·00
6	4	0·119	5·40	5·60	5·62	5·00	5·33
7	4	0·091	6·15	0·00	6·40	5·46	5·67
8	—	0·064	6·74	9·00	7·88	6·99	6·00
9	1	0·043	7·12	10·00	8·75	7·75	6·33
10	2	0·027	7·30	14·67	9·29	7·95	6·67
11	4	0·016	7·41	0·00	9·79	7·95	7·00
12	—	0·010	7·45	13·00	10·16	8·18	7·33
13	1	0·005	7·48	0·00	10·76	8·18	7·67
14	—	0·003	7·50	0·00	11·58	8·18	8·00
15	—	0·002	7·50	0·00	12·32	8·18	8·33
16	—	0·001	7·50	0·00	12·84	8·18	8·67
$W(\ )$			1·90	41·04	3·38	2·08	1·67

(c) Step-function approximation of G: three methods were employed for estimating $\lambda_1, \lambda_2, \lambda_3$ in G_3. The ML method requires maximizing

$$L_n(\boldsymbol{\lambda}) = \sum_x f_n(x) \ln \left\{ \tfrac{1}{3} \sum_{j=1}^{3} \exp(-\lambda_j)\lambda_j^x / x! \right\}$$

w.r.t. $\lambda_1 \leqslant \lambda_2 \leqslant \lambda_3$. The resulting estimates are

$$\hat{\lambda}_1 = 3{\cdot}57, \quad \hat{\lambda}_2 = 3{\cdot}58, \quad \hat{\lambda}_3 = 8{\cdot}14.$$

A minimum χ^2 method described in Maritz (1967), using grouping of the frequencies in Table 3.2, gave

$$\check{\lambda}_1 = 3{\cdot}3, \quad \check{\lambda}_2 = 3{\cdot}6, \quad \check{\lambda}_3 = 7{\cdot}5.$$

The method of moments was used, choosing the λ's so as to equate the first and second moments of the observed and fitted X_G-distributions while minimizing the absolute difference of third moments:

$$\bar{\lambda}_1 = 3{\cdot}50, \quad \bar{\lambda}_2 = 3{\cdot}59, \quad \bar{\lambda}_3 = 7{\cdot}87.$$

The results in column (c) of Table 3.2 are based on the minimum χ^2 estimates.

(d) The simple EBE

$$\delta_n(x) = (x+1)f_n(x+1)/\{1+f_n(x)\}.$$

(e) A smoothed version of (d), fitting a straight line by the weighted least squares method motivated by (3.4.13). Starting with an eye-fitted line

$$y^{(0)} = 2{\cdot}65 + 0{\cdot}465x$$

one iteration gave the result

$$y^{(1)} = 2{\cdot}15 + 0{\cdot}760x.$$

(f) A monotonized version of $\delta_n(x)$ according to the method of van Houwelingen (1977) described in section 3.4.5.

(g) Monotonic ordinate $\hat{\eta}_x$, $x = 0, 1, 2, \ldots$ fitted by the ML method suggested by (3.4.15): expressing $p^0(x_i)$ values in terms of trial ordinates η_x^0 and inspecting the partial derivatives of $\ln L$ w.r.t. the η's shows that η_x should be constant for all x such that $\sum_{r=0}^{x-1} f_n(r) = 0$ and $\sum_{r=x+1}^{\infty} f_n(r) = 0$. Then maximize $\ln L$ subject to $\hat{\eta}_0 \leqslant \hat{\eta}_1 \ldots$.

(h) Linear EB: using (3.6.2) the linear EBE is

$$\delta(\bar{w}_0, \bar{w}_1) = 2{\cdot}741 + 0{\cdot}452x,$$

Table 3.3 *Expected losses W( ) for EB estimators based on the data of Table 3.1 and for Bayes and best non-Bayes estimators*

Method	$W(\ )$
Parametric G: moments	1·77
: ML	1·74
G_3: moments	1·90
δ_n	41·04
δ_n smoothed: straight line	1·74
monotonized (van Houwelingen)	3·38
Non-decreasing ordinates: ML	2·08
Best non-Bayes: $T = x$	5·00
Bayes	1·67

which is the same as $\bar{\delta}_G(x)$, as it should be in this special case.

The values of some of the estimators are listed in Table 3.2. Unlisted values for other finite approximation types are close to those tabulated, while other unlisted values are readily calculated. The table also shows $p_G(x)$ and $\delta_G(x)$, to be used in computing $W(\cdot)$ values. The values of W for various estimators are given in Table 3.3.

All of the EBEs except δ_n have much better performance than T; indeed their $W(\cdot)$ values are remarkably close to $W(\delta_G)$.

Obtaining analytic results for $E_n W(\text{EB})$, where EB here stands for any EBE, in particular any one of (a)–(g), seems virtually impossible, except as approximations for n large. Even then formulae would be excessively unwieldy. One needs for each x an expression for $\text{var}_n(\text{EB})$, in terms of x, and then a summation of such terms after multiplication by $p_G(x)$. For example, if we take estimator (a), the gamma G case, we can write

$$\delta_G(x) = \frac{\mu_1'^2 + x(\mu_2' - \mu_1'^2 - \mu_1')}{(\mu_2' - \mu_1'^2)}, \qquad (3.7.1)$$

where $\mu_r' = E(X_G^r)$, $r = 1, 2$. The EBE $\bar{\delta}_g(x)$ is given by replacing μ_r' by m_r' where m_r' is the rth sample moment of X_G. Obviously, an expression in terms of α, β, n, x can be written down for $\text{var}\{\bar{\delta}_G(x)\}$, using standard methods of approximation. Suppose we put $\text{var}\{\bar{\delta}_G(x)\} = (1/n)V(\alpha, \beta, x)$. Then

$$E_n W(\bar{\delta}_G) \simeq W(\delta_G) + \frac{1}{n}\sum_x V(\alpha, \beta, x) p_G(x) \qquad (3.7.2)$$

and such a formula could be used to obtain numerical values for various α, β, n. The results given in Table 3.4 were, however, obtained by simulation because such a method is in any event needed for small n.

A general comment on Table 3.4 is that the Bayes estimator becomes, relative to T, less advantageous as the ratio $\text{var}(\Lambda)/\{E\,\text{var}(X|\Lambda)\}$ increases. In the Poisson–gamma case this ratio is $1/\alpha$. Of course, the same is true of EBEs. We may also note that the combination of G_3 and MM seems to provide a notably poor EBE for $n = 10$. Finally, allowing for the rather small $n = 10$ in Table 3.4 the order of magnitude of the ratios $[E_{10}\{W(\text{EB})\} - W(\delta_G)]/[E_{50}\{W(\text{EB})\} - W(\delta_G)]$ seems acceptably close to 5, as would be expected according to approximations such as given in (3.7.2).

Table 3.4 *Estimates of $E_n W(\cdot)$ for the following EB estimators in the Poisson-gamma (α, β) case:*
(a) Parametric G, moments
(b) Parametric G, ML
(c) Finite approximation, k = 3, moments
(c) Finite approximation, k = 3, ML
(g) Direct smoothing, non-decreasing ordinates

		$\alpha=2, \beta=10$	$\alpha=2, \beta=2$	$\alpha=5, \beta=25$	$\alpha=5, \beta=5$
$W(\delta_G)$		1·67	0·33	0·83	0·17
$W(T)$		5	1	5	1
	(a)	$1{\cdot}86 \pm 0{\cdot}02$	$0{\cdot}380 \pm 0{\cdot}005$	$0{\cdot}99 \pm 0{\cdot}02$	$0{\cdot}204 \pm 0{\cdot}004$
	(b)	$1{\cdot}86 \pm 0{\cdot}02$	$0{\cdot}377 \pm 0{\cdot}005$	$0{\cdot}99 \pm 0{\cdot}02$	$0{\cdot}205 \pm 0{\cdot}004$
$n=50$	(c)	$1{\cdot}99 \pm 0{\cdot}04$	$0{\cdot}427 \pm 0{\cdot}011$	$1{\cdot}02 \pm 0{\cdot}02$	$0{\cdot}213 \pm 0{\cdot}004$
	(c)	$2{\cdot}00 \pm 0{\cdot}04$	$0{\cdot}399 \pm 0{\cdot}005$	$1{\cdot}02 \pm 0{\cdot}02$	$0{\cdot}211 \pm 0{\cdot}004$
	(g)	$2{\cdot}15 \pm 0{\cdot}05$	$0{\cdot}422 \pm 0{\cdot}013$	$1{\cdot}11 \pm 0{\cdot}05$	$0{\cdot}218 \pm 0{\cdot}008$
	(a)	$2{\cdot}47 \pm 0{\cdot}09$	$0{\cdot}53 \pm 0{\cdot}03$	$1{\cdot}33 \pm 0{\cdot}07$	$0{\cdot}31 \pm 0{\cdot}02$
$n=10$	(b)	$2{\cdot}46 \pm 0{\cdot}09$	$0{\cdot}53 \pm 0{\cdot}03$	$1{\cdot}35 \pm 0{\cdot}07$	$0{\cdot}31 \pm 0{\cdot}02$
	(c)	$7{\cdot}11 \pm 0{\cdot}25$	$0{\cdot}85 \pm 0{\cdot}03$	$5{\cdot}02 \pm 0{\cdot}19$	$0{\cdot}57 \pm 0{\cdot}02$
	(c)	$2{\cdot}81 \pm 0{\cdot}12$	$0{\cdot}65 \pm 0{\cdot}04$	$1{\cdot}64 \pm 0{\cdot}12$	$0{\cdot}37 \pm 0{\cdot}02$

3.7.2 *The binomial distribution*

We consider estimation of λ in

$$p(x \mid \lambda) = \binom{m}{x} \lambda^x (1-\lambda)^{m-x}$$

for fixed m.

(a) Parametric G, known to be beta (p, q): we have

$$\delta_G(x) = (p+x)/(m+p+q)$$

and expressions for $E(X_G^r)$, $r = 1, 2$ are easily derived. Estimates of p and q by the MM can be obtained from

$$p = \mu_1'(\mu_2' - m\mu_1')/\{\mu_1'^2(m-1) + m(\mu_1' - \mu_2')\} \tag{3.7.3}$$
$$q = (m - \mu_1')(\mu_2' - m\mu_1')/\{\mu_1'^2(m-1) + m(\mu_1' - \mu_2')\}$$

where $\mu_r' = E(X_G^r)$, $r = 1, 2$. Substituting sample moments for μ_1' and μ_2' gives estimates of p and q; note that these estimates have to be truncated away from zero since $p, q > 0$ for a proper prior G. The estimates $\bar{p}, \bar{q}$ of p and q derived from (3.7.3) are clearly consistent and $\bar{\delta}_G(x) = (\bar{p} + x)/(m + \bar{p} + \bar{q})$ is a consistent estimate of $\delta_G(x)$.

Other methods of estimating p, q can be used, like ML, with the above mentioned truncation of the estimates.

(b) Simple EB estimation:

(i) Simple EB estimation of the odds ratio $\rho = \lambda/(1-\lambda)$ is straightforward, for

$$\delta_{\rho,G}(x) = \int \binom{m}{x} \frac{\lambda}{(1-\lambda)} \lambda^x (1-\lambda)^{m-x} dG(\lambda)/p_G(x)$$

$$= \left(\frac{x+1}{m-x}\right) p_G(x+1)/p_G(x). \tag{3.7.4}$$

This leads to the simple EBE of ρ according to (3.4.3). If an estimate of λ, rather than ρ is required, an approximation in the style of (3.4.11) could be used. It will be noted that estimation of the posterior variance of ρ is also straightforward since $E\{\theta^2/(1-\theta)^2 | x\} = [(x+2)(x+1)/\{(m-x)(m-x-1)\}]p_G(x+2)/p_G(x)$.

(ii) Although it is not possible to obtain a simple EBE of λ directly in the form (3.4.3) we can write the Bayes estimate $\delta_G(x) = \delta_m(G, \lambda; x)$ as

$$\delta_m(G, \lambda; x) = \left(\frac{x+1}{m+1}\right) \frac{p_{G,m+1}(x+1)}{p_{G,m}(x)}$$

where $p_{G,m}(x) = \int \binom{m}{x} \lambda^x (1-\lambda)^{m-x} dG(\lambda)$. To use this formula for EBE construction one would need past observations via binomial $(m+1, \lambda)$ variates. Alternatively one could write

$$\delta_{m-1}(G, \lambda; x) = \left(\frac{x+1}{m}\right) p_{G,m}(x+1)/p_{G,m-1}(x).$$

For EBE construction one could take x to be the number of successes in the first $m-1$ of the current trials; see Robbins (1955). Elaboration of this scheme is possible. For example, one could take all permutations of the current trials, produce an x-value for each of them and take the mean of the resulting simple EBEs.

(c) Linear empirical Bayes estimation: referring to formulae (3.6.1) and (1.12.2) we note that if we put $U_1(X) = X/m$, $U_2(X) = X(X-1)/((m-1)m)$ then

$$E\{U_1(X)|\lambda\} = \lambda, \qquad E\{U_2(X)|\lambda\} = \lambda E(X|\lambda).$$

Therefore we can estimate the r.h.s. elements of (1.12.1) by

$(1/m)\sum_{i=1}^{n} x_i/n$ and $(1/m)\sum_{i=1}^{m} x_i(x_i - 1)/n$ respectively, giving $(\bar{w}_0, \bar{w}_1)$ as the solution of

$$\begin{bmatrix} 1 & \bar{x} \\ \bar{x} & \overline{x^2} \end{bmatrix} \begin{bmatrix} \bar{w}_0 \\ \bar{w}_1 \end{bmatrix} = \frac{1}{m} \begin{bmatrix} \bar{x} \\ \overline{x^2} & -\bar{x} \end{bmatrix}$$

where $\overline{x^r} = \sum_{i=1}^{n} x_i^r/n$.

(d) Approximation of G by G_k: the method of obtaining an approximate EBE via approximation of G by a step-function G_k is essentially the same as outlined for the Poisson case.

The results of some studies of the performance of EBEs are summarized in Table 3.5. The results were obtained with beta(p, q) prior distributions and the method of moments was used in the estimation of parameters of the parametric G distribution and of G_k. In the parametric G case the correct form of prior, i.e. beta, was assumed.

3.7.3 *The normal distribution*

We take the X data distribution for given mean λ to be $N(\lambda, \sigma^2)$ with σ^2 known, and therefore without loss of generality $\sigma^2 = 1$.

Table 3.5 *Binomial data distribution, beta(p, q) prior, $E_n W(\cdot)$ values obtained by simulation, using 20 trials. The largest estimated coefficient of variation of any $E_n W$ for $n = 10$ was at $m = 5$, $p = 10$, $q = 9$, the estimated coefficient of variation being 12%. For $n = 50$ the largest estimated c.v. was 3.7% at $m = 5$, $p = 3$, $q = 18$.*

N	p	q	$W(T)$	$W(\delta_G)$	$E_n W(\check{\delta}_G)$	$E_n(\check{\phi}_3)$	$P_n(\check{\delta}_G)$	$P_n(\check{\phi}_3)$
					$n = 10$			
10	10	9	0·0247	0·0082	0·0126	0·0141	0·05	0·05
10	3	18	0·0117	0·0038	0·0067	0·0073	0·05	0·05
25	10	9	0·0094	0·0054	0·0073	0·0084	0·15	0·15
25	3	18	0·0047	0·0025	0·0043	0·0049	0·40	0·55
5	10	9	0·0474	0·0099	0·0184	0·0186	0·05	0·05
5	3	18	0·0234	0·0045	0·0069	0·0078	0·00	0·00
					$n = 50$			
10	10	9	0·0237	0·0082	0·0091	0·0094	0·00	0·00
10	3	18	0·0117	0·0038	0·0042	0·0046	0·00	0·00
25	10	9	0·0095	0·0054	0·0058	0·0069	0·00	0·05
25	3	18	0·0047	0·0025	0·0026	0·0031	0·00	0·00
5	10	9	0·0474	0·0099	0·0119	0·0121	0·00	0·00
5	3	18	0·0234	0·0045	0·0055	0·0058	0·00	0·00

(a) Parametric G known to be $N(\mu_G, \sigma_G^2)$: the Bayes estimate is

$$\delta_G(x) = (x + \mu_G/\sigma_G^2)/(1 + 1/\sigma_G^2)$$

and the marginal X-distribution $F_G(x)$ is $N(\mu_G, 1 + \sigma_G^2)$. Hence obvious estimates of μ_G and σ_G^2 are $\bar{x}$ and $\max(0, s^2 - 1)$ where $\bar{x}$ and s^2 are the sample mean and variance calculated from the past x-observations. More details for this case have already been given in Example 3.5.2.

(b) Approximation of G by G_k: some information on the goodness of such approximations, judged by values of $W(\phi_k)$, are given in section 3.3. Estimation of λ_j, $j = 1, \ldots, k$ can be by ML or by the method of moments, or otherwise. Implementation of the ML method is as indicated for the Poisson case, the method of moments requires expressions for the first k moments of F_{G_k} in terms of the λ's. For example, with $k = 3$ these moments are $\alpha_{31} = (\lambda_1 + \lambda_2 + \lambda_3)/3$, $\alpha_{32} = (\lambda_1^2 + \lambda_2^2 + \lambda_3^2)/3 + 1$, $\alpha_{33} = (\lambda_1^3 + \lambda_2^3 + \lambda_3^3)/3 + 3(\lambda_1 + \lambda_2 + \lambda_3)/3$. In practice estimates of λ_1, λ_2, λ_3 can be obtained by equating α_{31} and $\bar{x}$ and minimizing distances between α_{3r} and $\overline{x^r}$, $r = 2, 3$.

(c) Linear empirical Bayes estimation: let $U_1(X) = X$ and $U_2(X) = X^2 - 1$, then $E\{U_1(X)|\lambda\} = \lambda$, $E\{U_2(X)|\lambda)\} = \lambda^2 = \lambda E(X|\lambda)$. Thus the r.h.s. elements of (1.12.1) can be estimated by $\bar{x}$ and $\overline{x^2} - 1$ respectively, giving

$$\delta(\bar{w}_0, \bar{w}_1; x) = \bar{x}/s^2 + x(s^2 - 1)/s^2.$$

Table 3.6 *Data distribution $N(\lambda, 1)$, prior distribution $N(0, \sigma_G^2)$, values of W(Bayes), W(MLE), $E_n W(EB)$ for various EB estimators, and values of n. All $E_n W$ values were obtained by simulation; the largest s.e.'s of the tabulated values are for $E_{20}W$ (simple EBE d(ii) smoothed), 0·03, 0·04, 0·04. All other estimated $E_n W$ values are subject to s.e.'s $\leqslant$ 0·02*

		$\sigma_G^2 = 0{\cdot}1$	$\sigma_G^2 = 0{\cdot}5$	$\sigma_G^2 = 1{\cdot}0$
	W(Bayes)	0·09	0·33	0·50
	W(MLE)	1·00	1·00	1·00
E_{10}	W(Parametric EBE)	0·21	0·49	0·67
E_{20}	W(Parametric EBE)	0·16	0·42	0·60
E_{10}	W(G_3 approx. EBE)	0·25	0·55	0·82
E_{20}	W(G_3 approx. EBE)	0·16	0·47	0·69
E_{20}	W(simple EBE d(ii) smoothed)	0·29	0·74	1·01
E_{100}	W(simple EBE d(ii) smoothed)	0·17	0·45	0·65
E_{100}	W(simple EBE d(i) smoothed)	0·21	0·47	0·65

(d) Simple EB estimation: At least two approaches are possible.
(i) As in Example 1.3.7 we can write, noting $\sigma^2 = 1$,

$$\delta_G(x) = x + f'_G(x)/f_G(x). \tag{3.7.5}$$

Implementation of (3.7.5) for EB estimation requires estimation of $f_G(x)$ and $f'_G(x)$, a topic discussed in section 3.4.6.
(ii) Following the suggestions of section 3.4.4, one can put $T_1F(x|\lambda) = \int_{-\infty}^{x} u dF(x|\lambda)$, $T_2F(x|\lambda) = F(x|\lambda)$, leading to a fairly straightforward EBE obtained from the approximation (3.4.11). Details can be found in Maritz and Lwin (1975).

In Table 3.6 numerical results for smoothed versions of the simple EBEs of types (i) and (ii) are shown. The smoothing method is also described in Maritz and Lwin (1975).

The results of numerical case studies using the EB methods (a)–(d) are also shown in Table 3.6. They indicate that EB estimates can be considerably better than MLEs even with relatively small values of n.

3.8 EB estimation with multiple current observations: one parameter

3.8.1 Introduction and general considerations

For realistic applications of the EB approach the sampling scheme of one current observation and one past x_i at each λ_i seems severely restrictive. Let us now suppose that $m \geqslant 1$ independent observations $x_1, x_2, \ldots, x_m$ on X are made at the current value λ of Λ. Also assume that m_i independent x-observations $x_{i1}, x_{i2}, \ldots, x_{im_i}$ are made at the past realization λ_i of Λ, $i = 1, 2, \ldots, n$. One of the immediate consequences of this more general sampling scheme is that it enables one to deal with nuisance parameters. For example, in the case where $X \overset{d}{=} N(\lambda, \sigma^2)$ we need not assume σ^2 known, because it can be estimated using the multiple observations at each λ_i. At the same time derivation of EBEs can become more complicated, especially if the m_i values differ from each other.

When the m_i values do not vary, in particular, if $m_i = m$ for all i, it is sometimes possible to reduce the problem to the single-observation case. This happens when a one-dimensional sufficient statistic $T(\mathbf{x})$ for λ exists, in which case one can simply replace the current observations by the value t of the sufficient statistics, and similarly replace each set of past observations by the corresponding t_i, $i = 1, 2, \ldots, n$.

If reduction by sufficiency is not possible a sub-optimal **pseudo-Bayes** approach may be adopted whereby the current observations are replaced by an estimate $\hat{\lambda}$, such as the MLE, with distribution $F(\hat{\lambda}|\lambda, m)$. Then calculate the estimate

$$\delta_{G,m}(\hat{\lambda}) = \frac{\int \hat{\lambda} dF(\hat{\lambda}|\lambda, m) dG(\lambda)}{\int dF(\hat{\lambda}|\lambda, m) dG(\lambda)}. \tag{3.8.1}$$

This estimate is not the Bayes estimate, hence $W(\delta_{G,m}(\hat{\lambda})) \geqslant W(\text{Bayes})$. In general, calculation of the Bayes and pseudo-Bayes estimates, and of $W(\text{Bayes})$ and $W(\text{pseudo-Bayes})$, is rather complicated. Consequently it seems impossible to make an accurate statement of the loss in overall efficiency resulting from such a non-sufficient reduction. The following example gives an indication of the quantitative effects of non-sufficient reduction.

Example 3.8.1 Suppose that $X \stackrel{d}{=} N(\lambda, 1)$, $\Lambda \stackrel{d}{=} N(\mu_G, \sigma_G^2)$ and that $\hat{\lambda} = \text{median } (x_1, x_2, \ldots, x_m)$. Then, for m reasonably large $\hat{\lambda} \stackrel{d}{\simeq} N(\lambda, \pi/2m)$. The pseudo-Bayes estimate based on $\hat{\lambda}$ is

$$\delta_{G,m}(\hat{\lambda}) = \frac{2m\hat{\lambda}/\pi + \mu_G/\sigma_G^2}{2m/\pi + \mu_G/\sigma_G^2}$$

and $W(\text{pseudo-Bayes}) \simeq 1/(2m/\pi + 1/\sigma_G^2)$. In this case the Bayes estimate is

$$\delta_G(\mathbf{x}) = \frac{m\bar{x} + \mu_G/\sigma_G^2}{m + \mu_G/\sigma_G^2}$$

with $W(\text{Bayes}) = 1/(m + 1/\sigma_G^2)$. As we have remarked before, see for example section 3.1, we are really only concerned with cases where σ_G^2 is sufficiently small for EB methods to be potentially useful. So, for example, if $\sigma_G^2 = 1/m$ we see that $W(\text{pseudo-Bayes})/W(\text{Bayes}) \simeq 1{\cdot}22$.

If the data reduction is done by the ML method we have asymptotic equivalence of the pseudo-Bayes and Bayes estimates through the asymptotic sufficiency of the MLE as $m \to \infty$. For a discussion of asymptotic efficiency, see for example Cox and Hinkley (1974, p. 307). Asymptotic sufficiency of the MLE and asymptotic normality of its distribution can be useful in evaluating $W(\cdot)$. In what follows we shall usually assume that reduction by the MLE is done.

Multiple current observations also enable one to deal with nuisance parameters. Perhaps the most obvious case is that of the X-distribution being $N(\lambda, \sigma^2)$ with σ^2 unknown but fixed, while $\Lambda \stackrel{d}{=} N(\mu_G, \sigma_G^2)$; in other words the prior distribution of σ is taken to be degenerate. In this case σ^2 can be estimated by $\hat{\sigma}^2 = \sum_{i=1}^n \sum_{j=1}^{m_i} (x_{ij} - x_{i.})^2 / (\sum_{i=1}^n (m_i - 1))$, the usual within-group estimator. In formulae to do with the EBE of λ the parameter σ can then be replaced by $\hat{\sigma}^2$, appropriate allowance being made for this estimation of σ^2 in the calculation of $E_n W(\text{EB})$. Again, if all m_i are equal estimation of μ_G and σ_G^2 is straightforward.

In the case of multiple parameter estimation we shall also adopt the policy of reducing the observations to the MLEs. Both in the one-parameter and the multiple-parameter cases estimation of the prior distribution and construction of simple EBEs remains a non-trivial exercise.

3.8.2 *Unequal m_i: simple EB estimation*

The case studies of simple and other EBEs that have been reported indicate that the performance of the simple EBEs is relatively poor. Therefore, although many possibilities exist for constructing simple EBEs when $m_i \geqslant 1$, unequal m_i, only some of these will be discussed. Generally they lead to rather unwieldy calculations.

1. $m \leqslant$ all m_i: let the current observations be summarized in the MLE $\hat{\lambda}$ of λ. Select m observations from each past set and calculate $\hat{\lambda}_1, \ldots, \hat{\lambda}_n$. Then use one of the single observation methods of constructing a simple EBE. This procedure can, of course, be followed using every possible subset of size m of the observations at each past λ_j. Averaging the results is one way of obtaining a single EBE using all of the past observations. There are $\prod_{i=1}^n \binom{m_i}{m}$ possible estimates, hence the calculations could be tedious.
2. $m \nleqslant$ all m_i: a compromise method seems to be the only practical possibility, namely, instead of trying to estimate the Bayes estimate based on the m current observations, obtain EBEs based on subsets of the current observations, and average them. In order to proceed in a manner similar to that suggested in 1, the smallest subsets would have to be of size $\min(m_1, m_2, \ldots, m_n)$. Of course the

average of Bayes estimates from subsets of m observations is not the Bayes estimate itself, and the loss in efficiency may be great. This can be illustrated quite simply the considering the $N(\lambda, 1)$, $N(0, 1)$ case where the Bayes estimate is

$$\delta_G(\mathbf{x}) = \frac{m\bar{x}}{m+1}$$

and the average of m single observation Bayes estimates is $\overline{\delta_G}(x) = \bar{x}/2$. The respective expected losses are

$$W\left(\frac{\bar{x}}{2}\right) = \frac{1}{m+1} \qquad \to 0 \text{ as } m \to \infty$$

$$W\left(\frac{\bar{x}}{2}\right) = \frac{1}{4}\left(1 + \frac{1}{m}\right) \to \frac{1}{4} \text{ as } m \to \infty.$$

In the light of the difficulties mentioned under 1 and 2 above we shall concentrate on EBEs depending on estimation of G or an approximation to G.

3.8.3 *Unequal m_i: parametric G of known form, and approximation by G_k, using sufficient statistics*

Suppose that the estimate $t(x_1 \cdots x_m)$ is a sufficient statistic, and that $t_i(x_{i1} \cdots x_{im_i})$ are the corresponding sufficient statistics calculated from the past samples. The distribution of t_i depends on λ and on m_i, and its p.d.f. is $\mathscr{L}(t_i | \lambda, m_i)$. The prior distribution G is of known parametric form $G(\lambda; \boldsymbol{\theta})$, depending on parameters $\boldsymbol{\theta}$; the dimension of $\boldsymbol{\theta}$ is k.

The likelihood of the observations $t_1, t_2, \ldots, t_n$ is

$$L(\mathbf{t}; \boldsymbol{\theta}) = \prod_{i=1}^{n} \int \mathscr{L}(t_i | \lambda_i, m_i) dG(\lambda_i; \boldsymbol{\theta}) \qquad (3.8.2)$$

and estimation of the parameters $\boldsymbol{\theta}$ can now be done by maximizing $L(\mathbf{t}, \theta)$ w.r.t. $\boldsymbol{\theta}$.

In some cases it may be relatively easy to use the method of moments for estimating the parameters $\boldsymbol{\theta}$. Suppose that $E(t_i^r | \lambda_i) = \mu_r(\lambda_i, m_i)$, $r = 1, 2, \ldots, k$. Then

$$E\left(\sum_{i=1}^{n} T_i^r\right) = \sum_{i=1}^{n} \int \mu_r(\lambda; m_i) dG(\lambda; \boldsymbol{\theta}), \qquad r = 1, 2, \ldots, k, \quad (3.8.3)$$

and for certain distributions (3.8.3) has a fairly tractable form.

Replacing the l.h.s. of (3.8.3) by observed sums and solving the equations gives MM estimates of $\boldsymbol{\theta}$.

Example 3.8.2 Let $X \stackrel{d}{=}$ Poisson (λ) and put $T_i = (X_{i1} + \cdots + X_{im_i})/m_i$. Then $E(T_i|\lambda_i) = \lambda_i$, $E(T_i^2|\lambda_i) = \lambda_i^2 + \lambda_i/m_i$. If the prior distribution of Λ is gamma (α, β) with p.d.f. $\{1/\Gamma(\beta)\}\alpha^\beta \lambda^{\beta-1} e^{-\alpha\lambda}$, $\lambda > 0$ we have

$$E\left(\sum_{i=1}^{n} T_i\right) = n\beta/\alpha$$

$$E\left(\sum_{i=1}^{n} T_i^2\right) = n\beta(\beta+1)/\alpha^2 + (\beta/\alpha)\sum_{i=1}^{n} 1/m_i.$$

When the form of G is not given the approach of approximating G by G_k, a step-function, can be taken. In (3.8.2) and (3.8.3) $G(\lambda; \theta)$ is replaced by $G_k(\lambda; \theta_1, \theta_2, \ldots, \theta_k)$ where $\theta_1, \theta_2, \ldots, \theta_k$ are the points at which the function has jumps of size $1/k$.

3.8.4 *Linear EB using sufficient statistics*

In certain relatively simple cases the linear Bayes method of section 1.12 can be adapted for the present situation. Suppose that

$$E(T|\lambda) = \lambda$$

$$E(T^2|\lambda) = \lambda^2 + C\lambda/m, \text{ where } C \text{ is a constant.}$$

Then (1.12.2) can be written

$$\begin{bmatrix} 1 & E(\Lambda) \\ E(\Lambda) & E(\Lambda^2) + CE(\Lambda)/m \end{bmatrix} \begin{bmatrix} w_0 \\ w_1 \end{bmatrix} = \begin{bmatrix} E(\Lambda) \\ E(\Lambda^2) \end{bmatrix}. \tag{3.8.4}$$

From the past observations estimates of $E(\Lambda)$ and $E(\Lambda^2)$ can be obtained by noting that

$$\bar{t} = \frac{1}{n}\sum_{i=1}^{n} t_i \text{ is an estimate of } E(\Lambda);$$

$$\overline{t^2} = \frac{1}{n}\sum_{i=1}^{n} t_i^2 \text{ is an estimate of } E(\Lambda^2) + E(\Lambda)(C/n)\sum_{i=1}^{n} 1/m_i.$$

Example 3.8.3 Let $X \stackrel{d}{=}$ Poisson (λ) and put $t_i = \sum_{j=1}^{m_i} x_{ij}/m_i$. Then

$E(T_i|\lambda_i)=\lambda_i$, $E(T_i^2|\lambda_i)=\lambda_i^2+\lambda_i/m_i$, $i=1,2,\ldots,n$. Writing $\gamma_r=E(\Lambda^r)$, $r=1,2$, estimates of γ_1,γ_2 are given by

$$\bar{t}=\hat{\gamma}_1$$

$$\overline{t^2}=\hat{\gamma}_2+\hat{\gamma}_1\frac{1}{n}\sum(1/m_i),$$

which may be compared with the results in Example 3.8.2.

Somewhat more generally we may have $E(T|\lambda)=\lambda$, but $E(T^2|\lambda)=\lambda^2+q(\lambda)v$, where v is known, as is every v_i corresponding to the ith component. An obvious special case is $v_i=1/m_i$. Now $\overline{t^2}$ is an estimate of $E(\Lambda^2)+E\{q(\Lambda)\}(1/n)\sum_{i=1}^{n}v_i$. We need an estimate of $E(\Lambda^2)$, consequently we need to find $U_2(X)$ such that $E(U_2(X)|\lambda)=\lambda^2$ or $U_q(X)$ such that $E(U_q(X)|\lambda)=q(\lambda)$.

These may be non-trivial problems, but in many important cases the form of $q(\lambda)$ is such as to cause little difficulty. In particular, we may have $q(\lambda)$ not dependent on λ, as when the data distribution is $N(\lambda,\sigma^2)$. If σ^2 is known application of the linear EB method is straightforward in this case. Another interesting application arises in the design of a quality measurement plan as described in section 8.3.6. Here $T_i=I_i$ where the I_i are quality indices, the model being that $E(I_i|\lambda_i)=\lambda_i$ and $\operatorname{var}(I_i|\lambda_i)=\lambda_i/e_i$, where the e_i are known constants.

3.8.5 *Unequal m_i: reduction by ML estimation*

In the light of our earlier discussion on the use of MLEs, section 3.8.1, the methods to be employed here are essentially the same as those for the sufficient statistics. Typically difficulties will arise because the distributions of the MLEs will not be easily written down when they are not sufficient statistics. When the m_i are large enough, approximate normality of the MLEs could be used.

3.8.6 *Unequal m_i generally: parametric G and G_k approximations*

If G has the known parametric form $G(\lambda;\boldsymbol{\theta})$ it is, in principle, possible to estimate the parameters of G by the method of maximum likelihood, using the more general version of (3.8.2):

$$L(\mathbf{X};\boldsymbol{\theta})=\prod_{i=1}^{n}\int\left\{\prod_{j=1}^{m_i}f(x_{ij}|\lambda_i)dG(\lambda_i;\boldsymbol{\theta})\right\}. \tag{3.8.5}$$

Apart from a multiplicative factor not dependent on $\lambda_1, \lambda_2, \ldots, \lambda_m$, the likelihood in (3.8.5) reduces to the likelihood in (3.8.2) when t is a sufficient statistic. Finding MLEs of $\theta_1, \theta_2, \ldots, \theta_k$ from (3.8.3) is a straightforward computational problem. As usual however, calculation of Bayes and EB estimates, while straightforward, can be computationally difficult since multiple integration is required.

If the parametric form of G is not known and G_k is used to approximate G, all calculations are as outlined above with G replaced by G_k. In this case the EM algorithm can be particularly useful.

3.9 Unequal $m_i \geqslant 1$: application to particular distributions

3.9.1 *The Poisson distribution*

The statistic $T = (X_1 + \cdots + X_n)/m$ is sufficient for λ and since the distribution of $X_1 + \cdots + X_m$ is Poisson $(m\lambda)$ we have

$$E(T \mid \lambda) = \lambda$$
$$E(T^2 \mid \lambda) = \lambda^2 + \lambda/m;$$

see also Examples 3.8.2 and 3.8.3.

(a) Parametric G; $\Lambda \overset{d}{=} \text{gamma}(\alpha, \beta)$; *ML estimation of* α, β

In this example it is convenient to use $T^* = mT$ in the ML estimation because we have

$$P(T^* = r) = \int_0^\infty \frac{m^r}{\Gamma(\beta)} \alpha^\beta \frac{\lambda^{\beta + r - 1}}{r!} e^{-(\alpha + m)\lambda} d\lambda$$
$$= \frac{m^r \alpha^\beta}{(\alpha + m)^{\beta + r}} \frac{\Gamma(\beta + r)}{\Gamma(\beta) r!} \qquad (3.9.1)$$

and $L(T; \boldsymbol{\theta})$ in (3.8.2) becomes

$$L(\mathbf{T}^*; \boldsymbol{\theta}) = \prod_{i=1}^{n} \frac{\alpha^\beta}{(\alpha + m_i)^{\beta + t_i^*}} \frac{\Gamma(\beta + t_i^*)}{t_i^*! \Gamma(\beta)} m_i^{t_i^*}. \qquad (3.9.2)$$

It should be noted that the formula (3.9.2) can also be applied when m_i is not an integer. This can occur if only the values of t_i^* are reported and they are, for example, Poisson counts over time periods of variable lengths proportional to m_i.

From (3.8.5),

$$\frac{\partial \ln L}{\partial \alpha} = \frac{n\beta}{\alpha} - \sum_{i=1}^{n} \left(\frac{\beta + t_i^*}{\alpha + m_i} \right)$$

$$\frac{\partial \ln L}{\partial \beta} = n \ln \alpha - \sum_{i=1}^{n} \ln(\alpha + m_i) + \sum_{i=1}^{n} \Psi(\beta + t_i^*) - n\Psi(\beta), \quad (3.9.3)$$

where $\Psi(z) = \Gamma'(z)/\Gamma(z)$, the digamma function. Setting $S_1(\mathbf{t}^*; \alpha, \beta) = \partial \ln L/\partial\alpha = 0$ and $S_2(\mathbf{t}^*; \alpha, \beta) = \partial \ln L/\partial\beta = 0$ and solving the equations gives the MLEs $\hat{\alpha}, \hat{\beta}$ of α, β.

For the purposes of evaluating the performance of the EBEs the covariance matrix of $\hat{\alpha}, \hat{\beta}$ is useful. For more details we refer to section 3.11, but consider here the calculation of the relevant covariance matrix. The method used here is the two-parameter version of that described briefly in section 3.5.2.

The estimates of α and β are obtained by solving the two equations in a and b,

$$S_r(t^*; a, b) = 0. \quad (3.9.4)$$

An approximate, large n, formula for the covariance matrix $\mathbf{C}$ of $\hat{\alpha}, \hat{\beta}$ is

$$\mathbf{C} \simeq \boldsymbol{\delta}^{-1}\mathbf{V}(\boldsymbol{\delta}^{\mathrm{T}})^{-1},$$

where the elements of $\mathbf{V}$ are $\operatorname{cov}\{S_r, S_s\}$, $\quad r, s = 1, 2, \quad$ (3.9.5)

and the elements of $\boldsymbol{\delta}$ are $\left\{ \dfrac{\partial E S_r(\mathbf{T}^*; a, b)}{\partial c} \right\}_{a=\alpha, b=\beta}, \quad c = a, b.$

Expressions for $\operatorname{var}(S_1)$, etc. in terms of α, β can be obtained using (3.9.1) and the independence of the T_i^* statistics. An interesting, and practically possibly more realistic approach is to allow the m_i to be independent realizations of r.v. M. A considerable advantage of such an approach is that estimates of the elements of $\mathbf{V}$ and $\boldsymbol{\delta}$ can be calculated directly from the observed data. This is like using the observed information matrix in the usual ML estimation, a procedure advocated by many statisticians; see for example, Cox and Hinkley (1974, p. 302).

Write $S_1(\mathbf{T}^*; a, b) = \sum_{i=1}^{n} U_i(a, b)$ and $S_2(\mathbf{T}^*; a, b) = \sum_{i=1}^{n} V_i(a, b)$, where, from (3.9.3) we have

$$U_i(\alpha, \beta) = \beta/\alpha - (\beta + T_i^*)/(\alpha + M_i)$$

$$V_i(\alpha, \beta) = \ln \alpha - \ln(\alpha + M_i) + \Psi(\beta + T_i^*) - \Psi(\beta).$$

We shall simplify notation by putting $S_r(T^*:a,b) = S_r(a,b)$, $r = 1, 2$.

Now (U_i, V_i), $i = 1, 2, \ldots, n$ are independent realizations of the two-dimensional r.v. (U, V). Then

$$\hat{E}S_1(a,b) = \sum U_i(a,b)$$

$$\frac{\partial \hat{E}S_1(a,b)}{\partial a} = \sum \partial U_i/\partial a$$

and we estimate $\{\partial ES_1(a,b)/\partial a\}_{a=\alpha, b=\beta}$ by $\{\partial \hat{E}S_1(a,b)/\partial a\}_{a=\hat{\alpha}, b=\hat{\beta}}$, etc. We can estimate var $\{S_1(\alpha,\beta)\}$, var $\{S_2(\alpha,\beta)\}$, cov $\{S_1(\alpha,\beta), S_2(\alpha,\beta)\}$ by nS_u^2, nS_v^2, nS_{uv} where S_u^2, S_v^2, S_{uv} are the usual sample variances and covariance calculated using observed $U_i(\hat{\alpha},\hat{\beta})$, $U_i(\hat{\alpha},\hat{\beta})$ values. For example, let $U.(\hat{\alpha},\hat{\beta}) = \sum_{i=1}^n U_i(\hat{\alpha},\hat{\beta})/n$, then

$$S_u^2 = \sum_{i=1}^n \{U_i(\hat{\alpha},\hat{\beta}) - U.(\hat{\alpha},\hat{\beta})\}^2/(n-1).$$

(b) Parametric G: $\Lambda \stackrel{d}{=}$ gamma (α, β): *MM estimation of* α, β

Details of this case have already been given in Example 3.8.2. The MMEs $\bar{\alpha}$, $\bar{\beta}$ of α and β are obtained as the solutions of

$$M_1(\alpha,\beta) = \sum_{i=1}^n t_i - n\beta/\alpha = 0$$

$$M_2(\alpha,\beta) = \sum t_i^2 - n\beta(\beta M)/\alpha^2 - \beta/\alpha \sum_{i=1}^n (1/m_i) = 0. \tag{3.9.6}$$

Essentially the same techniques as those used in the MLE case are applicable here. In (3.9.5) $\hat{\alpha}, \hat{\beta}$ are replaced by $\bar{\alpha}, \bar{\beta}$; S_r by $M_r, r = 1, 2$, etc. Instead of $U_i(a,b)$, $V_i(a,b)$ we have $Y_i(a,b) = T_i - b/a$, $Z_i(a,b) = t_i^2 - b/am_i$.

(c) G_k *approximation of G*

We have to replace the gamma (α,β) distribution which leads to (3.8.4) by G_k giving

$$P(T^* = r) = \frac{1}{k}\sum_{j=1}^k e^{-m\lambda_j}(m\lambda_j)^r/r!. \tag{3.9.7}$$

Unfortunately this formula cannot be simplified, consequently the analogue of (3.8.5) becomes

$$L_k(\mathbf{T}^*; \lambda) = \prod_{i=1}^n \left\{\frac{1}{k}\sum_{j=1}^k e^{-m\lambda_j}(m\lambda_j)^{t_i^*}/t_i^*!\right\}$$

leading to somewhat unwieldy formulae for the derivatives of $\ln L_k$ w.r.t. λ_j. There is also the complication that one needs to put $\lambda_1 \leqslant \lambda_2 \cdots \leqslant \lambda_k$ in order to make computations feasible.

The MM estimation of $\lambda_1, \ldots, \lambda_k$ is also complicated in this case because, with $k \geqslant 3$ it is not necessarily possible to find $\bar{\lambda}$ such that the first k moment equations are satisfied exactly. The same difficulty was noted for $m_i = 1$.

(d) Linear EB

According to Example 3.8.3 estimates $\check{\gamma}_1$, $\check{\gamma}_2$ for $E(\Lambda)$, $E(\Lambda^2)$ are obtained as

$$\check{\gamma}_1 = \bar{t}$$

$$\check{\gamma}_2 = \bar{t}^2 - \bar{t}\frac{1}{n}\sum_{i=1}^{n}(1/m_i)$$

and estimates $\check{w}_0$, $\check{w}_1$ are obtained from (3.8.4) which becomes

$$\begin{pmatrix} 1 & \check{\gamma}_1 \\ \check{\gamma}_1 & \check{\gamma}_2 + \check{\gamma}_1/m \end{pmatrix}\begin{pmatrix} \check{w}_0 \\ \check{w}_1 \end{pmatrix} = \begin{pmatrix} \check{\gamma}_1 \\ \check{\gamma}_2 \end{pmatrix}.$$

Here also assessment of the covariance matrix of $\check{w}_0$, $\check{w}_1$ is relatively straightforward if we take m_i to be independent realizations of a r.v. M. Note that $\check{\gamma}_1$, $\check{\gamma}_2$ are solutions of

$$nM_1^*(\gamma_1, \gamma_2) = \sum_{i=1}^{n} t_i - n\gamma_1 = 0$$

$$nM_2^*(\gamma_1, \gamma_2) = \sum (t_i^2 - \gamma_1/m_i) - n\gamma_2 = 0$$

so that U_i, V_i in section (c) above can be replaced by

$$Y_i^*(\gamma_1, \gamma_2) = T_i - \gamma_1$$

$$Z_i^*(\gamma_1, \gamma_2) = T_i^2 - \gamma_1/m_i.$$

This enables one to calculate an estimate of the covariance matrix of $\hat{\gamma}_1\hat{\gamma}_2$ and hence of $\hat{w}_0, \hat{w}_1$.

3.9.2 *The binomial distribution*

We take λ to be the probability of success and $X_i = 1$ or 0 according as a success or a failure occurs. Then the distribution of $T^* =$

$X_1 + \cdots + X_m$ is binomial (m, λ), and for $T = T^*/m$ we have

$$E(T|\lambda) = \lambda$$
$$E(T^2|\lambda) = \lambda^2 + \lambda(1-\lambda)/m$$

The statistic T is sufficient for λ hence we can adopt the approach of section 3.8.3. In Bayes estimation of λ the natural conjugate prior distribution is beta(β_1, β_2), and we shall consider only this distribution for the parametric G case. As in the Poisson case this choice and limitation are defended on the grounds that the beta family is rich enough to cover many realistic situations. Details for other parametric G families can be worked out following the patterns for the beta family.

(a) Parametric G: $\Lambda \overset{d}{=} \text{beta}(\beta_1, \beta_2)$*: ML estimation of* β_1, β_2

The marginal T^* distribution is

$$P(T^* = r) = \frac{1}{B(\beta_1, \beta_2)} \int_0^1 \binom{m}{r} \lambda^{\beta_1 + r - 1}(1-\lambda)^{\beta_2 + m - r - 1} d\lambda$$
$$= \binom{m}{r} \frac{\Gamma(\beta_1 + r)\Gamma(\beta_2 + m - r)}{\Gamma(\beta_1 + \beta_2 + m)} \frac{\Gamma(\beta_1 + \beta_2)}{\Gamma(\beta_1)\Gamma(\beta_2)}$$

and the likelihood in (3.8.2) becomes

$$L(\mathbf{T}^*; \boldsymbol{\beta}) = \prod_{i=1}^{n} \binom{m_i}{t_i^*} \frac{\Gamma(\beta_1 + t_i^*)\Gamma(\beta_2 + m_i - t_i^*)}{\Gamma(\beta_1 + \beta_2 + m_i)} \frac{\Gamma(\beta_1 + \beta_2)}{\Gamma(\beta_1)\Gamma(\beta_2)}. \tag{3.9.8}$$

The MLEs of β_1, β_2 are the solutions $\hat{\beta}_1, \hat{\beta}_2$ of

$$S_1(\beta_1, \beta_2) = \sum_{i=1}^{n} \psi(\beta_1 + t_i^*) + n\psi(\beta_1 + \beta_2)$$
$$- n\psi(\beta_1 + \beta_2 + m_i) - n\psi(\beta_1) = 0$$
$$S_2(\beta_1, \beta_2) = \sum_{i=1}^{n} \psi(\beta_2 + m_i - t_i^*) + n\psi(\beta_1 + \beta_2)$$
$$- n\psi(\beta_1 + \beta_2 + m) - n\psi(\beta_2) = 0.$$

Estimation of the covariance matrix follows steps like those given for the Poisson distribution. Here too considerable simplification is achieved if $m_1, \ldots, m_n$ are taken to be independent realizations of a r.v.

M. Let $U_i(\beta_1, \beta_2)$ and $V_i(\beta_1, \beta_2)$ be the elements of the sums defining S_1 and S_2 respectively. Then the remaining calculations are like those in section 3.9.1(a).

To estimate β_1, β_2 by the method of moments we have to solve

$$\sum_{i=1}^{n} Y_i(\beta_1, \beta_2) = \sum_{i=1}^{n} \left\{ t_i - \frac{\beta_1}{\beta_1 + \beta_2} \right\} = 0$$

$$\sum_{i=1}^{n} Z_i(\beta_1, \beta_2) = \sum_{i=1}^{n} \left\{ t_i^2 - \frac{\beta_1(\beta_1 + 1)(1 - 1/m_i)}{(\beta_1 + \beta_2)(\beta_1 + \beta_2 + 1)} - \frac{\beta_1}{m_i(\beta_1 + \beta_2)} \right\} = 0.$$

Calculations of estimates of the elements of the covariance matrix of the MMEs $\bar{\beta}_1, \bar{\beta}_2$ are performed according to (3.9.5) with U_i, V_i replaced by Y_i, Z_i.

(b) G_k approximation of G

The details of this procedure run parallel to those for the Poisson distribution with the obvious modifications to (3.8.11) and the analogue of (3.8.5).

(c) Linear EB

Refer to the method described in section 3.8.3 and note that

$$E(T|\lambda) = \lambda$$
$$E(T^2|\lambda) = \lambda^2 + \lambda(1 - \lambda)/m$$

so that (3.8.4) becomes

$$\begin{bmatrix} 1 & \check{\gamma}_1 \\ \check{\gamma}_1 & \check{\gamma}_2 + (\check{\gamma}_1 - \check{\gamma}_2)/m \end{bmatrix} \begin{bmatrix} \check{w}_0 \\ \check{w}_1 \end{bmatrix} = \begin{bmatrix} \check{\gamma}_1 \\ \check{\gamma}_2 \end{bmatrix}$$

where $\gamma_r = E(\Lambda^r)$, $r = 1, 2$. Estimates $\check{\gamma}_r$ of γ_r, $r = 1, 2$ are obtained by solving

$$\sum_{i=1}^{n} Y_i^*(\gamma_1, \gamma_2) = \sum_{i=1}^{n} (t_i - \gamma_1) = 0$$

$$\sum_{i=1}^{n} Z_i^*(\gamma_1, \gamma_2) = \sum_{i=1}^{n} \{t_i^2 - \gamma_2 - (\gamma_1 - \gamma_2)/m_i\} = 0.$$

Estimates of the variances and covariance of the $\check{\gamma}$'s, and hence of the $\check{w}$'s are again obtainable through (3.9.5).

3.9.3 *The normal distribution*

We shall suppose that the distribution of X is $N(\lambda, \sigma^2)$ with σ^2 fixed and known. Since $m_i \geqslant 1$ it would be possible to compute estimates of an unknown σ^2 see also section 7.3.2(b). In the parametric G case the prior distribution of Λ will be assumed $N(\mu_G, \sigma_G^2)$*. For convenience below we shall write $\sigma_G^2 = v_G$. Details for other parametric G distributions will be similar, but usually more complicated algebraically.

(a) Parametric G: $N(\mu_G, v_G)$: ML and MM estimation

The marginal distribution of T is $N(\mu_G, v_G + \sigma^2/m)$ hence it follows that the MLEs of μ_G and v_G are the solutions of

$$\sum_{i=1}^{n} U_{1i}(\mu_G, v_G) = \sum_{i=1}^{n} (t_i - \mu_G)/(v_G + \sigma^2/m_i) \qquad = 0$$

$$\sum_{i=1}^{n} U_2(\mu_G, v_G) = \sum_{i=1}^{n} \left\{ \frac{-1}{(v_G + \sigma^2/m_i)} + \frac{(t_i - \mu_G)^2}{(v_G + \sigma^2/m_i)^2} \right\} = 0.$$

The solution is helped by noting that

$$\hat{\mu}_G = \sum_{i=1}^{n} t_i/(\hat{v}_G + \sigma^2/m_i) \Big/ \sum_{i=1}^{n} 1/(\hat{v}_G + \sigma^2/m_i).$$

In this case, also, the derivates needed in (3.9.5) are much easier to calculate than for the Poisson and binomial distributions.

For MM estimation we note that

$$E(T|\lambda) = \lambda; \qquad E(T) = \mu_G$$
$$E(T^2|\lambda) = \lambda^2 + \sigma^2/m; \qquad E(T^2) = \mu_G^2 + v_G + \sigma^2/m$$

so that we estimate μ_G and v_G by solving

$$\sum_{i=1}^{n} Y_i(\mu_G, v_G) = \sum_{i=1}^{n} (t_i - \mu_G) \qquad = 0$$

$$\sum_{i=1}^{n} Z_i(\mu_G, v_G) = \sum (t_i^2 - \mu_G^2 - v_G - \sigma^2/m_i) = 0,$$

giving

$$\tilde{\mu}_G = \bar{t} = (1/n) \sum_{i=1}^{n} t_i$$

$$\tilde{v}_G = \sum_{i=1}^{n} (t_i^2 - \bar{t}^2 - \sigma^2/m_i)/n.$$

In theory we should take $\bar{v}_G = \max\{0, \sum_{i=1}^{n}(t_i^2/n - \bar{t} - \sigma^2/m_i)/n\}$; with large n the truncation should rarely become operative.

(b) Approximation of G by G_k

The details are essentially the same as for the parametric G case, but they are computationally more complicated. Some simplification can be effected by assuming G to be symmetrical. If $k = 3$ this means that we have three equally spaced λ_j values to estimate.

(c) Linear EB

Equation (3.8.4) becomes

$$\begin{bmatrix} 1 & \gamma_1 \\ \gamma_1 & \gamma_2 + \sigma^2/m \end{bmatrix} \begin{bmatrix} w_0 \\ w_1 \end{bmatrix} = \begin{bmatrix} \gamma_1 \\ \gamma_2 \end{bmatrix}$$

where $\gamma_r = E(\Lambda^r)$, $r = 1, 2$. Estimates of γ_1 and γ_2 can be obtained as the solutions of

$$\sum_{i=1}^{n} Y_i^*(\gamma_1, \gamma_2) = \sum_{i=1}^{n} (t_i - \gamma_1)$$

$$\sum_{i=1}^{n} Z_i^*(\gamma_1, \gamma_2) = \sum_{i=1}^{n} (t_i^2 - \gamma_2 - \sigma^2/m_i),$$

giving

$$\check{\gamma}_1 = \bar{t}; \qquad \check{\gamma}_2 = \overline{t^2} - (i/n)\sigma^2 \sum_{i=1}^{n} (1/m_i) = \bar{t}^2 - \sigma^2\overline{(1/m)}$$

and

$$\check{w}_0 = (\check{\gamma}_1\sigma^2/m)/(\check{\gamma}_2 - \check{\gamma}_1^2 + \sigma^2/m)$$
$$\check{w}_1 = (\check{\gamma}_2 - \check{\gamma}_1^2)/(\check{\gamma}_2 - \check{\gamma}_1^2 + \sigma^2/m).$$

The covariance matrix of $\hat{\gamma}_1, \hat{\gamma}_2$ can be estimated according to (3.9.5) by substituting Y_i^*, Z_i^* for U_i, V_i, etc.

3.10 Nonparametric EB estimation

When we have multiple past and current observations an argument put forward by Johns (1957) shows that it is possible to develop yet another type of EB estimator. It has two great merits. First, G need not be estimated explicitly. Second, it is nonparametric in the sense that the exact mathematical form of $f(x|\lambda)$ need not be known. A disadvantage is the difficulty of application when X is continuous.

We first consider the simplest case where Johns's method can be applied. Let there be n pairs of past observations, and one current observation, denoted as follows:

$$\begin{array}{ll} \text{past} & \text{current} \\ \overbrace{x_{11}, x_{21}, \ldots, x_{n1}} & x \\ x_{12}, x_{22}, \ldots, x_{n2} & \end{array}$$

The two observations (x_{i1}, x_{i2}) are two independent realizations of the r.v. X_t which has the d.f. $F(x|\lambda_i)$, $i = 1, 2, \ldots, n$. We assume that $E(X|\lambda) = \lambda$; re-parametrization can be carried out if necessary. Now, every pair of past observations can also be looked upon as realizations of two independent r.v.s X_1 and X_2, each of which has the d.f. $F(x|\lambda)$ for a certain value of λ. Hence the joint density of X_1, X_2 and Λ is

$$f(x_1|\lambda)f(x_2|\lambda)dG(\lambda),$$

and

$$\begin{aligned} E(X_2|X_1 = x) &= \frac{\int_\lambda \int_{x_2} x_2 dF(x_2|\lambda)f(x|\lambda)dG(\lambda)}{\int_\lambda \int_{x_2} dF(x_2|\lambda)f(x|\lambda)dG(\lambda)} \\ &= \frac{\int \lambda f(x|\lambda)dG(\lambda)}{\int f(x|\lambda)dG(\lambda)} \\ &= \delta_G(x). \end{aligned} \tag{3.10.1}$$

Equation (3.10.1) suggests that, in the case of a discrete X, we can obtain an estimate of the Bayes estimator, i.e., an EB estimator as follows: first select all those pairs (x_{i1}, x_{i2}) in which $x_{i1} = x$. Let the number of these pairs be n_x, and denote the x_{i2} values for which $x_{i1} = x$ by $x_{j2}(x)$, $j = 1, 2, \ldots, n_x$. Then, if n_x is greater than 0, the EB estimator is the mean of the $x_{j2}(x)$. Formally, the EB estimator $\delta^*(x)$ is defined by

$$\delta^*(x) = \begin{cases} \dfrac{1}{n_x} \displaystyle\sum_{j=1}^{n_x} x_{j2}(x), & n_x > 0 \\ 0, & \text{otherwise.} \end{cases} \tag{3.10.2}$$

We observe that, owing to the independence of X_1 and X_2, every $\Delta_j = \lambda_j(x) - x_{j2}(x)$, where $\lambda_j(x)$ is the parameter value obtaining when the observation $x_{j2}(x)$ was generated, may be regarded as an independent realization of a r.v. with zero mean.

Hence,

$$\delta^*(x) = \frac{1}{n_x}\sum_{j=1}^{n_x} \lambda_j(x) - \frac{1}{n_x}\sum_{j=1}^{n_x} \Delta_j \to E(\Lambda|x) - 0 = \delta_G(x),$$

in probability, as $n \to \infty$, if $n_x \to \infty$, in probability, and $P(n_x > 0) \to 1$. The latter conditions are clearly satisfied in all cases of practical interest.

When n_x is small the estimate $\delta^*(x)$ of $E(\Lambda|x)$ defined by equation (3.10.2) may be too unreliable, and to counter this effect, Johns (1957) has suggested the following modification:

$$\begin{aligned} \delta_c(x) = \delta^*(x), &\quad n_x > c \\ x, &\quad n_x < c. \end{aligned} \tag{3.10.3}$$

Thus, for $n_x < c$, a 'conventional' unbiased estimator is used. It is easily verified that $\delta_c^*(x) \to \delta_G(x)$, (P), as $n \to \infty$.

Another modification concerns the assumption, above, that we have one current observation but pairs of past observations. In circumstances where EB procedures are likely to be used, it is also likely that there will be two current observations, x_1 and x_2, say. One way of utilizing both of these is to put

$$\bar{\delta}_c^*(x_1, x_2) = \tfrac{1}{2}\delta_c^*(x_1) + \tfrac{1}{2}\delta_c^*(x_2). \tag{3.10.4}$$

Denoting $\bar{\delta}_c^*(x_1, x_2)$ by $\bar{\delta}$, we have

$$W(\bar{\delta}_c^*(x_1, x_2)) = E(\bar{\delta} - \Lambda)^2, \tag{3.10.5}$$

by definition, where E denotes integration w.r.t. λ, x_1, x_2. The r.h.s. of (3.10.5) can be expanded to

$$\begin{aligned} &E(\Lambda)^2 + \tfrac{1}{4}\{E[\delta_c^*(x_1)]^2 + E[\delta_c^*(x_2)]^2\} + \tfrac{1}{2}E\{\delta_c^*(x_1)\cdot\delta_c^*(x_2)\} \\ &\quad - E\{\Lambda\cdot\delta_c^*(x_1) + \Lambda\delta_c^*(x_2)\}. \end{aligned}$$

Also,

$$W(\delta_c^*(x_1)) = E(\Lambda)^2 + E[\delta_c^*(x_1)]^2 - 2E[\Lambda\cdot\delta_c^*(x_1)],$$

and remembering that, since x_1 and x_2 are identically distributed,

$$\begin{aligned} E[\delta_c^*(x_1)]^2 &= E[\delta_c^*(x_2)]^2 \\ E[\Lambda\cdot\delta_c^*(x_1)] &= E[\Lambda\cdot\delta_c^*(x_2)], \end{aligned}$$

we find that

$$W(\delta_c^*(x_1)) - W(\bar{\delta}) = \tfrac{1}{4}E[\delta_c^*(x_1) - \delta_c^*(x_2)]^2 \geqslant 0. \tag{3.10.6}$$

By the same argument as before,

$$\bar{\delta} \to \tfrac{1}{2}(\delta_G(x_1) + \delta_G(x_2)) = \bar{\delta}_G(x_1, x_2).$$

In general

$$\tfrac{1}{2}(\delta_G(x_1) + \delta_G(x_2)) \neq \delta_G(x_1, x_2),$$

but by the same argument giving (3.8.6),

$$W(\bar{\delta}_G(x_1, x_2)) \leqslant W(\delta_G(x_1)).$$

This result, with (3.10.6), provides some justification for this particular method of combining both current results in an EB estimator.

Referring to section 3.2, we observe that, since $\delta_c^*(x) \to \delta_G(x)$, (P), in the case of a single current observation, $\delta_c^*(x)$ is a.o. under the condition (3.2.1). In a like manner, it can be shown that

$$W(\bar{\delta}) \to W(\bar{\delta}_G(x_1, x_2)), (P), \quad \text{and} \quad E_n W(\bar{\delta}) \to W(\bar{\delta}_G(x_1, x_2)).$$

Johns (1957) has computed upper and lower bounds for $E_n(\delta_1^*(x))$ and $E_n(\delta^*(x))$ in the case where $f(x|\lambda)$ is the Poisson distribution and $G(\lambda)$ is the Γ-distribution given in Example 1.3.2 with $\alpha = 2$, $\beta = 10$. These results are shown in Table 3.7.

Table 3.7 may be compared with the appropriate entries in Table 3.4. The Johns estimators appear to be markedly better than the Robbins estimators, but it must be remembered that the Johns estimators are based on rather more past data. The Robbins estimators are relatively inefficient because cognisance is taken only of the frequencies of occurrence of x and $x + 1$, the individual values of other observations not being used directly. The Johns estimators are somewhat less wasteful since the actual values of some of the x_{i2} are used. Another relevant point is that the Johns estimators could possibly be improved by smoothing.

Table 3.7 *The non-parametric EB estimator of Johns (1957). Bounds for $E_n W(\cdot)$ in the case of a Poisson kernel, gamma (2, 10) prior distribution*

n	*Bounds* *Lower*	*Upper*	$W(\delta_G)$	$W(T)$
15	10·70	12·51	1·67	5·00
60	4·46	5·54	1·67	5·00
120	3·20	3·87	1·67	5·00

The processes described above can be extended in an obvious manner to the case where we have $r-1$ independent current observations and n sets of r independent past observations. For example, let $r=3$ and let the current observations be (x_1, x_2), the past series being n triplets (x_{i1}, x_{i2}, x_{i3}), $i=1,\ldots,n$. By steps similar to those leading to equation (3.8.1) it can be shown that

$$\begin{aligned} E(X_3 \mid X_2 = x_2, X_1 = x_1) &= \frac{\int_{x_3}\int_\lambda x_3\, dF(x_3\mid\lambda) f(x_2\mid\lambda) f(x_1\mid\lambda)\, dG(\lambda)}{\int_{x_3}\int_\lambda dF(x_3\mid\lambda) f(x_2\mid\lambda) f(x_1\mid\lambda)\, dG(\lambda)} \\ &= \delta_G(x_1, x_2). \end{aligned} \tag{3.10.7}$$

If we now select from amongst the past triplets those for which (x_{i1}, x_{i2}) or $(x_{i2}, x_{i1}) = (x_1, x_2)$, and compute the average of the corresponding x_{i3}'s an EB estimator is obtained.

Formally, let

$$M_i(x_1, x_2) = \begin{cases} 1, & \text{if one of the permutations of } (x_1, x_2) = (x_{i1}, x_{i2}) \\ 0, & \text{otherwise,} \end{cases}$$

and

$$M(x_1, x_2) = \sum_{i=1}^{n} M_i(x_1, x_2).$$

Then the EB estimator is

$$\delta^*(x_1, x_2) = \begin{cases} \dfrac{1}{M(x_1, x_2)} \displaystyle\sum_{i=1}^{n} M_i(x_1, x_2) x_{i3}, & M(x_1, x_2) > 0 \\ 0, & \text{otherwise.} \end{cases}$$

The asymptotic properties of $\delta^*(x_1, x_2)$, as $n \to \infty$, are like those of $\delta^*(x)$. Modifications analogous to those of equations (3.10.3) and (3.10.4) can be made in an obvious manner. Further, extension of the definitions to current observational vectors with r elements, and past vectors with $r+1$ elements can be readily carried out. Johns (1957) describes this general case.

Returning to the case of a single current observation, and pairs of past observations, we introduce another modification to $\delta^*(x)$. Since the past observations are randomly generated for every parameter value, there is no reason for matching x with x_{i1} rather than x_{i2}. Therefore, let $(x_{i1}^{(s)}, x_{i2}^{(s)})$, $s = 1, 2$ be the two permutations of (x_{i1}, x_{i2});

we distinguish all permutations, even if the values of x_{i1} and x_{i2} are the same.

Put

$$M_{is}(x) = \begin{cases} 1, & \text{if } x_{i1}^{(s)} = x \\ 0, & \text{otherwise,} \end{cases}$$

and

$$M(x) = \sum_{i=1}^{n} \sum_{s=1}^{2} M_{is}(x).$$

Then define the EB estimator by

$$\begin{aligned} \gamma(x) &= \frac{1}{M(x)} \sum_{i=1}^{n} \sum_{s=1}^{2} M_{is}(x) x_{i2}^{(s)}, \quad && \text{for } M(x) > 0 \\ &= x, && \text{otherwise.} \end{aligned}$$

Example 3.10.1 illustrates the difference between $\delta^*(x)$ and $\gamma(x)$.

Example 3.10.1 Let the past and current observations be:

2 3 8 4 2 0 1 2

5 4 1 3 2 4 2

Then

$$\delta^*(2) = (5 + 2)/2 = 7/2$$
$$\gamma(2) = (5 + 2 + 2 + 1)/4 = 10/4.$$

The above modification can be easily extended to observational vectors with $r + 1$ and r elements, and alterations for the case of an $r + 1$ element current vector may be made as before.

Krutchkoff (1967) has proposed an extension of the Johns method applicable when, subsequent to evaluation of every current observation, further information about the current parameter becomes available. It is assumed to take the form of a statistic, Y, independent of X, the current observation, such that $E(Y|\lambda) = \lambda$. Thus, when the current observation is $X = x$, the available information is:

Past	current
$\overbrace{x_1, x_2, \ldots, x_n}$	x
$y_1, y_2, \ldots, y_n$	

The distribution of Y need not be the same as the distribution of X. Nevertheless, by essentially the same argument as that leading to equation (3.8.1),

$$E(Y|X=x)=E(\Lambda|X=x),$$

leading to the construction of an EB estimator as before. Put

$$M_i(x)=\begin{cases}1, & \text{when } x_i=x\\ 0, & \text{otherwise,}\end{cases}$$

and

$$M(x)=\sum_{i=1}^{n} M_i(x).$$

Then the EB estimator is

$$\psi_n(x)=\begin{cases}\dfrac{1}{M(x)}\sum_{i=1}^{n} M_i(x)y_i, & \text{for } M(x)>0\\ x, & \text{otherwise.}\end{cases}$$

Asymptotic optimality of $\psi_n(x)$ has been established by Krutchkoff (1967).

There is some difficulty in applying John's approach to a continuous X. In theory, the probability of a past value equalling x is zero. However, if all results are rounded off, or grouped, that is, if the r.v. X is effectively 'discretized', this difficulty can be overcome (Johns, 1957). Evidently, problems concerning the coarseness of the grouping arise, but Johns has shown that 'discretizing' can be carried out such that asymptotically satisfactory results are obtainable.

For any finite past sample the nonparametric EB estimators defined are clearly 'non-smooth' functions of x, in general. They share this property with the Robbins EB estimators for the Poisson distribution and other members of the exponential family of distributions. The possibility of direct smoothing of such estimators was discussed in section 3.4.5. Justification of smoothing by polynomial fitting, or similar devices, may be sought in the fact that substitution of a parametric prior d.f., G, usually leads to a 'well-behaved' function, $\delta_G(x)$, as the Bayes estimator.

Krutchkoff (1967) has considered smoothing of the 'supplementary sample' nonparametric EB estimators by fitting a straight line. If $E(\Lambda|x)=\alpha+\beta x$, then, from (3.10.1), $E(Y|x)=\alpha+\beta x$, and the proposed procedure is to estimate α and β by the usual method of least squares, treating the x_i as controlled variables, and the y_i as observations subject to error.

When smoothing any EB estimators, the argument used in section 3.4.5 suggests the use of a form of weighted least squares fitting, with weights proportional to $p_G(x)$, when X is discrete. In practice the $p_G(x)$ will be unknown, but estimates directly proportional to the $M(x)$ or n_x can be formed. Using these estimated weights will, in Krutchkoff's case, lead to the same procedure which he advocates. The choice of a smoothing function remains open, but it should be influenced by knowledge of certain general characteristics of the Bayes estimator. Thus, in examples of the type discussed above it would be appropriate to impose constraints which make the function non-decreasing.

Additionally the parameter may be restricted to a certain known interval. The geometric distribution (cf. Example 1.3.4) demands attention to both types of constraints. Krutchkoff (1967) has given the following example, which shows that smoothing can be effective, even when the fitted function is not strictly appropriate: the kernel distribution is binomial with index parameter $n = 5$, and the prior distribution is normal with mean 0·2 and s.d. 0·05. The best unbiased estimator $T = X/5$ has $W(T) = 0{\cdot}0315$, while $W(\delta_G) = 0{\cdot}0023$. With $n = 30$, $W(\psi_n) = 0{\cdot}0060$. Smoothing by fitting a straight line gives $\psi_n^*(x) = a + bx$, and with $n = 30$, $W(\psi_n^*) = 0{\cdot}0037$.

3.11 Assessing performance of EBEs in practice

Many studies of the performance of EBEs have been reported, some of them to do with finite and relatively small numbers, n, of past realizations of Λ, others with asymptotic properties as $n \to \infty$. On the whole they show that EBEs can perform very well by comparison with non-Bayes estimators. The relative goodness of EBEs depends largely on the dispersion of the prior distribution, and on n. These studies do not, however, give a direct answer to the following question that arises in every particular potential application of EB estimation. Is the expected performance of the EBE better than that of the MLE for these conditions and these past data? In order to answer such a question one needs estimates of W(Bayes), W(MLE), $E_n W$(EB), at least.

In this section we shall discuss estimation of the W values needed to make a judgement in practice of the relative merits of EB and other estimation methods. The most tractable solutions of the problem are obtainable in the parametric G situation, and in linear EB estimation.

3.11.1 *Parametric G, single past and current observations, $m_i = m = 1$*

If G is known to have parametric form $G(\lambda; \boldsymbol{\alpha})$, depending on parameters $\boldsymbol{\alpha}$, the mixed X-distribution is $F(x; \boldsymbol{\alpha})$. From the past observations $x_1, x_2, \ldots, x_n$ it is possible to estimate $\boldsymbol{\alpha}$ by one of the standard methods, for example maximum likelihood. Suppose $W(\text{MLE})$, $E_n W(\text{EB})$ are functions of $\boldsymbol{\alpha}$, and that they are estimated by the same functions of $\check{\boldsymbol{\alpha}}$. In particular $\check{\boldsymbol{\alpha}}$ may be $\hat{\boldsymbol{\alpha}}$, the MLE of $\boldsymbol{\alpha}$.

In order to decide, say, between EB and ML estimation we may need the estimate $\check{W}(\text{MLE})$ of $W(\text{MLE})$ and its standard error, and similarly for the other estimation methods. Refinements might be to ask for estimates of $W(\text{ML}) - W(\text{Bayes})$, with standard errors, and so on. Now, in principle estimation of $W(\text{ML})$, $W(\text{Bayes})$ and $W(\cdot)$ for any other non-Bayes method is straightforward. Estimation of $E_n W(\text{EB})$ is less straightforward, and we shall give some illustrations in the following examples.

(a) Poisson data distribution, gamma prior G

Suppose that the prior gamma density is written as in Example 1.3.7, $dG(\lambda) = \{1 | \Gamma(\beta)\} \alpha^{\beta} \lambda^{\beta-1} e^{-\alpha\lambda}$, $\lambda > 0$. Then the MLEs of α and β from the past data are the solutions of

$$\frac{\partial \ln L}{\partial \alpha} = \frac{n\beta}{\alpha} - \sum_{i=1}^{n} \frac{\beta + \alpha_i}{\alpha + 1} = 0$$

$$\frac{\partial \ln L}{\partial \beta} = n \ln \alpha - n \ln(\alpha + 1) + \sum_{i=1}^{n} \psi(\beta + x_i) - n\psi(\beta) = 0;$$

these equations are the special versions of (3.9.3) with $m_i = 1$. We shall follow the steps in section 3.9.1. Hence we put

$$U_i(a, b) = b/a - (b + x_i)/(a + 1)$$
$$V_i(a, b) = \ln a - \ln(a + 1) - \psi(b + x_i) + \psi(b)$$

and obtain

$$\partial U_i/\partial a = -b/a^2 + (b + x_i)/(a + 1)^2$$
$$\partial U_i/\partial b = 1/a - 1/(a + 1)$$
$$\partial V_i/\partial a = 1/a - 1/(a + 1)$$
$$\partial V_i/\partial b = -\psi'(b + x_i) - \psi'(b).$$

Using these results we can obtain an estimate covariance matrix for the MLEs $\hat{\alpha}$, $\hat{\beta}$.

For the EBE based on $\hat{\alpha}$, $\hat{\beta}$ we have

$$W(\text{EB}) - W(\text{Bayes}) = \left\{ \frac{\hat{\beta} + \beta/\alpha}{\hat{\alpha} + 1} - \frac{\beta + \beta/\alpha}{\alpha + 1} \right\}^2 + \frac{(\alpha - \hat{\alpha})^2 \beta}{(\hat{\alpha} + 1)^2 (\alpha + 1) \alpha^2}. \tag{3.11.1}$$

A fairly simple approximation for $E_n W(\text{EB}) - W(\text{Bayes})$ is obtained by replacing $\hat{\alpha}$ in the denominators in (3.11.1) by α before finally taking expectations. The result is

$$E_n W(\text{EB}) - W(\text{Bayes}) \simeq \frac{\text{var}(\hat{\beta}) + (\beta/\alpha)^2 \text{var}(\hat{\alpha}) - \partial(\beta/\alpha) \text{cov}(\hat{\alpha}, \hat{\beta})}{(\alpha + 1)^2} + \frac{\text{var}(\hat{\alpha})}{(\alpha + 1)^3} \frac{\beta}{\alpha^2}. \tag{3.11.2}$$

Example 3.11.1 We take the $n = 50$ observations to be those summarized in the column headed $f_n(x)$ of Table 3.2. Straightforward calculations give the estimated matrix $\mathbf{\Delta}$ as

$$\hat{\mathbf{\Delta}} = \begin{pmatrix} -82{\cdot}575 & 16{\cdot}53 \\ 16{\cdot}53 & -3{\cdot}4189 \end{pmatrix}$$

and

$$\hat{\mathbf{V}} = 50 \begin{pmatrix} 1{\cdot}7466 & -{\cdot}3528 \\ -{\cdot}3528 & 0{\cdot}0725 \end{pmatrix}$$

giving the estimated covariance matrix of $\hat{\alpha}$, $\hat{\beta}$ as

$$\hat{\mathbf{c}} = \begin{pmatrix} 0{\cdot}21 & 1{\cdot}01 \\ 1{\cdot}01 & 5{\cdot}15 \end{pmatrix}.$$

In the calculations above we used the MLEs $\hat{\alpha} = 1{\cdot}310$, $\hat{\beta} = 6{\cdot}548$.

The estimates of $W(\text{Bayes})$, $W(\text{ML})$ are

$$\hat{W}(\text{Bayes}) = \frac{\hat{\beta}}{\hat{\alpha}(\hat{\alpha} + 1)} = 2{\cdot}16$$

$$\hat{W}(\text{ML}) = \hat{\beta}/\hat{\alpha} = 5{\cdot}00$$

From formula (3.11.2) the estimate of $E_n W(\text{EB}) - W(\text{Bayes})$ is 0·12. Hence

$$\hat{E}_n W(\text{EB}) = 2{\cdot}28.$$

These results suggest that EB estimation should be preferred to ML in conditions generating the observed data. Recall that the particular set of data was generated with $\alpha = 2$, $\beta = 10$, a case studied in some detail in section 3.7.1. The estimate $\hat{E}_n W(\text{EB}) = 2{\cdot}28$ agrees quite well with the $E_n W(\text{EB})$ values reported in Table 3.4.

The estimates of $W(\text{Bayes})$, $W(\text{ML})$, $E_n W(\text{EB})$ obtained as in Example 3.11.1 have standard errors which could be calculated if a more formal decision between EB and ML is to be made. To obtain the s.e.s of the estimates of $W(\text{Bayes})$ and $W(\text{ML})$ is straightforward, for $E_n W(\text{EB})$ it can be more complicated. However every estimate of these quantities can, in some circumstances, be expressed in terms of $\hat{\alpha}$, $\hat{\beta}$, so that an estimate of the s.e. can be calculated. The circumstance in which this can be done quite easily is where it is appropriate to express $\mathbf{C}$ in (3.9.5) as the inverse information matrix, whose elements are functions of α and β. Then var $(\hat{\alpha})$, etc., in (3.11.2) can be replaced by these expressions, $E_n W(\text{EB}) - W(\text{Bayes})$ can be expressed as a function $H_n(\alpha, \beta)$, estimated by $H_n(\hat{\alpha}, \hat{\beta})$ whose s.e. can be estimated in the obvious way. This procedure will be illustrated in the case of the normal data distribution for which the actual manipulations are less complicated.

(b) Normal data distribution, normal prior G

Following the notation of section 3.7.3 we estimate μ_G by $\bar{x}$, the mean of the previous observations, and $\sigma_G^2 + 1$ by s^2, the sample variance of the previous observations. For the present purpose we shall take $\hat{\sigma}_G^2 = s^2 - 1$, not $\max(0, s^2 - 1)$, since this simplifies calculations, and introduces no complications to do with $\hat{\delta}_G$, $W(\hat{\delta}_G)$ or $E_n W(\hat{\delta}_G)$.

The EB estimate is

$$\hat{\delta}_G(x) = \frac{x(s^2 - 1) + \bar{x}}{s^2},$$

and from (3.5.3)

$$E_n W(\hat{\delta}_G) - W(\delta_G) = \left\{\frac{n-1}{n(n-3)} + \frac{2n+6}{(n-3)(n-5)}\right\} \Big/ (1 + \sigma_G^2).$$

$$W(\hat{\delta}_G) = \{\sigma_G^2 + (\bar{x} - \mu_G)^2 + (s^2 - 1)^2\}/s^4.$$

The estimate of $E_n W(\text{EB}) - W(\text{Bayes})$ is, therefore,

$$\hat{D}(\text{EB}) = \left\{\frac{n-1}{n(n-3)} + \frac{2n+6}{(n-3)(n-5)}\right\} \Big/ s^2$$

and the estimate of $W(\text{ML}) - W(\text{Bayes})$ is $\hat{D}(\text{ML}) = 1/s^2$.

We might choose between EB and ML on the basis of $\hat{D}(\text{ML, EB}) = \hat{D}(\text{ML}) - \hat{D}(\text{EB})$ for which we have

$$E\{\hat{D}(\text{ML, EB})\} = \left\{1 - \frac{n-1}{n(n-3)} - \frac{2n+6}{(n-3)(n-5)}\right\}\left(\frac{n-1}{n-3}\right)\frac{1}{(1+\sigma_G^2)}$$

$$\text{var}\,\{\hat{D}(\text{ML, EB})\} = \left\{1 - \frac{n-1}{n(n-3)} - \frac{2n+6}{(n-3)(n-5)}\right\}^2$$
$$\times \left\{\frac{(n-1)}{(n-3)(n-5)} - \frac{(n-1)^2}{(n-3)^2}\right\} \times \frac{1}{(1+\sigma_G^2)^2}.$$

Example 3.11.2 Suppose that $n = 20$ past observations give the result $s^2 = 1{\cdot}32$. Then the estimated value of $E\{\hat{D}(\text{ML, EB})\}$ is 0·65 with estimated s.e. $= 0{\cdot}26$. Such a result would indicate superiority of EB over ML.

3.11.2 Linear EB: $m_i = m = 1$

(a) Poisson data distribution

Following section 3.9.1 (d) with $m_i = 1, t = x$, the linear EB estimate is

$$\delta(x; \check{w}_0, \check{w}_1) = \check{w}_0 + \check{w}_1 x$$

where

$$\check{w}_0 = \bar{x}^2/(\overline{x^2} - (\bar{x})^2)$$
$$\check{w}_1 = \{\overline{x^2} - (\bar{x})^2 - \bar{x}\}/(\overline{x^2} - (\bar{x})^2).$$

Large n approximations for $E_n W(\text{linear EB})$ can be obtained by taking $\check{w}_0$, $\check{w}_1$ to be approximately unbiased for w_0, w_1, giving

$$E_n W(\text{linear EB}) - W(\text{linear Bayes}) \simeq \text{var}\,(\check{w}_0)$$
$$+ (\gamma_1^2 + \gamma_2)\,\text{var}\,(\check{w}_1) + 2\gamma_1\,\text{cov}\,(\check{w}_0, \check{w}_1).$$

Approximate values for $\text{var}\,(\check{w}_0)$, $\text{var}\,(\check{w}_1)$, $\text{cov}\,(\check{w}_0, \check{w}_1)$ can be obtained as suggested in section 3.9.1(d) or directly by using standard formulae for variances and covariances of moments.

Also

$$W(\text{linear Bayes}) = w_0^2 + (w_1 - 1)^2(\gamma_1^2 + \gamma_2 - \gamma_1)$$
$$+ \{2(w_1 - 1)w_0 + w_1^2\}\gamma_1$$

where $\gamma_r = E(\Lambda^r)$, $r = 1, 2$, and w_0, w_1 can be expressed in terms of

γ_1, γ_2. An estimate of W(linear Bayes) is obtained by noting that estimates of $\check{\gamma}_1$, $\check{\gamma}_2$ are $\check{\gamma}_1 = \bar{x}$, $\check{\gamma}_2 = \overline{x^2} - \bar{x}$.

(b) The normal data distribution

Taking the known $\sigma^2 = 1$ in section 3.9.3(c), and $m_i = m = 1$ the linear EBE becomes

$$\delta(x; \check{w}_0, \check{w}_0) = \check{w}_0 + \check{w}_1 x$$

with

$$\check{w}_0 = \bar{x}/s^2$$
$$\check{w}_1 = (s^2 - 1)/s^2.$$

We can express $E_n W$(linear EB) – W(linear Bayes) as

$$E_n\left\{\frac{\bar{x} - E(X_G)}{s^2}\right\}^2 + \operatorname{var}(X_G) E_n\left\{\frac{1}{s^2} - \frac{1}{\operatorname{var}(X_G)}\right\}^2.$$

A crude but useful approximation is obtained by putting s^2 in the denominators of the expression above equal to $\operatorname{var}(X_G)$, giving

$$\{1 + \mu_4(X_G) - \mu_2^2(X_G)\}/\{n\mu_2(X_G)\}$$

where $\mu_r(X_G)$ are the central moments of the marginal X-distribution. These can be estimated directly from the past observations.

CHAPTER 4

Empirical Bayes point estimation: vector parameters

4.1 Introduction

Most realistic estimation problems involve vector parameters. In univariate studies the simplest problems with more than one parameter tend to be those of location and scale, but many others occur where the data distribution F depends on more than one parameter. Multivariate data distributions usually have vector parameters of dimension greater than one.

The applications of EB methods discussed in Chapter 8 illustrate the statements made above. They also indicate that interest often centres on just one parameter or one function of several parameters. Keeping in mind that the general idea behind the use of EB methods is to improve the precision of individual estimates by using past results, one may well decide to apply an EB approach only to parameters of primary concern. We shall discuss this in more detail for the location-scale problem.

Vector parameter problems occur naturally in linear regression. In the usual general linear model it is reasonable to assume that the covariance matrix of the estimates of the parameters of interest is known apart from a multiplicative constant, the residual variance. This leads to a special class of EB problems, and in particular, those involving the multivariate normal distribution with known covariance matrix.

4.2 Location–scale estimation

We deal here with data distributions of the form $F\{(x-\lambda)/\sigma\}$; λ and σ are, respectively the location and scale parameters. Generally we may consider past random samples of sizes m_i, $i = 1, 2, \ldots, n$, obtained with past realizations (λ_i, σ_i) of r.v.s (Λ, Σ), from populations of the form F. However, a rather common type of statistical model is one in which σ does not vary, i.e. Σ is actually a constant σ. As an example, one can consider the one-way analysis of variance model with random location effects.

4.2.1 *Fixed unknown σ*

Every past sample with $m_i > 1$ provides an estimate of σ, and the natural steps are to pool these estimates into one estimate $\hat{\sigma}$ of σ. Then we replace σ by $\hat{\sigma}$ in whatever formulae arise for Bayes or EB estimates. Apart from this obvious change the methods for obtaining EB estimates are exactly like those described in Chapter 3.

Example 4.2.1 $F\{(x-\lambda)/\sigma\}$ is the $N(\lambda, \sigma^2)$ distribution, $m_i = m > 1$, the distribution of Λ is $N(\mu_G, \sigma_G^2)$. Then $\hat{\sigma}^2$ is the usual pooled sample variance, $\hat{\sigma}^2/m + \hat{\sigma}_G^2 = s^2$, the sample variance of the estimates $\bar{x}_i$ of the individual λ_i and the EBE of the current λ is

$$\hat{\delta}_G(\bar{x}) = (m\bar{x}\hat{\sigma}_G^2 + \hat{\mu}_G\hat{\sigma}^2)/(ms^2).$$

Calculations like those giving (3.5.2) lead to

$$E_n W(\text{EB}) = \frac{(m\sigma_G^2 + \sigma^2)}{nm} E_n\left(\frac{1}{ms^2}\right)^2 + \frac{m}{(m\sigma_G^2 + \sigma^2)} E_n\left\{\frac{\hat{\sigma}_G^2\sigma^2 - \sigma_G^2\hat{\sigma}^2}{ms^2}\right\}^2$$

which can be simplified somewhat by steps like those giving (3.5.3). The result is

$$E_n W(\text{EB}) \simeq \left\{\frac{1}{n}\frac{(n-1)^2}{(n-3)(n-5)} + \frac{\sigma^2}{m^2}\left(1 - 2\left(\frac{n-1}{n-3}\right) + \frac{n(m-1)+2}{n(m-1)}\frac{(n-1)^2}{(n-3)(n-5)}\right)\right\}\Big/(\sigma_G^2 + \sigma^2/m) + W(\text{Bayes}).$$

The details for non-normal $F((x-\lambda)/\sigma)$ are more complicated if one attempts to derive an EBE which is an estimate of the actual

Bayes estimate. Two compromises which have been discussed before are:

1. reducing the m sample values of the MLE $\hat{\lambda}$ of λ, treating $\hat{\lambda}$ as being normally distributed, i.e. $N(\lambda, c\sigma^2/m)$, where c is a constant determined by the form of F;
2. adopting a linear Bayes approach.

4.2.2 *Fixed unknown σ: linear Bayes*

Suppose that $\check{\lambda}$ is an estimate of λ, unbiased with $\text{var}(\check{\lambda}) = k\sigma^2/m$; the constant k, like c above, is determined by the form of F and the type of estimate $\check{\lambda}$. Then a linear Bayes estimate, based on $\check{\lambda}$, is $\delta(\check{\lambda}; w_0, w_1) = w_0 + w_1, \check{\lambda}$ with (w_0, w_1) given by

$$\begin{bmatrix} 1 & E(\Lambda) \\ E(\Lambda) & E(\Lambda^2) + k\sigma^2/m \end{bmatrix} \begin{bmatrix} w_0 \\ w_1 \end{bmatrix} = \begin{bmatrix} E(\Lambda) \\ E(\Lambda^2) \end{bmatrix}; \tag{4.2.1}$$

see also (1.12.2).

An empirical version of the linear Bayes estimate is obtained simply by replacing the elements of the l.h.s. matrix and r.h.s. vector in (4.2.1) by estimates derived from the past observations. Estimation of σ can be carried out in diverse ways, but generally one would take the final estimate $\check{\sigma}$ to be a pooled version of estimates derived from the n past samples, each of size m.

Example 4.2.2 Let $(1/\sigma) f((x - \lambda)/\sigma) = (1/2\sigma) \exp(-|(x - \lambda)/\sigma|)$, $-\infty < x < \infty$, the double exponential distribution. A simple estimate of λ_i, in the ith past sample is the sample mean $\bar{\lambda}_i$, and the ith sample variance, v_i, is an estimate of $2\sigma^2$. Taking the final estimate $\bar{\sigma}^2$ of σ^2 to be $\frac{1}{2}$(pooled sample variance over n past samples), (4.2.1) becomes

$$\begin{bmatrix} 1 & A_1 \\ A_1 & A_2 \end{bmatrix} \begin{bmatrix} \bar{w}_0 \\ \bar{w}_1 \end{bmatrix} = \begin{bmatrix} A_1 \\ A_2 - 2\bar{\sigma}^2/m \end{bmatrix}, \tag{4.2.2}$$

where A_1, A_2 and $\bar{\sigma}$ are the estimates obtained below.

From an actual data set the difference $E_n W(\text{linear EB}) - W(\text{linear Bayes})$ can be estimated by first finding approximate values for the variances and covariances of $A_1, A_2, \bar{\sigma}^2$. These can be obtained by the method discussed in sections 3.9 and 3.11. We obtain $A_1, A_2, \bar{\sigma}^2$ from the estimating equations

$$\left.\begin{aligned} \sum_{i=1}^{n} (\bar{\lambda}_i - A_1) &= \sum U_{1i} = 0 \\ \sum_{i=1}^{n} (\bar{\lambda}_i^2 - A_2) &= \sum U_{2i} = 0 \\ \sum_{i=1}^{n} (v_i/2 - \bar{\sigma}^2) &= \sum U_{3i} = 0 \end{aligned}\right\}. \tag{4.2.3}$$

Application of the previously described methods is now straightforward. The covariance matrix of $A_1, A_2, \bar{\sigma}^2$ is estimated by $\bar{\mathbf{C}}$ whose elements are $(1/n^2)\sum U_{pi}U_{qi}$, $p, q = 1, 2, 3$; note that the method (3.9.4) is followed and that $\mathbf{\Delta} = n\mathbf{I}$ in this case.

Since $\bar{w}_0, \bar{w}_1$ are obtained from the estimating equations (4.2.2), which can be written in the same form as (4.2.3), the approximate covariance matrix of $\bar{w}_0, \bar{w}_1$ is estimated by

$$\mathbf{P}^{-1}\mathbf{Q}\bar{\mathbf{C}}\mathbf{Q}'(\mathbf{P}^{-1})^{\mathrm{T}}$$

where $\mathbf{P} = \begin{bmatrix} 1 & A_1 \\ A_1 & A_2 \end{bmatrix}$, $\mathbf{Q} = \begin{bmatrix} \bar{w}_1 & 0 & 0 \\ \bar{w}_2 & \bar{w} & -2/m \end{bmatrix}$. The value of W(linear Bayes) is estimated by

$$\{\bar{w}_0 + (\bar{w}_1 - 1)A_1\}^2 + (\bar{w}_1 - 1)^2\{A_2 - 2\bar{\sigma}^2/m - A_1^2\} + \bar{w}_1^2(2\bar{\sigma}^2/m)$$

and $E_n W$(linear EB) by

$$W(\text{linear Bayes}) + \widehat{\text{var}}(\bar{w}_0) + 2A_1\,\widehat{\text{cov}}(\bar{w}_0, \bar{w}_1) + A_2\,\text{var}(\bar{w}_1) \tag{4.2.4}$$

Finally, for the non-Bayes estimate $\bar{\lambda}$, $W(\bar{\lambda})$ is estimated by $2\bar{\sigma}^2/m$.

Comments on Example 4.2.2:

1. The procedure outlined in the example applies with minor changes to many other location–scale distributions.
2. $E_n W$(linear EB) is given approximately by (4.2.4) with all estimates replaced by the corresponding parameter values.
3. Linear EB quantile estimation, a topic considered more generally in section 4.2.4, is straightforward here because we can express the p-quantile as $\xi_p = \lambda + \gamma_p\sigma$, and the linear EBE of ξ_p is $\delta(\bar{\lambda}; \bar{w}_0, \bar{w}_1) + \gamma_p\bar{\sigma}$, when γ_p is a known constant.

4.2.3 *Estimating both λ and σ: parametric G*

If it is agreed to estimate λ and σ by the appropriate means of the posterior joint distribution of Λ and Σ no new principle is involved. But calculations can be complicated, depending on the form of F. If a two-dimensional sufficient statistic for (λ, σ) does not exist choosing a suitable type of prior G is not straightforward. Moreover, as we have seen before, calculation of Bayes and EB estimates tends to be difficult in these cases, involving tedious integration. The only relatively simple cases seem to be the normal data distribution and the use, for whatever reason, of a finite joint distribution of λ and σ.

(a) The normal $N(\lambda, \sigma^2)$ data distribution

The joint sufficiency of the sample mean and variance for λ and σ^2 and the existence of a relatively tractable natural conjugate prior can be exploited in this case. For convenience we write $\beta = \sigma^2$, $\bar{x}$ for the usual sample mean and $b = \sum_{i=1}^{m}(x_i - \bar{x})^2/(m-1)$. Then the joint p.d.f. of $\bar{x}$ and b is

$$f(\bar{x}, b|\lambda, \beta) = (\text{const.})\beta^{-(m-1)/2} b^{m/2-3/2} \times \exp[-m\{(m-1)b/m + (\bar{x}-\lambda)^2\}/2\beta]. \tag{4.2.5}$$

Following Raiffa and Schlaifer (1961) and Evans (1964), the natural conjugate prior distribution of λ and β has the joint p.d.f.

$$g(\lambda, \beta) = (\text{const.})\beta^{-(1+(1/2)\nu)} \exp[-\{A + Z(\lambda - D)^2\}/2\beta], \tag{4.2.6}$$

where ν, A, $Z > 0$ and $-\infty < D < +\infty$ are constants.

While we may, in general, be interested in estimation of a variety of functions of λ and β, we shall here consider only estimation of λ, β and the p-quantile $\lambda + \gamma_p\sqrt{\beta}$. In every case we shall use the quadratic loss function, so that the Bayes estimate is the posterior mean of the function being estimated.

For λ we have the Bayes estimator

$$E(\lambda|\bar{x}, b) = \frac{\iint \lambda f(\bar{x}, b|\lambda, \beta) g(\lambda, \beta) d\lambda d\beta}{\iint f(\bar{x}, b|\lambda, \beta) g(\lambda, \beta) d\lambda d\beta},$$

and noting that

$$m(\bar{x}-\lambda)^2 + Z(\lambda - D)^2 = (m+Z)\left[\lambda - \frac{m\bar{x} + ZD}{(m+Z)}\right]^2 + \frac{mZ}{(m+Z)}(\bar{x} - D)^2, \tag{4.2.7}$$

it follows easily that

$$E(\lambda|\bar{x}, b) = \frac{m\bar{x} + ZD}{m + Z}. \tag{4.2.8}$$

Again making use of (4.2.7), the Bayes estimator of β is

$$E(\beta|\bar{x}, b) = \frac{(m-1)b + A + mZ(\bar{x} - D)^2/(m+Z)}{(m + v - 3)},$$

a result given by Evans (1964). By the same technique the Bayes estimator of $\sqrt{\beta}$ is

$$E(\sqrt{\beta}|\bar{x}, b) = \left\{\frac{(m-1)b + A + mZ(\bar{x} - D)^2/(m+Z)}{2}\right\}^{1/2} \times \frac{\Gamma[\frac{1}{2}(m+v) - 1]}{\Gamma[\frac{1}{2}(m+v) - \frac{1}{2}]}.$$

To check that the estimator of $\sqrt{\beta}$ is of the 'proper' form, we note that, as m becomes large,

$$\frac{\Gamma[\frac{1}{2}(m+v) - \frac{1}{2}}{\Gamma[\frac{1}{2}(m+v) - 1]} \sim \left(\frac{m+v}{2} - 1\right)^{1/2}\left(1 - \frac{1}{2m + 2v - 1}\right),$$

so that $E(\sqrt{\beta}|\bar{x}, b) \sim \sqrt{b}$. Finally, the Bayes estimator of $\lambda + \gamma_p\sqrt{\beta}$ is

$$E(\lambda|\bar{x}, b) + \gamma_p E(\sqrt{\beta}|\bar{x}, b).$$

Let us now suppose that we have an EB situation, n past estimates $(\bar{x}_i, b_i)$, $i = 1, 2, \ldots, n$, being available, generated in the same way as $\bar{x}$ and b. In the notation of section 3.8 we are taking $m_i = m$; the variable m_i case is slightly more complicated. They may be regarded as n pairs of observations drawn at random from the bivariate population whose p.d.f. is

$$f_G(\bar{x}, b) = \iint f(\bar{x}, b|\lambda, \beta)g(\lambda, \beta)d\lambda d\beta$$

$$= K\frac{b^{1/2(m-3)}}{[(m-1)b + A + mZ(\bar{x} - D)^2/(m+Z)]^{1/2(v+m)-1}}.$$

The method of maximum likelihood can be employed to estimate the parameters A, Z, D, v, but the method of moments is more tractable. To find integrals of the type $\iint \bar{x}^r b^s f_G(\bar{x}, b)d\bar{x}$ it is easier to use the form $\mu'_{rs} = \iiiint \bar{x}^r b^s f(\bar{x}, b|\lambda, \beta)d\bar{x}dbd\lambda d\beta$, integrating first w.r.t. $\bar{x}$ and b. The following results are obtained:

$$\left.\begin{aligned} \mu'_{10} &= D \\ \mu'_{01} &= \frac{A}{\nu-3} \\ \mu'_{20} &= \left(\frac{A}{\nu-3}\right)\left(\frac{1}{m}+\frac{1}{Z}\right)+D^2 \\ \mu'_{02} &= \left(\frac{m+1}{m-1}\right)\left(\frac{A}{\nu-3}\right)\left(\frac{A}{\nu-5}\right). \end{aligned}\right\} \tag{4.2.9}$$

It will be noticed that we require $\nu > 5$ for these moments to exist. Estimates of the μ'_{rs} from past data are just the first and second sample moments of the $\bar{x}_i$ and b_i. For example, μ'_{10} is estimated by $\breve{\mu}'_{10} = (\bar{x}_1 + \bar{x}_2 + \cdots + \bar{x}_n)/n$. Equations (4.2.9) can be solved to express A, Z, D, ν in terms of $\mu'_{10}, \ldots, \mu'_{02}$, and replacing μ'_{rs} by $\breve{\mu}'_{rs}$, estimates $\breve{A}$, $\breve{Z}$, $\breve{D}$, $\breve{\nu}$ are found. The restriction that $\nu > 5$ must be observed, but since $\nu = 5 + 2k\mu'_{01}/(\mu'_{02} - k\mu'^2_{01})$, where $k = (m+1)/(m-1)$, this poses no difficulty unless $\breve{\mu}'_{02} < k\breve{\mu}'^2_{01}$. Since $\nu > 5$ implies that $\mu'_{02}/\mu'^2_{01} > k$, decreasing as ν increases, we adopt the convention of letting $\breve{\nu} = +\infty$ when $\breve{\mu}'_{02} < k\breve{\mu}'^2_{01}$. This means that $\breve{A} \to +\infty$ as $\breve{\nu} \to +\infty$ since $\breve{\mu}'_{01}$ will be finite and positive. An estimate of Z is obtained from

$$(1/Z) + (1/m) = \mu_{20}/\mu'_{01},$$

and we use the convention that $\breve{Z} = +\infty$ when $(\mu_{20}/\mu'_{01} - 1/m) \leqslant 0$. In practical computations it is convenient to replace $\breve{Z}$ and $\breve{\nu}$ by large positive numbers when the conventions suggest the value $+\infty$.

The EB estimators are, as usual, obtained on replacing A, Z, D, ν, by their estimates in the expressions for the Bayes estimators; for example, denoting the EB estimators by

$$\breve{E}(\lambda \mid \bar{x}, b), \qquad \breve{E}(\beta \mid \bar{x}, b), \qquad \breve{E}(\sqrt{\beta} \mid \bar{x}, b),$$

we have

$$\breve{E}(\lambda \mid \bar{x}, b) = (m\bar{x} + \breve{Z}\breve{D})/(m + \breve{Z}), \text{ etc.}$$

Since

$$\left(\frac{Z}{m+Z}\right) = \left(\frac{A}{\nu-3}\right)\Big/(m\mu'_{20}),$$

the effect of letting $\breve{\nu} \to +\infty$ while $[\breve{A}/(\breve{\nu}-3)]$ remains finite, is that

$$\breve{E}(\beta \mid \bar{x}, b) \to \breve{\mu}'_{01}$$

and

$$\check{E}(\sqrt{\beta}\,|\,\bar{x}, b) \rightarrow (\check{\mu}'_{01})^{1/2}.$$

When we do not have all $m_i = m$ the estimating equations for A, D, Z, v, derived from (4.2.9) become

$$\left.\begin{aligned}
\check{\mu}_{10} &= \check{D} \\
\check{\mu}'_{01} &= \check{A}/(\check{v} - 3) \\
\check{\mu}_{20} &= \{\check{A}/(\check{v} - 3)\}\left\{\left(\sum_{i=1}^{n} 1/m_i\right) + 1/\check{Z}\right\} + \check{D}^2 \\
\mu_{02} &= \check{A}^2\left\{\sum_{i=1}^{n} (m_i + 1)/(m_i - 1)\right\}\Big/\{(\check{v} - 3)(\check{v} - 5)\}
\end{aligned}\right\} \qquad (4.2.10)$$

The sample values $\check{\mu}_{10}$, etc. are defined as before.

Performance of the EBEs. Consider first the EBE of λ. It can be expressed as

$$\check{E}(\lambda\,|\,\bar{x}, b) = \check{\omega}_0 + \check{\omega}_1 \bar{x}$$

with $\check{\omega}_0 = \check{Z}\check{D}/(m + \check{Z})$, $\check{\omega}_1 = m/(m + \check{Z})$. For this estimate of λ we have

$$\begin{aligned}
W(\check{\omega}_0 + \check{\omega}_1 \bar{x}) &= \{(\check{\omega}_1 - 1)D + \check{\omega}_0\}^2 \\
&\quad + \{(\check{\omega}_1 - 1)^2/Z + \check{\omega}_1^2/m\}A/(v - 3) \qquad (4.2.11) \\
W(\omega_0 + \omega_1 \bar{x}) &= A/\{(m + Z)(v - 3)\} \\
W(\bar{x}) &= A/\{m(v - 3)\}.
\end{aligned}$$

From (4.2.11), assuming $\check{\omega}_0, \check{\omega}_1$ to be approximately unbiased for ω_0, ω_1,

$$\begin{aligned}
E_n W(\text{EB}) \simeq{} & D^2 \operatorname{var}(\check{\omega}_1) + \operatorname{var}(\check{\omega}_0) + 2D \operatorname{cov}(\check{\omega}_1, \check{\omega}_0) \\
& + \operatorname{var}(\check{\omega}_1)(m + Z)A/\{(v - 3)mZ\} + W(\text{Bayes}).
\end{aligned}$$

Given a particular data set the covariance matrix of $\check{\omega}_0, \check{\omega}_1$ can be estimated by the technique discussed in Example 4.2.2, using the estimating equations (4.2.10). An approximate expression for $E_n W(\text{EB})$ in terms of the parameters A, Z, D, v can be obtained on replacing $\bar{\mathbf{C}}$ by the approximate matrix of theoretical moments.

We can write the EBE of β as

$$\check{E}(\beta\,|\,\bar{x}, b) = \check{h}_0 + \check{h}_1 b + \check{h}_2(\bar{x} - \check{D})^2$$

from which expressions for the relevant expected losses $W(\cdot)$ can be derived. They are somewhat more complicated than the corresponding results for λ. Expressions for the relevant quantities when estimating $\sqrt{\beta}$ are even more difficult. Estimation of quantiles therefore presents a rather difficult problem, even in the apparently simple case of the normal data distribution. We defer further discussion to section 4.3 where an alternative approach to estimation of quantiles is considered. Finally we note that estimation of the various parameters could be performed by the ML method instead of the method of moments as was done above.

(b) Finite G

By finite G is meant a distribution with a finite number of discrete mass points. The marginal distributions of both λ and β are, therefore, finite step-functions. As an example we shall consider a distribution with mass points of equal weight 1/6 at $(\beta_1, D \pm \delta_1)$, $(\beta_2, D \pm \delta_2)$, $(\beta_3, D \pm \delta_3)$ with $0 \leqslant \beta_1 \leqslant \beta_2 \leqslant \beta_3$, $0 \leqslant \delta_1 \leqslant \delta_2 \leqslant \delta_3$. This distribution has some of the features of the distribution (4.2.6); and many other configurations are possible.

For given $\boldsymbol{\beta}$, D, $\boldsymbol{\delta}$ calculation of the Bayes estimate of λ straightforward according to

$$E(\Lambda|\mathbf{x})$$

$$= \frac{\sum_{i=1}^{3} \left\{(D-\delta_i)\prod_{j=1}^{m} f\left(\frac{x_j - D + \delta_i}{\beta_i}\right) + (D+\delta_i)\prod_{j=1}^{m} f\left(\frac{x_j - D - \delta_i}{\beta_i}\right)\right\}}{\sum_{i=1}^{3} \left\{\prod_{j=1}^{m} f\left(\frac{x_j - D + \delta_i}{\beta_i}\right) + \prod_{j=1}^{m} f\left(\frac{x_j - D - \delta_i}{\beta_i}\right)\right\}}$$

but evaluation of W(Bayes), etc., is very tedious.

Estimation of $\boldsymbol{\beta}$, D, $\boldsymbol{\delta}$ under the EB sampling scheme is easiest by the method of moments; the details are similar to those for linear EB estimation, as given in section 4.2.4. Alternatively, one may fix the above mesh points and assign probability masses θ_i to them. The θ_i can then be estimated by application of the EM algorithm.

4.2.4 *Estimating both λ and σ: linear Bayes and EB*

There are several ways in which the idea of linear Bayes estimation can be applied in the present context. The simplest seems to be to begin with conventional estimates $\check{\lambda}$ and $\check{\sigma}$ of λ and σ; typically they may be linear functions of order statistics, unbiased with minimum variance. Now consider estimation of a linear function $A_\lambda\lambda + A_\sigma\sigma$, A_λ and A_σ being given constants. By choosing A_λ, A_σ suitably this includes estimation of λ, σ or a p-quantile.

Suppose that we estimate $A_\lambda\lambda + A_\sigma\sigma$ by

$$\delta(\check{\lambda}, \check{\sigma}; \boldsymbol{\omega}) = \omega_0 + \omega_\lambda\check{\lambda} + \omega_\sigma\check{\sigma}$$

where ω_0, ω_λ, ω_σ are chosen to minimize

$$\iint(\omega_0 + \omega_\lambda\check{\lambda} + \omega_\sigma\check{\sigma} - A_\lambda\lambda - A_\sigma\sigma)^2 dF(\check{\lambda}, \check{\sigma}|\lambda, \sigma)dG(\lambda, \sigma).$$

Let $\gamma_{rs} = \int\lambda^r\sigma^s\, dG(\lambda, \sigma)$, $\operatorname{var}(\check{\lambda}|\lambda, \sigma) = \sigma^2 V_{11}(m)$, $\operatorname{var}(\check{\sigma}|\lambda, \sigma) = \sigma^2 V_{22}(m)$, $\operatorname{cov}(\check{\lambda}, \check{\sigma}|\lambda, \sigma) = \sigma^2 V_{12}(m)$, $E_G(\check{\lambda}^r\check{\sigma}^s) = \iint\check{\lambda}^r\check{\sigma}^s\, dF(\check{\lambda}, \check{\sigma}|\lambda, \sigma)\cdot dG(\lambda, \sigma)$. Then the optimum, i.e. linear Bayes $\boldsymbol{\omega}$ is given by

$$\begin{bmatrix} 1 & E_G(\check{\lambda}) \\ E_G(\check{\lambda}) & E_G(\check{\lambda}^2) \\ E_G(\check{\sigma}) & E_G(\check{\lambda}\check{\sigma}) \end{bmatrix} \begin{bmatrix} E_G(\check{\sigma}) \\ E_G(\check{\lambda}\check{\sigma}) \\ E_G(\check{\sigma}^2) \end{bmatrix} \begin{bmatrix} \omega_0 \\ \omega_\lambda \\ \omega_\sigma \end{bmatrix} = \begin{bmatrix} A_\lambda\gamma_{10} + A_\sigma\gamma_{01} \\ A_\lambda\gamma_{20} + A_\sigma\gamma_{11} \\ A_\lambda\gamma_{11} + A_\sigma\gamma_{02} \end{bmatrix} \tag{4.2.12}$$

which we can write as

$$\mathbf{M}_G\boldsymbol{\omega} = \boldsymbol{\Gamma}\mathbf{A}.$$

The expected loss, W(linear Bayes) is

$$\mathbf{A}^{\mathrm{T}}\{\boldsymbol{\Gamma}_* - \boldsymbol{\Gamma}^{\mathrm{T}}\mathbf{M}_G^{-1}\boldsymbol{\Gamma}\}\mathbf{A}$$

where

$$\boldsymbol{\Gamma}_* = \begin{pmatrix} \gamma_{10} & \gamma_{11} \\ \gamma_{11} & \gamma_{02} \end{pmatrix}.$$

Under the EB sampling scheme estimation of the elements of $\mathbf{M}_G$ and $\boldsymbol{\Gamma}$ can proceed as follows, estimates being written $\bar{E}_G(\check{\lambda})$, $\bar{\gamma}_{10}$, etc., and note that

$$E_G(\check{\lambda}^2) = \gamma_{20} + V_{11}(m)\gamma_{02}, \quad E_G(\check{\lambda}, \check{\sigma}) = \gamma_{11} + V_{12}(m)\gamma_{02},$$
$$E_G(\check{\sigma}^2) = \gamma_{02}(1 + V_{22}(m)).$$

Let $\check{\lambda}_i$, $\check{\sigma}_i$ be the conventional unbiased estimates of λ, σ at the ith past realization of Λ, σ, based on m_i observations. Then

$$\left.\begin{aligned}
\bar{E}_G(\check{\lambda}) &= \bar{\gamma}_{10} = (1/n)\sum_{i=1}^{n} \check{\lambda}_i \\
\bar{E}_G(\check{\sigma}) &= \bar{\gamma}_{01} = (1/n)\sum_{i=1}^{n} \check{\sigma}_i \\
n\bar{\gamma}_{20} + \bar{\gamma}_{02}\sum_{i=1}^{n} V_{11}(m_i) &= \sum_{i=1}^{n} \check{\lambda}_i^2 \\
n\bar{\gamma}_{11} + \bar{\gamma}_{02}\sum_{i=1}^{n} V_{12}(m_i) &= \sum_{i=1}^{n} \check{\lambda}_i\check{\sigma}_i \\
\left\{n + \sum_{i=1}^{n} V_{22}(m_i)\right\}\bar{\gamma}_{02} &= \sum_{i=1}^{n} \check{\sigma}_i^2.
\end{aligned}\right\} \tag{4.2.13}$$

Estimation of $E_n W$(linear EB), etc., can be carried by straightforward modification of the procedures described for Example 4.2.2.

An alternative approach to linear Bayes and EB estmation in the location–scale family of data distributions is not to reduce the m independent observations to the estimates $\check{\lambda}$ and $\check{\sigma}$, but to base the estimates directly on order statistics. Thus, estimate $A_\lambda\lambda + A_\sigma\sigma$ by $\delta(\mathbf{x};\boldsymbol{\omega}) = \omega_0(m) + \sum_{j=1}^{m}\omega_j(m)x_{(j)}$ choosing $\boldsymbol{\omega}$ so as minimize

$$\iint \{\delta(\mathbf{x};\boldsymbol{\omega}) - A_\lambda\lambda - A_\sigma\sigma\}^2\, dF(\mathbf{x}|\lambda,\sigma)dG(\lambda,\sigma).$$

The details are similar to those given above and are shown more fully in Lwin (1976).

4.3 Quantile estimation

The p-quantile of the continuous distribution $F(x)$ is ξ_p and it is defined by the relation $F(\xi_p) = p$. If F is of the location–scale type as discussed in section 4.2 we can express ξ_p as a linear function of λ and σ, i.e. $\xi_p = \lambda + \gamma_p\sigma$. Thus, for location–scale type F this problem can be seen as a particular application of the ideas developed in section 4.2. However, the question of quantile estimation need not be addressed in only this restricted setting. It does arise quite naturally in the nonparametric context where the exact form of F may be unspecified. Since the linear Bayes and EB methods are less demanding about the form of underlying distribution than the true Bayes method, this approach seems well suited to the problem of quantile estimation.

4.3.1 *Linear Bayes and EB estimation*

Let G be the joint prior distribution function of the relevant parameters. Suppose that $\hat{\xi}_p$ is a conventional estimate of ξ_p based on m independent realizations of X in the current component problem, and that its variance is $\text{var}(\hat{\xi}_p|\xi_p) = v_p(m)$. We shall assume that the bias of $\hat{\xi}_p$ is negligible, and that to a satisfactory degree of approximation $v_p = \omega_p/m$. This latter assumption is not strictly needed but it is a useful simplification. In certain special cases more accurate statements about the first two moments of the conventional estimate can be made; an example is that of X having a normal distribution. In the notation just introduced G is the joint prior distribution of Ξ_p and Ω_p. It will also be useful to write E_G for expectation w.r.t. G, $E_{X|G}$ for expectation w.r.t. the observations conditional on fixed values of the parameters, and so on. Variances may be similary subscripted. Where there is no confusion subscripts may be omitted. Thus $\text{var}(\hat{\xi}_p|\xi_p) = \omega_p/m = \text{var}_{X|G}(\hat{\xi}_p|\xi_p)$.

In the manner of section 4.2.2 we define the linear Bayes estimate of ξ_p as $\delta_G(\beta_0, \beta_1; \hat{\xi} - p) = \beta_0 + \beta_1 \hat{\xi}_p$, with β_1 and β_2 determined so as to minimize the expected squared error $E_G E_{X|G}(\beta_0 + \beta_1 \hat{\xi}_p - \Xi_p)^2$. The appropriate values of β_0 and β_1 are given as the solutions of

$$\begin{pmatrix} 1 & E_G(\Xi_p) \\ E_G(\Xi_p) & E_G(\Xi_p^2) + E_G(\Omega_p)/m \end{pmatrix} \begin{pmatrix} \beta_0 \\ \beta_1 \end{pmatrix} = \begin{pmatrix} E_G(\Xi_p) \\ E_G(\Xi_p^2) \end{pmatrix}. \quad (4.3.1)$$

Solving this equation we find that the linear Bayes estimate can be expressed as

$$\delta_G(\beta_0, \beta_1; \hat{\xi}_p) = (1 - \beta_1) E_G(\Xi_p) + \beta_1 \hat{\xi}_p,$$

where

$$\beta_1 = \text{var}_G(\Xi_p) / \{\text{var}_G(\Xi_p) + E_G(\Omega_p)/m\}.$$

Also,

$$W(\text{linear Bayes}) = 1/[1/\{\text{var}_G(\Xi_p)\} + 1/\{E_G(\Omega_p)/m\}],$$

$$W(\hat{\xi}_p) = E_G(\Omega_p)/m.$$

For the empirical version of the linear Bayes estimate we need estimates of the elements of the matrix on the left side of (4.3.1) and of the vector on the right side. We shall assume that in each component problem we can calculate a conventional estimate $\hat{\xi}_{pi}$ of the component ξ_{pi}, and also a conventional estimate $\hat{\omega}_{pi}$ of the component ω_{pi}. We shall also assume that these estimates are unbiased or that their biases are negligible. Then estimates of $E_G(\Xi_p)$, $E_G(\Xi_p^2)$ and $E_G(\Omega_p)$ are given by

$$\hat{E}_G(\Xi_p) = (1/n)\sum_{i=1}^{n} \hat{\xi}_{pi}$$

$$\hat{E}_G(\Xi_{pi}^2) = (1/n)\sum_{i=1}^{n} (\hat{\xi}_{pi}^2 - \hat{\omega}_{pi}) \tag{4.3.2}$$

$$\hat{E}_G(\Omega_p) = (1/n)\sum_{i=1}^{n} \hat{\omega}_{pi}.$$

Replacing the expectations in (4.3.2) by their estimates gives estimates of the β's, and hence the linear empirical Bayes estimate $\delta(\hat{\beta}_0, \hat{\beta}_1; \hat{\xi}_p)$ whose expected loss is

$$W(\text{linear EB}) = \hat{\beta}_0^2 + (\hat{\beta}_1 - 1)^2 E_G(\Xi_p^2) + 2\hat{\beta}_0(\hat{\beta}_1 - 1)E_G(\Xi_p) + \hat{\beta}_1^2 E_G(\Omega_p/m). \tag{4.3.3}$$

Using formula (4.3.3) it is relatively straightforward to evaluate $E_n W(\text{linear EB})$; see also Maritz (1989).

4.3.2 *Location–scale distributions of known form*

Suppose that the distribution function of X is $F\{(x-\lambda)/\sigma\}$ where the form of F is given. Typically λ, σ will be estimated by the ML method, as will be $\xi_p = \lambda + \gamma_p \theta$. The constant γ_p will be known and we also have $\omega_p = k_p \sigma^2$ where k_p is another known constant. For example, if F is $N(\lambda, \sigma^2)$ and $p = 0{\cdot}75$ we have $\gamma_p = 0{\cdot}67449$ and $k_p = 1{\cdot}23$. In this example we then have $\hat{\xi}_{pi} = \bar{x}_i + 0{\cdot}67449 s_i$ and $\hat{\omega}_{pi} = 1{\cdot}23 s_i^2$.

4.3.3 *Distributions of unknown form*

When the form of the distribution of X is not specified the natural distribution-free estimate of ξ_p is the sample p-quantile. In this case $\operatorname{var}(\bar{\xi}_p | \xi_p) = p(1-p)/\{mf^2(\xi_p)\}$, approximately, where $f(\xi_p)$ is the density of the distribution of X at $x = \xi_p$. Thus, in this context, $\omega_p = p(1-p)/f^2(\xi_p)$.

We can now apply the theory of section 4.3.1, as was done in section 4.3.2, the only change of note being that the method of estimating var $(\bar{\xi}_p|\xi_p)$ is different. Various possibilities are open here, depending on the sample sizes m_i. With moderately large sample sizes one may begin with a kernel density estimate $\bar{f}_p$ of $f(\xi_p)$ and then estimate ω_p in the ith component by $\bar{\omega}_{pi} = p(1-p)/\bar{f}^2_{pi}$.

An important aspect of these calculations is that no specific assumptions are made about the form of the underlying F. The only parameters of consequence are $E_G(\Xi_p)$, $E_G(\Xi_p^2)$, and $E_G(\Omega_p)$. Therefore, in postulating a sequence of realizations $\{\xi_{pi}, \omega_{pi}\}$, $i = 1, 2, \ldots, n$ it is not necessary to suppose that they are generated by the same form of distribution F with varying parameters. The form of F itself could vary.

To conclude this section we consider estimation of W(linear Bayes) and $E_n W$(linear EB) from a given set of data. A bootstrap approach seems natural in the given conditions. We begin with the estimated values of the expectations appearing in relations (4.3.2); they are $\bar{E}_G(\Xi_p)$, etc., and note that $\overline{\text{var}}_G(\Xi_p) = \max[0, \bar{E}_G(\Xi_p^2) - \{\bar{E}_G(\Xi_p)\}^2]$. The estimated W(linear Bayes) is

$$\bar{W}(\text{linear Bayes}) = 1/[1/\{\overline{\text{var}}_G(\Xi_p)\} + 1/\{\bar{E}_G(\Omega_p)/m\}].$$

In the following bootstrap calculations the estimated moments and $\bar{\beta}_0$, $\bar{\beta}_1$ are treated as if they are the true values. The steps are:

1. Select one of the component data sets at random.
2. Select an m_i' value at random using the empirical distribution of component sample sizes.
3. Generate m_i' random observations from the empirical distribution function of the data in the set selected in step 1.
4. Do steps 1–3 n times.
5. Calculate $(\bar{\xi}'_{pi}, \bar{\omega}'_{pi})$, $i = 1, 2, \ldots, n$, and $\bar{\beta}_0$, $\bar{\beta}_1$ using the data generated in steps 1–4.
6. Calculate $W(\bar{\beta}'_0 + \bar{\beta}'\bar{\xi}'_p)$ by formula (4.3.3) treating $\bar{\beta}_0, \bar{\omega}_1$, etc., as the true values.
7. Do steps 1–6 a number, N, of times and find the mean of the W values obtained by step 6. It is an estimate of $E_n W$(linear EB).

4.4 The multivariate normal distribution

Study of the multivariate normal can be defended on the grounds that essentially multivariate data often are collected and analysed, and the EB sampling scheme may well apply to such collections. Here we would have to assume that both the mean vector and the covariance matrix, in general, have non-degenerate prior distributions. Estimation of multivariate normal means also arises quite naturally in applications of general linear models. An example of this sort is discussed in detail by Hui and Berger (1983); it is also discussed in Chapter 8. In this type of example it is sometimes not unrealistic to assume that the covariance matrix changes from component to component but is known, apart possibly from a multiplicative constant. This is analogous to the univariate normal case with variable numbers m_i of observations at the component problems, but with the variance remaining constant from component to component.

The notation we shall use is: the data distribution is $N(\boldsymbol{\lambda}, \boldsymbol{\Sigma})$ and the prior distribution of $(\boldsymbol{\lambda}, \boldsymbol{\Sigma})$ is $G(\boldsymbol{\lambda}, \boldsymbol{\Sigma})$ where the matrices are $k \times k$ and the vectors are $k \times i$.

4.4.1 *Known* $\boldsymbol{\Sigma}$, $\boldsymbol{\Lambda}$ *distributed* $N(\boldsymbol{\mu}_G, \boldsymbol{\Sigma}_G)$

Standard theory of the multivariate normal distribution gives the result that the posterior distribution of $\boldsymbol{\Lambda}|\mathbf{x}$ is multivariate normal with mean vector

$$E(\boldsymbol{\Lambda}|\mathbf{x}) = (\boldsymbol{\Sigma}^{-1} + \boldsymbol{\Sigma}_G^{-1})^{-1}(\boldsymbol{\Sigma}^{-1}\mathbf{x} + \boldsymbol{\Sigma}_G^{-1}\boldsymbol{\mu}_G) \qquad (4.4.1)$$

and covariance matrix $\mathbf{M} = (\boldsymbol{\Sigma}^{-1} + \boldsymbol{\Sigma}_G^{-1})^{-1}$. The marginal $\mathbf{x}$-distribution is $N(\boldsymbol{\mu}_G, \boldsymbol{\Sigma} + \boldsymbol{\Sigma}_G)$.

Recall from section 1.6 that we may choose to take the loss when estimating $\boldsymbol{\lambda}$ by $\boldsymbol{\delta}$ as $L(\boldsymbol{\delta}, \boldsymbol{\lambda}) = (\boldsymbol{\delta} - \check{\mathrm{l}})^r \mathbf{A}(\boldsymbol{\delta} - \boldsymbol{\lambda})$, this being a natural generalization of squared error loss for single parameter estimation. The Bayes estimate, i.e. that $\boldsymbol{\delta}$ which minimizes the expected loss, does not depend on $\mathbf{A}$, but the actual expected loss does. In order to simplify the following exposition $\mathbf{A}$ will be set equal to the unit matrix so that $L(\boldsymbol{\delta}, \boldsymbol{\lambda})$ is just a sum of squared errors.

With $\mathbf{A} = \mathbf{I}$ the expected squared error of the Bayes estimate is

$$W(\text{Bayes}) = \operatorname{tr}\{\mathbf{M}\boldsymbol{\Sigma}^{-1}\mathbf{M}^{\mathrm{T}}\} + \operatorname{tr}\{\mathbf{M}\boldsymbol{\Sigma}_G^{-1}\mathbf{M}^{\mathrm{T}}\} \qquad (4.4.2)$$

$$= \operatorname{tr}(\mathbf{M}^{\mathrm{T}}).$$

In the EB setting the past observations are $\mathbf{x}_i$, $i = 1, 2, \ldots, n$ and we shall assume that the operative k-variate normal data distribution at the ith component has mean $\boldsymbol{\lambda}_i$ and known covariance matrix $\boldsymbol{\Sigma}_i$. The subscript i to $\boldsymbol{\Sigma}$ indicates that we are considering component problems that are not necessarily identical. Then we can estimate $\boldsymbol{\mu}_G$ by $\bar{\mathbf{x}} = (1/n)\sum_{i=1}^{n}\mathbf{x}_i$. To estimate $\boldsymbol{\Sigma}_G$, let $\mathbf{S}$ be the matrix of second moments about the origin calculated from the past $\mathbf{x}_i$ vectors, thus

$$\mathbf{S}=(1/n)\sum_{i=1}^{n}\mathbf{x}_i\mathbf{x}_i^{\mathrm{T}}.$$

Now, since $E(\mathbf{x}_i\mathbf{x}_i^{\mathrm{T}}|\boldsymbol{\lambda}_i)=\boldsymbol{\lambda}_i\boldsymbol{\lambda}_i^{\mathrm{T}}+\boldsymbol{\Sigma}_i$ an estimate $\hat{\boldsymbol{\Sigma}}_G$ of $\boldsymbol{\Sigma}_G$ is given by

$$\mathbf{S}=\bar{\mathbf{x}}\bar{\mathbf{x}}^{\mathrm{T}}+\hat{\boldsymbol{\Sigma}}_G+(1/n)\sum_{i=1}^{n}\boldsymbol{\Sigma}_i. \tag{4.4.3}$$

This estimation is by the method of moments. ML estimation of $\boldsymbol{\mu}_G$ and $\boldsymbol{\Sigma}_G$ is possible but more complicated; see the analogous univariate normal case in section 3.9.3.

Example 4.4.1 In longitudinal health studies it is common to measure a response variable Y at times $t_{1i}<t_{2i}\cdots<t_{mi}$, say, on subject i, obtaining results $y_{1i},y_{2i},\ldots,y_{mi}$. Although analyses are often simpler if the t-values are the same for each subject, arranging such a data set in practice is usually not possible. Let us suppose that the data set for the ith subject is summarized in the intercept a_i and slope b_i of a Y on t regression line fitted to the data by the method of least squares.

Under the usual assumptions a_i and b_i are unbiased estimates of α_i and β_i the parameter values characterizing subject i, and the joint distribution of a_i, b_i is bivariate

$$N\left\{\begin{pmatrix}\alpha_i\\ \beta_i\end{pmatrix},\sigma_i^2\begin{pmatrix}m_i & \sum t_{ji}\\ \sum t_{ji} & \sum t_{ji}^2\end{pmatrix}^{-1}\right\}$$

where $\sum t_{ji}^r=\sum_{j=1}^{m_i}t_{ji}^r$. Often it is reasonable to take the residual variance σ_i^2, as fixed at σ^2 for every subject. We assume this to be so for our present illustration.

In this example (a_i,b_i) takes the place of (x_{1i},x_{2i}) in the preceding theory. The covariance matrix of (a_i,b_i) for given (α_i,β_i) is known except for σ^2 which can be estimated by a pooled residual variance. So,

$$\boldsymbol{\Sigma}_i=\sigma^2\begin{pmatrix}m_i & \sum t_{ji}\\ \sum t_{ji} & \sum t_{ji}^2\end{pmatrix},$$

$$\mathbf{S}-\bar{\mathbf{x}}\bar{\mathbf{x}}^{\mathrm{T}}=\begin{pmatrix}\frac{1}{n}\sum a_i^2-\bar{a}^2 & \frac{1}{n}\sum a_ib_i-\bar{a}\bar{b}\\ \frac{1}{n}\sum a_ib_i-\bar{a}\bar{b} & \frac{1}{n}\sum b_i^2-\bar{b}^2\end{pmatrix}$$

and

$$\hat{\boldsymbol{\mu}}_G = \begin{pmatrix} \bar{a} \\ \bar{b} \end{pmatrix} \quad \text{is an estimate of } \hat{\boldsymbol{\mu}}_G.$$

The empirical Bayes estimate given by $\hat{\boldsymbol{\mu}}_G = \bar{\mathbf{x}}$, and $\boldsymbol{\Sigma}_G$ obtained from (4.4.3), is

$$\hat{E}(\boldsymbol{\Lambda}|\mathbf{x}) = (\boldsymbol{\Sigma}^{-1} + \hat{\boldsymbol{\Sigma}}_G^{-1})^{-1}(\boldsymbol{\Sigma}^{-1}\mathbf{x} + \hat{\boldsymbol{\Sigma}}_G^{-1}\hat{\boldsymbol{\mu}}_G) = \hat{\mathbf{M}}(\boldsymbol{\Sigma}^{-1}\mathbf{x} + \hat{\boldsymbol{\Sigma}}_G^{-1}\hat{\boldsymbol{\mu}}_G)$$

and its expected loss, $W(\text{EB})$ is

$$\begin{aligned}(\hat{\mathbf{M}}\boldsymbol{\Sigma}^{-1}\boldsymbol{\mu}_G + \hat{\mathbf{M}}\hat{\boldsymbol{\Sigma}}_G^{-1}\hat{\boldsymbol{\mu}}_G - \boldsymbol{\mu}_G)^{\mathrm{T}}(\hat{\mathbf{M}}\boldsymbol{\Sigma}^{-1}\boldsymbol{\mu}_G + \hat{\mathbf{M}}\hat{\boldsymbol{\Sigma}}_G^{-1}\hat{\boldsymbol{\mu}}_G - \boldsymbol{\mu}_G)\\ + \operatorname{tr}(\hat{\mathbf{M}}\boldsymbol{\Sigma}^{-1} - \mathbf{I})\boldsymbol{\Sigma}_G(\hat{\mathbf{M}}\boldsymbol{\Sigma}^{-1} - \mathbf{I}) + \operatorname{tr}(\hat{\mathbf{M}}\boldsymbol{\Sigma}^{-1}\hat{\mathbf{M}}).\end{aligned}$$

The difference $W(\text{EB}) - W(\text{Bayes})$ can be expressed as

$$\begin{aligned}\{\hat{\mathbf{M}}\hat{\boldsymbol{\Sigma}}_G^{-1}(\hat{\boldsymbol{\mu}}_G - \boldsymbol{\mu}_G)\}^{\mathrm{T}}\{\hat{\mathbf{M}}\hat{\boldsymbol{\Sigma}}_G^{-1}(\hat{\boldsymbol{\mu}}_G - \boldsymbol{\mu}_G)\}\\ + \operatorname{tr}\{(\hat{\mathbf{M}} - \mathbf{M})\boldsymbol{\Sigma}^{-1}(\boldsymbol{\Sigma} + \boldsymbol{\Sigma}_G)\boldsymbol{\Sigma}^{-1}(\hat{\mathbf{M}} - \mathbf{M})\}.\end{aligned}$$

The expression for $E_n W(\text{EB}) - W(\text{Bayes})$ simplifies slightly through independence of $\hat{\mathbf{M}}$, $\hat{\boldsymbol{\Sigma}}_G$ and $\hat{\boldsymbol{\mu}}_G$ to

$$\begin{aligned}\frac{1}{n}E_n \operatorname{tr}\{(\mathbf{I} - \hat{\mathbf{M}}\boldsymbol{\Sigma}^{-1})(\boldsymbol{\Sigma} + \boldsymbol{\Sigma}_G)(\mathbf{I} - \hat{\mathbf{M}}\boldsymbol{\Sigma}^{-1})\}\\ + E_n \operatorname{tr}\{(\hat{\mathbf{M}} - \mathbf{M})\boldsymbol{\Sigma}^{-1}(\boldsymbol{\Sigma} + \boldsymbol{\Sigma}_G)\boldsymbol{\Sigma}^{-1}(\hat{\mathbf{M}} - \mathbf{M})\}. \qquad (4.4.4)\end{aligned}$$

Asymptotic optimality of the EBE is seen to follow quite readily from (4.4.4), but despite the distribution of the elements of $\hat{\mathbf{M}}$ being known, evaluation of the expression is not straightforward.

4.4.2 Linear Bayes and EB

(a) Known $\boldsymbol{\Sigma}$

We estimate $\boldsymbol{\lambda}$ by

$$\delta(\mathbf{x}; \mathbf{a}, \mathbf{B}) = \mathbf{a} + \mathbf{B}\mathbf{x} \qquad (4.4.5)$$

where $\mathbf{a}$ is $k \times 1$ vector and $\mathbf{B}$ is a $k \times k$ matrix, $\mathbf{a}$ and $\mathbf{B}$ chosen so as to minimize the expected loss

$$\begin{aligned}W(\mathbf{a} + \mathbf{B}\mathbf{x}) &= \int\int (\mathbf{a} + \mathbf{B}\mathbf{x} - \boldsymbol{\lambda})^{\mathrm{T}}(\mathbf{a} + \mathbf{B}\mathbf{x} - \boldsymbol{\lambda})\, dF(\mathbf{x}|\boldsymbol{\lambda}, \boldsymbol{\Sigma})\, dG(\boldsymbol{\lambda})\\ &= \{\mathbf{a} + (\mathbf{B} - \mathbf{I})\boldsymbol{\mu}_G\}^{\mathrm{T}}\{\mathbf{a} + (\mathbf{B} - \mathbf{I})\boldsymbol{\mu}_G\} + \operatorname{tr}(\mathbf{B}\boldsymbol{\Sigma}_G\mathbf{B}^{\mathrm{T}})\\ &\quad + \operatorname{tr}(\mathbf{B}\boldsymbol{\Sigma}\mathbf{B}^{\mathrm{T}}) \qquad (4.4.6)\end{aligned}$$

where F is the multivariate normal distribution with mean vector $\boldsymbol{\lambda}$ and known, fixed covariance matrix $\boldsymbol{\Sigma}'$.

Differentiating w.r.t. the elements of **a** and **B** the following equations are obtained for **a**, **B**:

$$\left.\begin{aligned} &\mathbf{a} + \mathbf{B}\boldsymbol{\mu}_G = \boldsymbol{\mu}_G \\ &\mathbf{a}\boldsymbol{\mu}_G^{\mathrm{T}} + \mathbf{B}(\boldsymbol{\mu}_G\boldsymbol{\mu}_G^{\mathrm{T}} + \boldsymbol{\Sigma}_G + \boldsymbol{\Sigma}) = \boldsymbol{\mu}_G\boldsymbol{\mu}_G^{\mathrm{T}} + \boldsymbol{\Sigma}_G \end{aligned}\right\} \tag{4.4.7}$$

in obvious analogy with earlier univariate results. From (4.4.5) we obtain, after multiplying the first equation by $\boldsymbol{\mu}_G^{\mathrm{T}}$,

$$\left.\begin{aligned} &\mathbf{B} = \boldsymbol{\Sigma}_G(\boldsymbol{\Sigma}_G + \boldsymbol{\Sigma})^{-1} = (\boldsymbol{\Sigma}_G^{-1} + \boldsymbol{\Sigma}^{-1})^{-1}\boldsymbol{\Sigma}^{-1} \\ &\mathbf{a} = \boldsymbol{\Sigma}(\boldsymbol{\Sigma}_G + \boldsymbol{\Sigma})^{-1}\boldsymbol{\mu}_G = (\boldsymbol{\Sigma}_G^{-1} + \boldsymbol{\Sigma}^{-1})^{-1}\boldsymbol{\Sigma}_G^{-1}\boldsymbol{\mu}_G. \end{aligned}\right\} \tag{4.4.8}$$

The results (4.4.8) agree with (4.4.1).

In order to construct an empirical version of the linear Bayes estimate we need estimates of $\boldsymbol{\Sigma}_G$ and $\boldsymbol{\mu}_G$, and they can be obtained by the method of moments. First we note the expectation of the marginal X_j is μ_{Gj}. Therefore we can estimate $\boldsymbol{\mu}_G$ by $\bar{\mathbf{x}}$ as before. Then, for the marginal variates X_r, X_s, we have

$$E(X_r, X_s) = E_G E_F(X_r X_s) = \sigma_{Grs} + \sigma_{rs} + \mu_{Gr}\mu_{Gs},$$

so that the estimate of $\boldsymbol{\Sigma}_G$ is again given by formula (4.4.3).

(c) $\boldsymbol{\Sigma} = \sigma^2\boldsymbol{\Gamma}$, $\boldsymbol{\Gamma}$ *known*

In Example 4.4.1 we took the residual variance $\sigma^2 = \omega$ to be the same in each component problem. In that case it seems obvious that one should estimate it by a pooled residual variance, and if n is large replacing σ^2 by such an estimate would have little effect on W(linear EB); a study of the effect of estimating ω in this context is reported by Martz and Krutchkoff (1969). One of the advantages of the linear Bayes and EB approach is that ω can be allowed not only to be unknown, but also to vary randomly from component to component. Thus we have a model where $\boldsymbol{\Sigma}_i = \omega_i\boldsymbol{\Gamma}_i$, $\boldsymbol{\Gamma}_i$ being known at each component, but ω_i, $i = 1, 2, \ldots, n$ being independent realizations of a non-degenerate random variable.

Under this model the linear Bayes estimate of the same form as in (4.4.5) is given by **a** and **B** derived from

$$\left.\begin{aligned} &\mathbf{a} + \mathbf{B}\boldsymbol{\mu}_G = \boldsymbol{\mu}_G \\ &\mathbf{a}\boldsymbol{\mu}_G^{\mathrm{T}} + \mathbf{B}(\boldsymbol{\mu}_G\boldsymbol{\mu}_G^{\mathrm{T}} + \boldsymbol{\Sigma}_G + \omega_G\boldsymbol{\Sigma}) = \boldsymbol{\mu}_G\boldsymbol{\mu}_G^{\mathrm{T}} + \boldsymbol{\Sigma}_G \end{aligned}\right\} \tag{4.4.9}$$

where $\omega_G = \int \omega \, dG(\lambda, \omega) = E_G(\omega)$. Note also that $W(\mathbf{a} + \mathbf{B}\mathbf{x})$ is of the same form as (4.4.6) with $\mathbf{\Sigma}$ replaced by $\omega_G \mathbf{\Sigma}$.

To derive an empirical version of the linear Bayes estimate we need estimates of $\boldsymbol{\mu}_G, \mathbf{\Sigma}_G, \omega_G$. Suppose that $\hat{\omega}_i$ is an unbiased estimate of ω_i at the ith component. Then we can estimate ω_G by $\bar{\omega} = (\omega_1 + \omega_2 + \cdots + \omega_n)/n$, $\boldsymbol{\mu}_G$ by $\bar{\mathbf{x}}$ as before and $\mathbf{\Sigma}_G$ by $\hat{\mathbf{\Sigma}}_G = \mathbf{S} - \bar{\mathbf{x}}\bar{\mathbf{x}}^{\mathrm{T}} - \bar{\omega}(1/n)\sum_{i=1}^{n} \mathbf{\Gamma}$. The linear EBE is of the same form as the linear Bayes estimate in (4.4.5) with $\mathbf{a}, \mathbf{B}$ replaced by $\hat{\mathbf{a}}, \hat{\mathbf{B}}$, the solution of (4.4.9) with $\boldsymbol{\mu}_G, \mathbf{\Sigma}_G, \omega_G$ replaced by their estimated values.

Example 4.4.2 In studies of the type discussed in Example 4.4.1 it is sometimes more realistic to assume that ω_i varies from subject to subject. Typically the sets of t_{ij} values will also differ from subject to subject. The data on $n = 50$ subjects summarized in this example were generated to resemble data actually collected in a longitudinal study of lung function of factory workers. A typical individual data set is the following for subject $i = 6$; the notation is as for Example 4.4.1:

t_{ij}	3	6	7	8	10
y_{ij}	3437·20	3704·00	4010·74	3918·82	4175·15

giving $a_6 = 3128{\cdot}42$, $b_6 = 105{\cdot}994$, $\hat{\omega}_6 = 8995$

$$\mathbf{\Sigma}_6 = \begin{pmatrix} 1{\cdot}9254 & -0{\cdot}2537 \\ -0{\cdot}2537 & 0{\cdot}03731 \end{pmatrix}$$

$$\begin{pmatrix} \bar{a} \\ \bar{b} \end{pmatrix} = \begin{pmatrix} 3976{\cdot}7 \\ 30{\cdot}66 \end{pmatrix}; \qquad (1/50) \sum_{i=1}^{50} \mathbf{\Sigma}_i = \begin{pmatrix} 0{\cdot}7635 & -0{\cdot}1092 \\ -0{\cdot}1092 & 0{\cdot}01986 \end{pmatrix}$$

$$\bar{\omega} = 56288$$

$$\mathbf{S} = \begin{pmatrix} 70278 & -5773{\cdot}8 \\ -5773{\cdot}8 & 1291{\cdot}7 \end{pmatrix}$$

$$\hat{\mathbf{\Sigma}}_G = \begin{pmatrix} 27302 & 372{\cdot}8 \\ 372{\cdot}8 & 173{\cdot}8 \end{pmatrix}.$$

The empirical Bayes estimate for subject $i = 6$ is obtained from

$$\hat{\mathbf{B}}_6 = \hat{\mathbf{\Sigma}}_G(\hat{\mathbf{\Sigma}}_G + \bar{\omega}\mathbf{\Sigma}_6)^{-1} = \begin{pmatrix} 0{\cdot}5842 & 3{\cdot}7372 \\ 0{\cdot}02836 & 0{\cdot}2497 \end{pmatrix}$$

and

$$\delta(\mathbf{x}; \hat{\mathbf{a}}, \hat{\mathbf{B}}) = \hat{\mathbf{B}}_6\left(\begin{pmatrix} a_6 \\ b_6 \end{pmatrix} - \begin{pmatrix} \bar{a} \\ \bar{b} \end{pmatrix}\right) + \begin{pmatrix} \bar{a} \\ \bar{b} \end{pmatrix} = \begin{pmatrix} 3762{\cdot}5 \\ 25{\cdot}42 \end{pmatrix}.$$

(c) General variable $\boldsymbol{\Sigma}$

In (a) and (b) we have dealt with rather special structures for the covariance matrix of the data distribution, although it has to be said that examples like Example 4.4.2 are common enough. More generally we may consider the multivariate analogue of the case treated in section 4.2.3(a), where $\boldsymbol{\Sigma}$ has a distribution with expectation $\int \boldsymbol{\Sigma}\, dG(\lambda; \boldsymbol{\Sigma}) = E_G(\boldsymbol{\Sigma})$.

With the linear Bayes estimate defined as before the equations for **a** and **B** are

$$\left.\begin{aligned} \mathbf{a} + \mathbf{B}\boldsymbol{\mu}_G &= \boldsymbol{\mu}_G \\ \mathbf{a}\boldsymbol{\mu}_G^{\mathrm{T}} + \mathbf{B}(\boldsymbol{\mu}_G\boldsymbol{\mu}_G^{\mathrm{T}} + \boldsymbol{\Sigma}_G + E_G\boldsymbol{\Sigma}) &= \boldsymbol{\mu}_G\boldsymbol{\mu}_G^{\mathrm{T}} + \boldsymbol{\Sigma}_G \end{aligned}\right\} \tag{4.4.10}$$

and note that $W(\mathbf{a} + \mathbf{B}\mathbf{x})$ is given by (4.4.6) with $\boldsymbol{\Sigma}$ replaced by $E_G\boldsymbol{\Sigma}$.

For an empirical linear Bayes estimate we need an estimate of $E_G\boldsymbol{\Sigma}$. Suppose that $\hat{\boldsymbol{\Sigma}}_i$ is an unbiased estimate of $\boldsymbol{\Sigma}_i$ at the ith component. Then we estimate $E_G\boldsymbol{\Sigma}$ by $(1/n)\sum_{i=1}^{n}\hat{\boldsymbol{\Sigma}}_i$, and estimation of the other unknowns in (4.4.10) proceeds as in cases (a) and (b).

4.4.3 *Simple EB estimation*

Here we shall take the covariance matrix of the data distribution, $\boldsymbol{\Sigma}$, as known, or known apart from a multiplicative constant and fixed from component to component. We are, as in previous sections, dealing with the multivariate normal distribution, concerned only with estimating the mean vector $\boldsymbol{\lambda}$.

Let $\mathbf{C} = \boldsymbol{\Sigma}^{-1}$, then the joint density of $\mathbf{x}$ for given $\boldsymbol{\lambda}$ is

$$f(\mathbf{x}|\boldsymbol{\lambda}) = (\text{const.})\exp\{-\tfrac{1}{2}(\mathbf{x} - \boldsymbol{\lambda})^{\mathrm{T}}\mathbf{C}(\mathbf{x} - \boldsymbol{\lambda})\} \tag{4.4.11}$$

and

$$\frac{\partial f(\mathbf{x}|\boldsymbol{\lambda})}{\partial x_r} = -f(\mathbf{x}|\boldsymbol{\lambda})\sum_{j=1}^{k}(x_j - \lambda_j)C_{rj}, \tag{4.4.12}$$

$r = 1, 2, \ldots, k$.

Integrating both sides of (4.4.11) w.r.t. $G(\boldsymbol{\lambda})$ and dividing by $f_G(\mathbf{x})$ we obtain

$$\frac{\partial f_G(\mathbf{x})/\partial x_r}{f_G(\mathbf{x})} + \sum_{j=1}^{k} x_j C_{rj} = \sum_{j=1}^{k} E(\Lambda_j|\mathbf{x})C_{rj}, \tag{4.4.13}$$

$r = 1, 2, \ldots, k$. Let $\rho_r = \{\partial f_G(\mathbf{x})|\partial x_r\}/f_G(\mathbf{x})$, $r = 1, 2, \ldots, k$; $\delta_{Gj} = E(\Lambda_j|\mathbf{x})$, $j = 1, 2, \ldots, k$. Then we can write (4.4.12) as

$$\boldsymbol{\rho} + \mathbf{C}\mathbf{x} = \mathbf{C}\boldsymbol{\delta}_G. \tag{4.4.14}$$

From (4.4.14) a simple EBE of $\boldsymbol{\lambda}$ can be calculated if $\boldsymbol{\rho}$ is replaced by an estimate $\hat{\boldsymbol{\rho}}$ of $\boldsymbol{\rho}$. Such an estimate requires estimation of the multivariate density $f_G(x)$ and its derivatives. Methods for estimating these quantities exist and can be applied to the observed vectors $\mathbf{x}_i$, $i = 1,2,\ldots,n$ which can be regarded as independent observations from a population with density $f_G(\mathbf{x})$. For multivariate density estimation, see, for example, Cacoullos (1966). A derivation and a study of simple EB estimation in multiple regression is given by Martz and Krutchkoff (1969). Smoothing of EBEs of this sort is discussed by Bennett and Martz (1972) and Lemon and Krutchkoff (1969).

4.4.4 *Performance of EBEs*

In the somewhat general, but clearly realistic setting of non-identical components, it is difficult to calculate quantities like $E_n W(\mathrm{EB})$. Numerical results for isolated cases seem to be less useful here because of the additional arbitrary element in the loss calculations represented by the choice of matrix $\mathbf{A}$ in $\mathscr{L}(\boldsymbol{\delta}, \boldsymbol{\lambda})$ as defined in sections 4.4.1 and 1.6. Even if $\mathbf{A}$ is a diagonal matrix it means that the total loss is a sum of weighted squared errors, and in a particular problem the overall relative values of $W(\text{Bayes})$, $W(\text{ML})$, $E_n W(\mathrm{EB})$ will depend on those weights. At the same time the expected squared errors of estimates of individual parameters are still minimized, these corresponding to obvious special choices of $\mathbf{A}$.

To illustrate some calculations that can be made and some problems, we reconsider Example 4.4.2.

Example 4.4.3 Take the data of Example 4.4.2 and consider a current component with t_j configuration giving the covariance matrix of (a, b) as $\sigma^2\boldsymbol{\Sigma}$. Then the linear EBE of the current α, β is given by (4.4.5) with $\mathbf{B}$ replaced by

$$\hat{\mathbf{B}} = \hat{\boldsymbol{\Sigma}}_G(\hat{\boldsymbol{\Sigma}}_G + \bar{\omega}\boldsymbol{\Sigma})^{-1}$$

and $\mathbf{a}$ replaced by

$$\hat{\mathbf{a}} = \begin{pmatrix} \bar{a} \\ \bar{b} \end{pmatrix} - \hat{\mathbf{B}}\begin{pmatrix} \bar{a} \\ \bar{b} \end{pmatrix}.$$

For this estimate

$$W(\mathrm{EB}) = (\hat{\mathbf{a}} + \hat{\mathbf{B}}\boldsymbol{\mu}_G - \boldsymbol{\mu}_G)^{\mathrm{T}}(\hat{\mathbf{a}} + \hat{\mathbf{B}}\boldsymbol{\mu}_G - \boldsymbol{\mu}_G) + \mathrm{tr}\,(\hat{\mathbf{B}}\boldsymbol{\Sigma}_G\hat{\mathbf{B}}^{\mathrm{T}}) + \omega_G\,\mathrm{tr}\,(\hat{\mathbf{B}}\boldsymbol{\Sigma}\mathbf{B}^{\mathrm{T}});$$

see (4.4.6). The data of Example 4.4.2 were generated according to a model in which the joint prior distribution of α, β is

$$N\left\{\begin{pmatrix}4000\\30\end{pmatrix}, \begin{pmatrix}40000 & 0\\0 & 9\end{pmatrix}\right\},$$

$\sqrt{\omega}$ is independently distributed $N(250, 625)$ and a random number m_i of t-values is selected from the integers $1, 2, \ldots, 10$. The distribution of M is uniform $(4, 10)$. Thus if we let the t-configuration be $(3, 6, 7, 8, 10)$, as in the case $i = 6$, we obtain

$$\hat{\mathbf{B}} = \begin{pmatrix}0{\cdot}5842 & 3{\cdot}7372\\0{\cdot}02836 & 0{\cdot}2497\end{pmatrix} \quad \hat{\mathbf{a}} = \begin{pmatrix}1538{\cdot}9\\-89{\cdot}78\end{pmatrix}$$

$$W(\text{EB}) = \begin{pmatrix}-12{\cdot}2\\1{\cdot}15\end{pmatrix}^{\mathrm{T}}\begin{pmatrix}-12{\cdot}2\\1{\cdot}15\end{pmatrix} + \operatorname{tr}(\hat{\mathbf{B}}\Sigma_G - \hat{\mathbf{B}}^{\mathrm{T}}) + W_G - \operatorname{tr}(\hat{\mathbf{B}}\Sigma\mathbf{B}^{\mathrm{T}})$$

$$= 18369 + 52.$$

The first term is $W(\text{EB})$ for estimating the intercept, the second is $W(\text{EB})$ for the slope estimate.

In the non-identical component setting of this example it seems virtually impossible to obtain an analogous result for $E_n W(\text{EB})$. In order to obtain an estimate of $E_n W(\text{EB})$ oue could have to generate $n = 50$ sets of observations and calculate $\hat{\mathbf{a}}$, $\hat{\mathbf{B}}$ for each of them, then $W(\text{EB})$, and finally average the results.

An alternative (quicker?), and perhaps more realistic calculation is to obtain the mean of the actual losses in the components of the realized observations. This mean is an estimate of the average expected squared error in $n = 50$ components. It can be regarded as an overall measure of performance of the estimation procedure in the EB sampling scheme. It is not an estimate of $E_n W(\text{EB})$ for a particular current component such as that exhibited above. For the one $n = 50$ component realization referred to in Example 4.4.2 the following mean squared errors were obtained.

Intercept estimate	ML	:	39888
	Bayes	:	9014
	EB	:	10858
Slope estimate	ML	:	1268
	Bayes	:	12
	EB	:	67

Finally, we consider assessing the relative performance of ML, linear Bayes and linear EB estimates from the realized data set. The linear Bayes estimate has

$$W(\text{linear Bayes}) = \text{tr}(\mathbf{B}\boldsymbol{\Sigma}_G\mathbf{B}^{\text{T}}) + \omega_G(\mathbf{B}\boldsymbol{\Sigma}\mathbf{B}^{\text{T}})$$

and the difference $E_n W(\text{linear EB}) - W(\text{linear Bayes})$ is

$$E_n\{(\hat{\mathbf{a}} - \mathbf{a}) + (\hat{\mathbf{B}} - \mathbf{B})\boldsymbol{\mu}_G\}^{\text{T}}\{(\hat{\mathbf{a}} - \mathbf{a}) + (\hat{\mathbf{B}} - \mathbf{B})\boldsymbol{\mu}_G\}$$
$$+ E_n \,\text{tr}(\hat{\mathbf{B}} - \mathbf{B})(\boldsymbol{\Sigma}_G + \omega_G\boldsymbol{\Sigma})(\hat{\mathbf{B}} - \mathbf{B})^{\text{T}}. \tag{4.4.15}$$

Assuming the estimates of **a** and **B** to be approximately unbiased the latter expression can be rewritten in terms of the variances and covariances of the estimates of the elements of **a** and **B**. The first term in (4.4.15) becomes

$$\sum_{j=1}^{2} \{\text{var}(\hat{a}_j) + 2\mu_{G1}\,\text{cov}(\hat{a}_j, \hat{b}_{j1}) + 2\mu_{G2}\,\text{cov}(a_j, \hat{b}_{j2})$$
$$+ \mu_{G1}^2\,\text{var}(\hat{b}_{j1}) + 2\mu_{G1}\mu_{G2}\,\text{cov}(\hat{b}_{j1}, \hat{b}_{j2}) + \mu_{G2}^2\,\text{var}(\hat{b}_{j2})\}.$$

The second term becomes

$$\omega_{11}\{\text{var}(\hat{b}_{11}) + \text{var}(\hat{b}_{21})\} + \omega_{22}\{\text{var}(\hat{b}_{12}) + \text{var}(\hat{b}_{22})\}$$
$$+ 2\omega_{12}\{\text{cov}(\hat{b}_{11}, \hat{b}_{12}) + \text{cov}(\hat{b}_{21}, \hat{b}_{22})\}$$

where ω_{ij} are the elements of $\mathbf{W} = \boldsymbol{\Sigma}_G' + \omega_G\boldsymbol{\Sigma}$.

The value of W(linear Bayes) can be estimated by substituting estimates for the unknown quantities. To estimate $E_n W(\text{linear EB}) - W(\text{linear Bayes})$ we need estimates of the variances and covariances of $\hat{a}_{ij}$, $\hat{b}_{ij}$, $i, j = 1, 2$. Since the estimates are obtained as the solutions of six estimating equations summarized in

$$\hat{\mathbf{a}} + \hat{\mathbf{B}}\bar{\mathbf{x}} = \bar{\mathbf{x}}$$

$$\hat{\mathbf{a}}\bar{\mathbf{x}}^{\text{T}} + \hat{\mathbf{B}}\left\{\mathbf{S} - \bar{\omega}\left(\frac{1}{n}\right)\sum_{i=1}^{n} \boldsymbol{\Gamma}_i + \bar{\omega}\boldsymbol{\Gamma}\right\} = \mathbf{S} - \bar{\omega}(1/n)\sum_{i=1}^{n} \boldsymbol{\Gamma}_i$$

the steps given in section 3.11 can be followed to obtain estimates of the required variances and covariances.

4.5 The multinomial distribution

We write the multinomial distribution as

$$P(\mathbf{X}:\mathbf{x}\,|\,\boldsymbol{\theta}) = P(X_1 = x_1, X_2 = x_2, \ldots, X_p = x_p | \theta_1, \theta_2, \ldots, \theta_p)$$
$$= \frac{m!}{\prod_{j=1}^{p} x_j!}\theta_1^{x_1}\theta_2^{x_2}\cdots\theta_p^{x_p} \tag{4.5.1}$$

where $x_1 + x_2 + \cdots + x_p = m$, $\theta_1 + \theta_2 + \cdots + \theta_p = 1$ and $0 \leqslant x_j \leqslant m$

for every j. It is fundamental in the analyses of categorical data, typical cases which we shall consider being the following:

1. Each of a number of subjects is given m questions or propositions to each of which $p = 3$ mutually exclusive responses is possible. For example, the responses might be strong agreement, strong disagreement, and neutral. Then each subject can be regarded as having probabilities $\theta_1, \theta_2, \theta_3$ of registering a response in the three categories respectively, and the observed numbers of strong agreement, etc., responses are multinomial observations. If $\boldsymbol{\theta} = (\theta_1, \theta_2, \theta_3)$ varies randomly from subject to subject we have, for n subjects, a sequence $\boldsymbol{\theta}_i$, $i = 1, 2, \ldots, n$ of parameters, and observations $\mathbf{x}_i$, $i = 1, 2, \ldots, n$ in a typical EB sampling scheme.
2. Cross-tabulated data in contingency tables. The simplest of these is a 2×2 contingency table. Such tables arise in a great diversity of applications. As an example, consider that a randomly selected group of subjects is randomly partitioned into two subgroups, one of which is treated with a certain drug, the other being a control group. The observed response is recovery or non-recovery from a certain condition. The results of such a trial can be summarized in a 2×2 contingency table. In a drug screening study one may have a sequence of drugs which could be regarded as having been sampled from a population of drugs, the trial of each drug giving rise to a 2×2 contingency table, thus again providing an EB setting for the multinomial distribution.
3. Univariate data summarized in a grouped frequency distribution, where the observed frequencies in the groups can be regarded as a multinomial observation: an approach to Bayes and EB estimation of distribution functions or quantiles is possible with the above setting as starting point.
4. Two-way contingency tables with n rows and p columns. If the rows can be regarded as corresponding to realizations of a random effect the results of each row can be taken as a multinomial observation, with randomly varying $\boldsymbol{\theta}$.

4.5.1 Estimation with Dirichlet priors

The Dirichlet prior density for $\theta_1, \theta_2, \ldots, \theta_p$ is

$$g(\boldsymbol{\theta}) = \frac{\Gamma(\alpha_1 + \alpha_2 + \cdots + \alpha_p)}{\Gamma(\alpha_1)\Gamma(\alpha_2)\cdots\Gamma(\alpha_p)} \theta_1^{\alpha_1 - 1}\theta_2^{\alpha_2 - 1}\cdots\theta_p^{\alpha_p - 1}, \qquad (4.5.2)$$

the conjugate prior for the multinomial distribution. Straightforward calculation gives the posterior distribution of $\boldsymbol{\theta}$ given $\mathbf{x}$ as

$$g(\boldsymbol{\theta}|\mathbf{x}) = \Gamma(A+m)\prod_{i=1}^{p}\frac{\theta_i^{\alpha_i+x_i-1}}{\Gamma(\alpha_i+x_i)}, \tag{4.5.3}$$

where $A = \alpha_1 + \alpha_2 + \cdots + \alpha_p$. It is also a Dirichlet distribution. Also

$$E(\theta_r|\mathbf{x}) = \frac{\alpha_r + x_r}{A+m}, \qquad r = 1, 2, \ldots, p \tag{4.5.4}$$

and the expected squared error of this estimate of θ_r is $\alpha_r(A-\alpha_r)/\{A(A+1)(A+m)\}$. If we take $W(\text{Bayes})$ to be the sum of the squared errors we have

$$W(\text{Bayes}) = \frac{A^2 - \sum_{i=1}^{p}\alpha_i^2}{A(A+1)(A+m)}. \tag{4.5.5}$$

In order to estimate the parameters $\alpha_1, \alpha_2, \ldots, \alpha_p$ of the prior distribution in the empirical Bayes setting, recall that we consider n past realizations $\boldsymbol{\theta}_1, \boldsymbol{\theta}_2, \ldots, \boldsymbol{\theta}_p$ of $\boldsymbol{\theta}$, unobserved, and the corresponding observations $\mathbf{x}_1, \mathbf{x}_2, \ldots, \mathbf{x}_n$. Estimation by the method of maximum likelihood is straightforward in principle, but we shall here deal only with estimation by the method of moments. In the marginal $\mathbf{X}$-distribution we have

$$E(X_j) = m\alpha_j/A$$

$$E\{X_j(X_j-1)\} = \frac{m(m-1)\alpha_j(\alpha_j+1)}{A(A+1)},$$

$j = 1, 2, \ldots, p$. Using the first moment equations we can let the MM estimates $\bar{\alpha}_1, \bar{\alpha}_2, \ldots, \bar{\alpha}_p$ of $\alpha_1, \alpha_2, \ldots, \alpha_p$ satisfy

$$\bar{x}_j = m\bar{\alpha}_j/\bar{A} = m\bar{\beta}_j, \qquad j = 1, 2, \ldots, p. \tag{4.5.6}$$

Also, using the second moment equations we let

$$Q = \frac{1}{n}\sum_{i=1}^{n}\sum_{j=1}^{p} x_{ji}(x_{ji}-1) = \frac{m(m-1)(\sum_{j=1}^{p}\bar{\alpha}_j^2 + \bar{A})}{\bar{A}(\bar{A}+1)}. \tag{4.5.7}$$

Writing $B = \bar{\beta}_1^2 + \bar{\beta}_2^2 + \cdots + \bar{\beta}_p^2$, the estimates are

$$\begin{aligned} \bar{A} &= (1-Q)/(Q-B), \qquad \text{for } Q > B, \\ \bar{\alpha}_j &= \bar{A}\bar{\beta}_j, \qquad j = 1, 2, \ldots, p. \end{aligned} \tag{4.5.8}$$

When $Q \leqslant B$ the prior distribution is estimated as being degenerate at

$(\bar{\beta}_1, \bar{\beta}_2, \ldots, \bar{\beta}_p)$. This means that if each component is in turn treated as the current realization, every one will have the same estimated θ_j; this is just the limiting case of 'shrinkage' estimation.

4.5.2 *Linear Bayes and EB estimation*

In previous discussions of multiple-parameter Bayes estimation we have seen that the loss can be taken as a sum of squared errors in deriving the Bayes estimate. This means, effectively, that we need consider only the estimation of individual parameters. To illustrate derivation of linear Bayes estimates for the multinomial case we consider the trinomial distribution; the more general case can be treated a similar way.

Let θ_1 be estimated by $a_1 + b_{11}(x_1/m) + b_{12}(x_2/m)$. Then the expected squared error of estimation is

$$\int_\theta \sum_{x_1, x_2} \left(a_1 + b_{11}\frac{x_1}{m} + b_{12}\frac{x_2}{m} - \theta_1 \right)^2 dG(\theta_1, \theta_2)$$

and we minimize it w.r.t. a_1, b_{11}, b_{12}. The first and second moments of $G(\theta_1, \theta_2)$ are $E(\theta_1) = \mu_{1G}$, $E(\theta_2) = \mu_{2G}$, $\operatorname{var}(\theta_1) = \sigma_{1G}^2$, $\operatorname{var}(\theta_2) = \sigma_{2G}^2$, $\operatorname{cov}(\theta_1, \theta_2) = \sigma_{12G}$, and differentiating w.r.t. a_1, b_{11}, b_{12} the following equations are obtained for the optimal a_1, b_{11}, b_{12}, after some manipulation.

$$a_1 + b_{11}\mu_{1G} + b_{12}\mu_{2G} = \mu_{1G}$$

$$\times \begin{pmatrix} \sigma_{1G}^2 - \dfrac{1}{m}(\sigma_{1G}^2 + \mu_{1G}^2) + \dfrac{1}{m}\mu_{1G} & \sigma_{12G} - \dfrac{1}{m}(\sigma_{12G} + \mu_{1G}\mu_{2G}) \\ \sigma_{12G} - \dfrac{1}{m}(\sigma_{12G} + \mu_{1G}\mu_{2G}) & \sigma_{2G}^2 - \dfrac{1}{m}(\sigma_{2G}^2 + \mu_{2G}^2) + \dfrac{1}{m}\mu_{2G} \end{pmatrix}$$

$$\times \begin{pmatrix} b_{11} \\ b_{12} \end{pmatrix} = \begin{pmatrix} \sigma_{1G}^2 \\ \sigma_{12G} \end{pmatrix}. \qquad (4.5.9)$$

If the prior distribution is Dirichlet as given in (4.5.2) $\mu_{1G} = \alpha_1/A$, $\mu_{2G} = \alpha_2/A$, $\sigma_{1G}^2 = \alpha_1(A - \alpha_1)/\{A^2(A+1)\}$, $\sigma_{2G}^2 = \alpha_2(A - \alpha_2)/\{A^2(A + 1)\}$, $\sigma_{12G} = -\alpha_1\alpha_2/\{A^2(A+1)\}$. Substituting these values in (4.5.9) gives $b_{12} = 0$, $b_{11} = m/(A+m)$, $a_1 = \alpha_1/(A+m)$, in agreement with (4.5.4).

In the marginal **X**-distribution we have

$$\begin{aligned} E(X_r) &= m\mu_{rG}, \qquad r=1,2,\\ \operatorname{var}(X_r) &= m(m-1)\sigma_{rG}^2 + m\mu_{rG}(1-\mu_{rG}), \qquad r=1,2, \\ \operatorname{cov}(X_1X_2) &= m(m-1)\sigma_{12G} - m\mu_{1G}\mu_{2G}. \end{aligned} \tag{4.5.10}$$

Replacing the left sides of (4.5.10) by the empirical moments calculated from n past realizations we obtain equations from which estimates of μ_{rG}, σ_{rG}^2, $r=1,2$ and σ_{12G} can be calculated.

4.5.3 *Non-identical components*

In many realistic applications of the ideas of sections 4.5.1 and 4.5.2 one may expect m to vary from one component to another; thus, at past component i we have m_i instead of m. The formulae for Bayes estimation at the current component do not change, but allowance has to be made for the variable m in estimating relevant parameters for EB estimation.

Dirichlet prior: for the ith component we can write the marginal expectations as

$$\begin{aligned} E(X_j^{(i)}) &= m_i\alpha_j/A \\ E\{X_j^{(i)}(X_j^{(i)}-1)\} &= m_i(m_i-1)\alpha_j(\alpha_j+1)/\{A(A+1)\}. \end{aligned}$$

The estimating equations (4.5.6) and (4.5.7) become

$$\bar{x}_j = \bar{\alpha}_j\bar{m}/\bar{A}$$

$$\frac{1}{n}\sum_{i=1}^{n}\sum_{j=1}^{p} x_{ji}(x_{ji}-1) = \left\{\frac{1}{n}\sum_{i=1}^{n} m_i(m_i-1)\right\} \times\left(\sum_{j=1}^{p}\bar{\alpha}_j^2+\bar{A}\right)\bigg/\{\bar{A}(\bar{A}+1)\}, \tag{4.5.11}$$

where $\bar{x}$ is the mean of the m_i values.

An alternative form of (4.5.11) is

$$\frac{1}{n}\sum_{i=1}^{n} x_{ji}/m_i = \bar{\alpha}_j/\bar{A} = \bar{\beta}_j, \qquad j=1,2,\ldots,p,$$

$$\begin{aligned} Q &= \frac{1}{n}\sum_{j=1}^{p}\sum_{i=1}^{n} x_{ji}(x_j-1)/\{m_i(m_i-1)\} \\ &= \left(\sum_{j=1}^{p}\bar{\alpha}_j^2+\bar{A}\right)\bigg/\{\bar{A}(\bar{A}+1)\}. \end{aligned} \tag{4.5.12}$$

Putting $B = \bar{\beta}_1^2 + \bar{\beta}_2^2 + \cdots + \bar{\beta}_p^2$ as before, the estimates are given again by equations (4.5.9).

An example of the application of these methods is given in section 8.4. It concerns a contingency table studied by Laird (1978) using a different model.

Linear EB: instead of the relations (4.5.10) we use

$$E(X_r^{(i)}) = m_i \mu_{rG}, \qquad r = 1, 2$$
$$E\{X_r^{(i)}(X_r^{(i)} - 1)\} = m_i(m_i - 1)(\sigma_{rG}^2 + \mu_{rG}^2), \qquad r = 1, 2$$
$$E(X_1^{(i)} X_2^{(i)}) = m_i(m_i - 1)(\sigma_{12G} + \mu_{1G}\mu_{2G}).$$

Replacing the l.h.s. expectations by appropriate mean values we obtain

$$(1/n)\sum_{i=1}^{n} x_{ri} = \bar{\mu}_{rG}\bar{m}, \qquad r = 1, 2$$

$$(1/n)\sum_{i=1}^{n} x_{ri}(x_{ri} - 1) = (\bar{\sigma}_{rG}^2 + \bar{\mu}_{rG}^2)(1/n)\sum_{i=1}^{n} m_i(m_i - 1) \qquad (4.5.13)$$

$$(1/n)\sum_{i=1}^{n} x_{1i}x_{2i} = (\bar{\sigma}_{12G} + \bar{\mu}_{1G}\bar{\mu}_{2G})(1/n)\sum_{i=1}^{n} m_i(m_i - 1)$$

from which to calculate estimates of the parameters in (4.5.9).

4.5.4 *Simple EB estimation*

For this discussion it will be convenient to use the notation $\mathbf{x}^{(m)}$ for the multinomial vector resulting from m trials. Also, let J_i be an operator such that

$$J_i \mathscr{L}(x_1, x_2, \ldots, x_p) = \mathscr{L}(x_1, x_2, \ldots, x_i + 1, \ldots, x_p).$$

Then the Bayes estimate of θ_i can be expressed as

$$E(\theta_i|\mathbf{x}) = \left(\frac{x_i^{(m+1)} + 1}{m + 1}\right)\frac{J_i p_{G,m+1}(\mathbf{x}^{(m+1)})}{p_{G,m}(\mathbf{x}^{(m)})}, \qquad (4.5.14)$$

where

$$p_{G,m}(\mathbf{x}^{(m)}) = \frac{m!}{x_1! x_2! \cdots x_p!}\int \theta_1^{x_1}\theta_2^{x_2}\cdots\theta_p^{x_p}\,dG(\boldsymbol{\theta}),$$

the marginal $\mathbf{x}$-distribution. Since we actually observe the outcomes of m trials only we have to use a version of (4.5.14) with m replaced by $(m - 1)$, when application to EB estimation is considered. Under the

EB sampling scheme, for fixed m, a direct estimate of $p_{G,m}(\mathbf{x}^{(m)})$ can be obtained. By omitting one of m observations in every component a direct estimate of $p_{G,m-1}(\mathbf{x}^{m-1})$ is obtainable, and schemes such as averaging estimates generated by successively omitting every observation in turn can be considered.

An alternative approach, analogous to that of section 3.4, is to let, for $i \neq p$,

$$K_i \mathscr{L}(x_1, x_2, \ldots, x_p) = \mathscr{L}(x_1, x_2, \ldots, x_i + 1, \ldots, x_p - 1).$$

Also let B be the posterior distribution function of $\boldsymbol{\theta}$, and E_B expectation w.r.t. B, i.e. posterior expectation.

Then

$$K_i p_{G,m}(\mathbf{x}^{(m)}) = \frac{x_p^{(m)}}{x_i^{(m)} + 1} E_B\left\{\frac{\theta_i}{\theta_p}\right\}. \tag{4.5.15}$$

Formula (4.5.15) shows that a simple EBE of the ratio θ_i/θ_p can be derived from the observations $\mathbf{x}_j^{(m)}, j = 1, 2, \ldots, n$ on the marginal $\mathbf{X}$-distribution.

An approximation for $E_B(\theta_i)$ can be derived, following Maritz and Lwin (1975). We need relations

$$K_i K_j p_{G,m}(\mathbf{x}^{(m)}) = \frac{x_p^{(m)}(x_p^{(m)} - 1)}{(x_i^{(m)} + 1)(x_j^{(m)} + 1)} E_B\left\{\frac{\theta_i \theta_j}{\theta_p^2}\right\}. \tag{4.5.16}$$

for $i, j = 1, 2, \ldots, p - 1$.

Write $\Lambda_i = \theta_i/\theta_p$, $i = 1, 2, \ldots, p$, so that $\theta_i = \Lambda_i/(\Lambda_1 + \cdots + \Lambda_p)$. From (4.5.15) and (4.5.16) we can get expressions for the posterior moments of order 1 and 2 of the Λ_i, and can derive approximations for the posterior expectations of the θ_i by the usual Taylor expansion technique.

4.6 Linear Bayes and EB, and subsets of parameters

Linear Bayes and EB estimation has been considered in the special cases dealt with in earlier sections of this chapter. Now we look at it in a slightly more general setting.

Suppose that m_i observations on the vector r.v. $\mathbf{X}$ at the ith component yield unbiased estimates t_i of parameters θ_i, $i = 1, 2, \ldots, p$ and also estimates ω of parameters $\boldsymbol{\omega}$. It is convenient to think of $\boldsymbol{\theta}$ as the parameters of primary interest, and $\boldsymbol{\omega}$ as nuisance parameters. For example, in the case of the multivariate normal distribution, $\boldsymbol{\theta}$ might be the vector of means, $\boldsymbol{\omega}$ the collection of dispersion parameters.

A linear Bayes estimate of $\boldsymbol{\theta}$, derived from $\mathbf{t}$, is obtained by minimizing the expectation

$$\int_{\theta}\int_{\mathbf{t}} (\mathbf{a}+\mathbf{Bx}-\boldsymbol{\theta})^{\mathrm{T}}(\mathbf{a}+\mathbf{Bx}-\boldsymbol{\theta})dF(\mathbf{t}\,|\,\boldsymbol{\theta})dG(\boldsymbol{\theta},\boldsymbol{\omega}).$$

Let $\boldsymbol{\Sigma}(\boldsymbol{\theta},\boldsymbol{\omega})$ be the covariance matrix of $\mathbf{x}$ for given $\boldsymbol{\theta}, \boldsymbol{\omega}$, $\boldsymbol{\Sigma}_G$ the prior covariance matrix of $\boldsymbol{\theta}$, and $\boldsymbol{\mu}_G$ the prior expectation of $\boldsymbol{\theta}$. Then the optimal $\mathbf{a}$ and $\mathbf{B}$ are given by

$$\left.\begin{aligned}&\mathbf{a}+\mathbf{B}\boldsymbol{\mu}_G=\boldsymbol{\mu}_G\\&\mathbf{a}\boldsymbol{\mu}_G^{\mathrm{T}}+B\{\boldsymbol{\mu}_G\boldsymbol{\mu}_G^{\mathrm{T}}+\boldsymbol{\Sigma}_G+E_G\boldsymbol{\Sigma}(\boldsymbol{\theta},\boldsymbol{\omega})\}=\boldsymbol{\mu}_G\boldsymbol{\mu}_G^{\mathrm{T}}+\boldsymbol{\Sigma}_G\end{aligned}\right\}\tag{4.6.1}$$

These equations are essentially like (4.4.10). Eliminating $\mathbf{a}$ from these equations, $\mathbf{B}$ is given by

$$\mathbf{B}\{\boldsymbol{\Sigma}_G+E_G\boldsymbol{\Sigma}(\boldsymbol{\theta},\boldsymbol{\omega})\}=\boldsymbol{\Sigma}_G.\tag{4.6.2}$$

The l.h.s. matrix in braces is the marginal $\mathbf{X}$ covariance matrix. Therefore, in the EB sampling scheme, where the conditional covariance matrix at the ith component is $\boldsymbol{\Sigma}_i(\boldsymbol{\theta}_i,\boldsymbol{\omega}_i)$ we can estimate $\boldsymbol{\Sigma}_G$ by

$$\bar{\boldsymbol{\Sigma}}_G=\mathbf{S}-\frac{1}{n}\sum_{i=1}^{n}\hat{\boldsymbol{\Sigma}}_i$$

where

$$\mathbf{S}=\frac{1}{n}\sum_{i=1}^{n}\mathbf{x}_i^{\mathrm{T}}\mathbf{x}_i-\bar{\mathbf{x}}^{\mathrm{T}}\mathbf{x}$$

and $\hat{\boldsymbol{\Sigma}}_i$ is an estimate of the conditional covariance matrix at the ith component. The mean vector $\boldsymbol{\mu}_G$ is estimated by $\bar{\mathbf{x}}$, assuming that the parametrization is appropriate.

In general the attraction of linear EB estimation is its relative simplicity, but there is the drawback that the number of elements of $\boldsymbol{\mu}_G$ and $\boldsymbol{\Sigma}_G$ to be estimated increases rapidly with p. Referring to Example 4.5.1, with $n=16$, $p=11$ the number of parameters in $\boldsymbol{\mu}_G$ and $\boldsymbol{\Sigma}_G$ is 66, a seemingly excessively large number given the size of the data set. A question which arises is whether much is lost if the dimensionality of the problem is reduced. The following examples illustrate the effect of reducing dimensionality.

Example 4.6.1 Consider the multinomial Dirichlet prior case. The Bayes estimate of θ_1 is given by (4.5.4) as

$$E(\theta_1|\mathbf{x}) = (\alpha_1 + x_1)/(A + m).$$

If we pool the classes $i = r, r+1, \ldots, p, r \geqslant 2$ simple calculations show that the Bayes estimate of θ_1 remains unchanged.

Example 4.6.2 $\begin{pmatrix} X_1 \\ X_2 \end{pmatrix}$ distributed $N\left\{\begin{pmatrix} \theta_1 \\ \theta_2 \end{pmatrix}, \begin{pmatrix} 1/3 & 0 \\ 0 & 1/2 \end{pmatrix}\right\}$, prior distribution of $\begin{pmatrix} \theta_1 \\ \theta_2 \end{pmatrix}$ is $N\left\{\begin{pmatrix} 0 \\ 1 \end{pmatrix}, \begin{pmatrix} 1/15 & 1/18 \\ 1/18 & 1/10 \end{pmatrix}\right\}$. Then the Bayes estimate of θ_1 is

$$E(\theta_1|\mathbf{x}) = -0{\cdot}0782 + 0{\cdot}1558x_1 + 0{\cdot}0782x_2$$

and its expected squared error is 0·0519.

Ignoring information about θ_1 supplied by observation on X_2, i.e. simply using the fact that $X \overset{\text{d}}{=} N(\theta_1, 1/3)$, $\theta_1 \overset{\text{d}}{=} N(0, 1/15)$, we have $E(\theta_1|x_1) = x_1/6$ with expected squared error 0·0555. The expected squared error of x_1, the MLE of θ_1, is 1/3. Thus the loss in precision by reducing the dimensionality of the problem is negligible in this example.

Example 4.6.3 A slightly more general version of the previous example is to take

$$\begin{pmatrix} X_1 \\ X_2 \end{pmatrix} \overset{\text{d}}{=} N\left\{\begin{pmatrix} \theta_1 \\ \theta_2 \end{pmatrix}, \begin{pmatrix} \sigma_1^2 & 0 \\ 0 & \sigma_2^2 \end{pmatrix}\right\}$$

and

$$\begin{pmatrix} \theta_1 \\ \theta_2 \end{pmatrix} \overset{\text{d}}{=} N\left\{\begin{pmatrix} \mu_{1G} \\ \mu_{2G} \end{pmatrix}, \begin{pmatrix} \sigma_{1G}^2 & \rho\sigma_{1G}\sigma_{2G} \\ \rho\sigma_{1G}\sigma_{2G} & \sigma_{2G}^2 \end{pmatrix}\right\}.$$

Then the expected squared error of the Bayes estimate of θ_1 is

$$\frac{1/\sigma_{G2}^2 + (1-\rho^2)/\sigma_2^2}{\left\{\dfrac{1}{\sigma_{G1}^2\sigma_{G2}^2} + \dfrac{1}{\sigma_{G2}^2\sigma_1^2} + \dfrac{1}{\sigma_{G1}^2\sigma_2^2} + \dfrac{(1-\rho^2)}{\sigma_1^2\sigma_2^2}\right\}}.$$

At $\rho = 0$ we have the usual one-parameter result $(1/\sigma_{G1}^2 + 1/\sigma_1^2)^{-1}$, and this is the maximum w.r.t. ρ. The minimum is at $|\rho| = 1$ and is

$$\frac{1}{\{1/\sigma_{G1}^2 + 1/\sigma_1^2 + \sigma_{G2}^2/(\sigma_{G1}^2\sigma_2^2)\}}.$$

This formula shows that a Bayes estimate of θ_1 involving x_2 as well as x_1 will only be appreciably better than the Bayes estimate involving x_1 alone if $\sigma_{G2}^2/(\sigma_{G1}^2\sigma_2^2)$ is large. In Example 4.6.2 $\sigma_{G2}^2/(\sigma_{G1}^2\sigma_2^2) = 3$, compared with $1/\sigma_{G1}^2 + 1/\sigma_1^2 = 18$ so that the largest and smallest possible expected squared error values of the Bayes estimates are 0·05 and 0·0476.

The discussions in the immediately preceding examples are not conclusive, but they do suggest that EB estimation with reduced dimensionality, thus requiring estimation of fewer parameters, may be better than EB estimation in higher dimensions. In other words, if data are available in an EB sampling scheme one may do better to use non-EB estimates of nuisance parameters. Making general qualitative statements does not seem possible; individual cases may have to be examined with careful analysis of available data.

4.7 Concomitant variables

In practical cases where application of EB methods might be considered appropriate there will often be concomitant information about the parameter values. Specifically, recall the EB sampling scheme where we have observations $(x_1, x_2, \ldots, x_n)$ when the parameter values are $(\lambda_1, \lambda_2, \ldots, \lambda_n)$. Every x_i is usually thought of as an estimate of the corresponding λ_i. Now, it may happen that we also have associated with every x_i an observation c_i on a concomitant variable C. Every c_i is not necessarily directly an estimate of λ_i, but C and Λ may not be independent, so that taking account of the observed c should improve the estimate of λ. However, the emphasis is still on estimating individual λ values, and not on exploring the relation between Λ and C, as one might do by examining the regression of Λ on C. The paper by Tsutakawa, Shoop and Marienfeld (1985) pays some attention to concomitant variables in the EB context. Other examples involving concomitant variables are discussed by Raudenbush and Bryk (1985) and Fay and Herriot (1979). More details of these examples are given in Chapter 8. In the discussion that follows we consider just a one-dimensional concomitant variable C; extension to vector $\mathbf{C}$ is not difficult.

The details of incorporating concomitant information are relatively straightforward if the joint distribution of Λ, X and C is normal. Retaining our earlier notation, the model is:

$$\Lambda \overset{d}{=} N(\mu_G, \sigma_G^2)$$

$$X|\lambda \overset{d}{=} N(\lambda, \sigma_G^2)$$

$$\Lambda|C \overset{d}{=} N\left\{\rho_{\Lambda C}\frac{\sigma_G}{\sigma_C}(c-\mu_C)+\mu_G, \sigma_G^2(1-\rho_{\Lambda C}^2)\right\}.$$

In the formulae above μ_C is the mean of C, σ_C^2 its variance, and $\rho_{\Lambda C}$ is the correlation of Λ and C. From these specifications the joint distribution of Λ, X and C has mean vector $(\mu_G, \mu_G, \mu_C)^T$ and covariance matrix as follows:

$$\begin{pmatrix} \sigma_G^2 & \sigma_G^2 & \rho_{\Lambda C}\sigma_G\sigma_C \\ \sigma_G^2 & \sigma^2+\sigma_G^2 & \rho_{\Lambda C}\sigma_G\sigma_C \\ \rho_{\Lambda C}\sigma_G\sigma_C & \rho_{\Lambda C}\sigma_G\sigma_C & \sigma_C^2 \end{pmatrix}$$

The Bayes estimate of λ, i.e. $E(\Lambda|x,c)$ is now

$$E(\Lambda|x,c) = \mu_G + (1/D)(\sigma_G^2 \quad \rho_{\Lambda C}\sigma_G\sigma_C)$$

$$\times\begin{pmatrix} \sigma_C^2 & -\rho_{\Lambda C}\sigma_G\sigma_C \\ -\rho_{\Lambda C}\sigma_G\sigma_C & \sigma^2+\sigma_G^2 \end{pmatrix}\begin{pmatrix} x-\mu_G \\ c-\mu_C \end{pmatrix},$$

where $D = \sigma_C^2\sigma^2 + \sigma_C^2\sigma_G^2(1-\rho_{\Lambda C}^2)$. Also,

$$\text{var}(\Lambda|x,c) = \sigma^2\sigma_G^2\sigma_C^2(1-\rho_{\Lambda C}^2)/D.$$

When $\rho_{\Lambda C}=0$ these formulae reduce to the previously established forms for $E(\Lambda|x)$ and $\text{var}(\Lambda|x)$.

From the point of view of EB estimation it is useful to rewrite the formulae for $E(\Lambda|x,c)$ and $\text{var}(\Lambda|x,c)$. First we note that

$$E(X|c) = \rho_{\Lambda C}(\sigma_G/\sigma_C)(c-\mu_C)+\mu_G = \gamma_0+\gamma_1,$$

and

$$\text{var}(X|c) = \sigma^2 + (1-\rho_{\Lambda C}^2)\sigma_G^2 = \sigma^2+\tau^2.$$

Then we have

$$E(\Lambda|x,c) = \frac{\tau^2}{(\sigma^2+\tau^2)}x + \frac{\sigma^2}{(\sigma^2+\tau^2)}E(X|c). \qquad (4.7.1)$$

Both $E(X|c)$ and $\text{var}(X|c)$ depend on the parameters of G, and when Λ and C are independent, equation (4.7.1) reduces to the usual formula for the Bayes estimate of λ.

For EB estimation the form of (4.7.1) suggests that one should look for estimates of γ_0, γ_1, σ^2 and τ^2. We consider the unequal components case where x_i is replaced by the mean $\bar{x}_i$, the mean of m_i independent observations at component i with $\text{var}(\bar{X}_i|\lambda_i) = \sigma^2/m_i = v_i$. We shall assume that σ^2 is known. If it is not, an estimate can be obtained as the usual within-groups sample variance. Now

$$E(T_i|c_i) = \gamma_0 + \gamma_1 c_i$$
$$\text{var}(T_i|c_i) = v_i + \tau^2,$$

and we write $v_i + \tau^2 = 1/\omega_i$. Conditioning on the observed c_i values and maximizing the likelihood of the observed $\bar{x}_i$ values we obtain the following equations in γ_0, γ_1 and τ^2 to be solved for their ML estimates:

$$\begin{pmatrix} \sum\omega_i & \sum\omega_i c_i \\ \sum\omega_i c_i & \sum\omega_i c_i^2 \end{pmatrix} \begin{pmatrix} \hat{\gamma}_0 \\ \hat{\gamma}_1 \end{pmatrix} = \begin{pmatrix} \sum\omega_i \bar{x}_i \\ \sum\omega_i c_i \bar{x}_i \end{pmatrix}, \tag{4.7.2}$$

and

$$\sum \frac{1}{(v_i + \tau^2)} = \sum \frac{(\bar{x}_i - \gamma_0 - \gamma_1 c_i)^2}{(v_i + \tau^2)^2}. \tag{4.7.3}$$

In the equations above every $\sum$ should be read as $\sum_{i=1}^{n}$. Solving them iteratively can be accomplished by starting with a trial value for τ^2, then solving the two linear equations for the trial values of the γ's, then checking equality of the left and right sides of the third equation and adjusting the trial τ^2 appropriately.

Finally we may note that equations like (4.7.2) and (4.7.3) for estimates of the parameters γ_0, γ_1 and τ^2 can be derived using a weighted least squares approach without appealing to normality of the underlying distributions.

CHAPTER 5

Testing of hypotheses

5.1 Introduction

In the non-Bayes approach to hypothesis testing a sharp distinction is usually made between the null hypothesis and alternative hypotheses. The null hypothesis holds a special place, a notion which is reinforced by the terminology 'testing of hypotheses'. The Bayes approach, or at least the Bayes decision theoretic approach, is different, emphasis being on the choice between hypotheses rather than one being singled out for special attention. There seems to be no natural Bayesian counterpart to the testing of a null hypothesis against a vague alternative which is just its negation. For a discussion of this and related matters see Cox and Hinkley (1974, p. 392).

From the point of view of decision theory the essential difference between point estimation and choosing between hypotheses is in the form of the loss function. Commonly a '0–1' loss function is used in the latter context, i.e. the loss is taken as 0 if the correct choice is made, otherwise it is 1. Other loss structures are of course also used, and in section 5.4 we shall discuss one which has found favour in EB theory.

The simplest types of hypothesis testing problems have to do with choice between k simple hypotheses. They can also be regarded as problems of point estimation where the prior distribution actually is discrete, having atoms of probability at $\lambda_1, \lambda_2, \ldots, \lambda_n$, say. However, here the loss is 0–1 and we do not take the mean of the posterior distribution as the point estimate.

Whatever loss structure is considered, the EB approach to hypothesis testing is in principle the same as to point estimation. From past data an estimate of the prior distribution is made, and it replaces the actual prior distribution in the Bayes decision rule, thus producing an EB decision rule. With a special loss structure such as that discussed briefly in section 1.5 it is possible to avoid the process of finding an explicit estimate of the prior distribution G. In other words, a 'simple' EB approach to hypothesis testing is possible as in point estimation. We shall take this up again in section 5.4, but in other parts of this chapter we shall use 0–1 loss unless another structure is specified.

5.2 Two simple hypotheses, one-parameter problems

5.2.1 *Single past and current observations:* $m_i = m = 1$

The random variable X has distribution function $F(x|\lambda)$ and λ is known to have one of two given values λ_1 or λ_2; $\lambda_1 < \lambda_2$. The prior probabilities of λ_1 and λ_2 are θ_1 and $\theta_2 = 1 - \theta_1$. With the 0–1 loss function the Bayes rule is to choose λ_1 when the observation x is such that $\theta_2 f(x|\lambda_2) < \theta_1 f(x|\lambda_1)$, where $f(x|\lambda)$ is the p.d.f. of X. Modifications for discrete X are obvious. If the ratio $f(x|\lambda_1)/f(x|\lambda_2)$ is monotonic in x, the Bayes rule is: choose λ_1 if $x < \xi_G$ where $x = \xi_G$ is the solution of

$$\theta_2 f(x|\lambda_2) = \theta_1 f(x|\lambda_1); \qquad (5.2.1)$$

see also section 1.4.

In this case, estimation of the prior G in order to derive an EB rule reduces to estimating θ_1. Letting $x_1, x_2, \ldots, x_n$ be the past observations, and noting that the marginal p.d.f. of X is $\theta_1 f(x|\lambda_1) + \theta_2 f(x|\lambda_2)$ one can estimate θ_1 by maximizing the likelihood

$$L_n = \sum_{j=1}^{n} \ln\,[\theta_1 f(x_j|\lambda_1) + (1-\theta_1) f(x_j|\lambda_2)]$$

w.r.t. θ_1, subject to $0 \leqslant \theta_1 \leqslant 1$. Thus the estimate $\hat{\theta}_1$ of θ_1 can be obtained as the solution of

$$\sum_{i=1}^{n} \frac{f(x_i|\lambda_1) - f(x_i|\lambda_2)}{\theta_1 f(x_i|\lambda_1) + (1-\theta_1) f(x_i|\lambda_2)} = 0, \qquad (5.2.2)$$

if the solution lies between 0 and 1; otherwise it is taken as 0 or 1, whichever gives the greater L_n.

Finding the solution of (5.2.2) is usually rather awkward, and simpler alternative methods of estimating θ_1 have been used. For example, suppose that $E(X|\lambda) = \lambda$. Then the mean of the mixed distribution $F_G(x)$ is

$$\int x\,dF_G(x) = \iint x\,dF(x|\lambda)\,dG(\lambda) = \lambda_1\theta_1 + \lambda_2\theta_2. \qquad (5.2.3)$$

Thus if $\bar{x}$ is the mean of the past observations, (5.2.3) suggests estimation of θ_1 by $\bar{\theta}_1$ from

$$\bar{x} = \bar{\theta}_1\lambda_1 + (1 - \bar{\theta}_1)\lambda_2,$$

giving

$$\bar{\theta}_1 = \begin{cases} (\lambda_2 - \bar{x})/(\lambda_2 - \lambda_1), & \text{for } \lambda_1 < \bar{x} < \lambda_2 \\ 0, & \text{for } \bar{x} \geqslant \lambda_2 \\ 1, & \text{for } \bar{x} \leqslant \lambda_1. \end{cases} \tag{5.2.4}$$

(a) An example: the case of normal F

Suppose that the distribution of X for given λ is $N(\lambda, 1)$ as in Example 1.4.1. Then ξ_G is given by (1.4.2), and using the estimate of θ_1 given by (5.2.4), the EB rule is: choose λ_1 or λ_2 according as $\bar{x} < \bar{\xi}_G$ or $\bar{x} \geqslant \bar{\xi}_G$, where

$$\bar{\xi}_G = \begin{cases} \dfrac{\lambda_1 + \lambda_2}{2} - \dfrac{1}{(\lambda_2 - \lambda_1)} \ln\left(\dfrac{\bar{x} - \lambda_1}{\lambda_2 - x}\right), & \lambda_1 < \bar{x} < \lambda_2 \\ -\infty, & \bar{x} \geqslant \lambda_2 \\ +\infty, & \bar{x} \leqslant \lambda_1. \end{cases} \tag{5.2.5}$$

The expected loss incurred by using the rule (5.2.5) is $W(\bar{\xi}_G)$ given by

$$W(\bar{\xi}_G) = \theta_1\{1 - \Phi(\bar{\xi}_G - \lambda_1)\} + \theta_2\Phi(\bar{\xi}_G - \lambda_2), \tag{5.2.6}$$

where $\Phi(u)$ is the standard normal distribution function.

In order to evaluate $E_n W(\bar{\xi}_G)$ we need the distribution of $\bar{\xi}_G$. Now, the distribution of $\bar{x}$ will be well approximated by a normal distribution for quite small values of n, because the marginal X-distribution is itself reasonably close to normal when $\lambda_2 - \lambda_1$ is not large by comparison with $\text{var}(X|\lambda)$, which equals 1 in this case. For example, with $\lambda_2 = +1$, $\lambda_1 = -1$, $\theta_1 = 0{\cdot}8$, the coefficients of skewness and kurtosis of the marginal X-distribution are 0·069 and 3·038.

The values of $E_n W(\bar{\xi}_G)$ in Table 5.1 were calculated using a normal approximation for the distribution of $\bar{x}$, the expression (5.2.5) of $\bar{\xi}$ in terms of $\bar{x}$ and numerical integration. Various values of θ_1 and of the difference between λ_1 and λ_2 were used. Two factors influenced these choices. When $\theta_1 = \theta_2$ the Bayes and 'best' non-Bayes rules coincide, so that the EB rule must necessarily give a worse result than T. As $\theta_1 \to 0$ or 1 the Bayes rule becomes relatively more effective than T. Also, for fixed θ_1, the Bayes rule tends to be relatively less effective

Table 5.1 *Two simple hypotheses, $H_1: \lambda = \lambda_1$, $H_2: \lambda = \lambda_2$ with prior probabilities θ_1 and $1 - \theta_1$*

θ_1	$W(T)$	$W(\xi_G)$	$E_n W(\check{\xi}_G)$ $n = 10$	$n = 50$	$n = 100$
(i) $\lambda_1 = -1, \lambda_2 = +1$					
0·5	0·159	0·159	0·195	0·164	0·162
0·6	0·159	0·154	0·180	0·160	0·157
0·7	0·159	0·139	0·173	0·145	0·142
0·8	0·159	0·112	0·142	0·120	0·116
0·9	0·159	0·069	0·091	0·078	0·075
(ii) $\lambda_1 = -2$, $\lambda_2 = +2$					
0·5	0·023	0·023	0·026	0·023	0·023
0·6	0·023	0·022	0·029	0·023	0·022
0·7	0·023	0·020	0·032	0·021	0·021
0·8	0·023	0·017	0·035	0·019	0·018
0·9	0·023	0·012	0·032	0·016	0·013

as $\lambda_2 - \lambda_1 \to \infty$. When θ_1 is not close to 0·5 considerable advantage may be gained by the EB approach. This emphasizes the need, when contemplating using an EB method, of having some idea as the spread of the prior distribution; in the present case the need is for preliminary information on θ_1.

When a situation of 'least favourable' prior distribution is encountered, for example $\theta_1 = 0{\cdot}5$, the EB method is less 'good' than the best conventional method. On the other hand, the results of Table 5.1 indicate that the possible gain in using the EB method for favourable values of θ_1 is much greater than the loss in using it when θ_1 is unsuitable. For example for $n = 100$, $\theta_1 = 0{\cdot}5$, $E_n W(\xi_G) - W(T) \simeq 0{\cdot}162 - 0{\cdot}1587 = 0{\cdot}0033$, and the '%loss' is $100\,(0{\cdot}0033/0{\cdot}1587) = 2{\cdot}1\%$. But when $\theta_1 = 0{\cdot}9$, $W(T) - E_n W(\xi_G) \simeq 0{\cdot}1587 - 0{\cdot}75 = 0{\cdot}0837$, and the '%gain' is $100\,(0{\cdot}0837/0{\cdot}1587) = 52{\cdot}7\%$. These results make a strong case for the use of EB methods in the present context.

The preceding discussion raises the question of judging the effectiveness in practice of the EB rule relative to non-Bayes procedures, i.e. when a set of past observations is given. Here, estimates of $E_n W(\xi_G)$, $W(\xi_G)$ and $W(\text{Bayes})$ are needed. All of these quantities are functions of θ for given n, λ_1, λ_2. Thus, if $\bar{\theta}_1$ is an estimate of $\bar{\theta}_1$, the corresponding estimates of $W(\text{Bayes})$ and $E_n W(\bar{\xi}_G)$ are obtained by replacing θ by $\bar{\theta}$, in the appropriate calculation. If

$\lambda_1 = -1$, $\lambda_2 = 1$ the tabulated values of the two functions given in Table 5.1(i) can be used, with interpolation, if necessary.

As an example, suppose that $n = 100$, $\lambda_1 = -1$, $\lambda_2 = +1$, $\bar{x} = -0{\cdot}63$. Then $\bar{\theta}_1 = 1{\cdot}63/2 = 0{\cdot}815$, and the estimates of $W(\bar{\xi}_G)$ and $E_n W(\bar{\xi}_G)$ are respectively 0·107 and 0·111.

Confidence limits for $W(\bar{\xi}_G)$ and $E_n W(\bar{\xi}_G)$ can be obtained by first getting confidence limits for θ_1. Approximate normality of the distribution of $\bar{x}$ and formulae

$$E(\bar{x}) = \lambda_2 + (\lambda_2 - \lambda_1)\theta_1$$
$$\operatorname{var}(\bar{x}) = \{1 + (\lambda_2 - \lambda_1)^2\theta_1(1 - \theta_1)\}/n$$

are used in straightforward calculations to find the confidence limits for θ_1. Then interpolation in Table 5.1 can again be done to find the other confidence limits.

For the example above, two-sided 80% confidence limits for the parameters listed below are

$$\begin{aligned} E(\bar{x}):&\quad -0{\cdot}468, \quad -0{\cdot}792 \\ \theta_1:&\quad 0{\cdot}734, \quad 0{\cdot}896 \\ W(\xi_G):&\quad 0{\cdot}072, \quad 0{\cdot}131 \\ E_n W(\xi_G):&\quad 0{\cdot}076, \quad 0{\cdot}133 \end{aligned}$$

Confidence limits for the particular $W(\xi_G)$ can be found more easily by substituting $\bar{\xi}_G$ and the two limit values for θ_1 in (5.2.6). In the example above the results are 0·078, 0·138. The similarity of the limits for $W(\bar{\xi}_G)$, $E_n W(\bar{\xi}_G)$, $W(\xi_G)$ in this example is accidental. In general it is possible for the upper limit for $W(\bar{\xi}_G)$ to be substantially greater than $W(T)$, but by its definition the upper limit for $W(\xi_G)$ cannot exceed $W(T)$.

Finally, by following steps like those above, we can find a point estimate of and confidence limits for the difference $E_n W(\bar{\xi}_G) - W(\xi_G)$. In our example the results are: point estimate = 0·004; two-sided 80% confidence limits 0·003, 0·006.

5.2.2 *$m, m_i \geqslant 1$ not necessarily equal*

If m current observations $\mathbf{x}^{\mathrm{T}} = (x_1, x_2, \ldots, x_m)$ are made independently on X, λ_1 being chosen if $\mathbf{x} \in A_1$, λ_2 if $\mathbf{x} \in A_2$, the expected loss is

$$\theta_1 \int_{\mathbf{x} \in A_2} f(\mathbf{x}|\lambda_1) d\mathbf{x} + \theta_2 \int_{\mathbf{x} \in A_1} f(\mathbf{x}|\lambda_2) d\mathbf{x},$$

where $f(\mathbf{x}|\lambda_j)=\prod_{i=1}^n f(x_i|\lambda_j)$, $j=1,2$. The Bayes rule $\delta_G(x)$ chooses λ_1 if $\theta_2 f(\mathbf{x}|\lambda_2)<\theta_1 f(\mathbf{x}|\lambda_1)$, and λ_2 otherwise. If a one-dimensional sufficient statistic, t, exists this rule reduces to choice of λ_1 if $\theta_2 f(t|\lambda_2,m)<\theta_1 f(t|\lambda_1,m)$ in an obvious notation.

Estimation of θ_1 and θ_2 remains a problem. Suppose that the observations at the ith past component are x_{ij}, $j=1,2,\ldots,m_i$. Then the likelihood of the entire collection of past observations can be expressed as

$$\prod_{i=1}^{n}\left\{\theta_1\prod_{j=1}^{m_i} f(x_{ij}|\lambda_1)+\theta_2\prod_{j=1}^{m_i} f(x_{ij}|\lambda_2)\right\},$$

which simplifies to

$$C(x)\prod_{i=1}^{n}\{\theta_1 f(t_i|\lambda_1,m_i)+\theta_2 f(t_i|\lambda_2,m_i)\}$$

when sufficiency of t for λ holds. This leads to an equation similar to (5.2.2) for finding the MLE of θ_1.

Again, it may be simpler to use the method of moments. For example, if $E(X|\lambda)=\lambda$, suppose that $\bar{x}_i$ is the mean of the observations at the ith component and let $\bar{x}_{\cdot}=\sum_{i=1}^n x_i/n$. Then we can take as an estimate of θ_1 the solution of

$$\bar{x}_{\cdot}=\bar{\theta}_1\lambda_1+(1-\bar{\theta}_1)\lambda_2, \tag{5.2.7}$$

truncated at 0 and 1, as appropriate; see also (5.2.4). Refer also to Example 2.5.1 which gives an illustration of the implementation of this method.

If no sufficient statistic exists, simplification through the use of an estimate of λ can be considered, in much the same way as was done in point estimation. The idea is to base the decision about λ on the observed value of an estimate of λ. A difficulty here is that the distribution of an estimator, for example the MLE, is usually not known exactly. If m is sufficiently large it will often be possible to approximate the distribution of the estimate by a normal distribution, leading to relatively straightforward calculations.

5.2.3 *Bayes cut-off rules*

Suppose that $\hat{\lambda}$ is an estimate of λ and that we consider rules of the type:

choose λ_1 if $\hat{\lambda}\leqslant\xi$, and λ_2 if $\hat{\lambda}>\xi$.

Such a rule may be referred to as a cut-off rule, ξ being the cut-off point. The expected 0–1 loss under this rule is

$$W(\xi) = \theta_1\{1 - \Psi(\xi|\lambda_1, m)\} + \theta_2\Psi(\xi|\lambda_2, m),$$

where $\Psi(\hat{\lambda}|\lambda, m)$ is the distribution function of $\hat{\lambda}$. The Bayes cut-off value is defined as the value of ξ which minimizes the expected loss. Under suitable conditions it can be obtained by differentiation of $W(\xi)$ w.r.t. ξ, so that it is the solution of

$$-\theta_1\Psi(\xi|\lambda_1, m) + \theta_2\Psi(\xi|\lambda_2, m) = 0. \tag{5.2.8}$$

In certain special cases the Bayes cut-off rule is the actual Bayes rule, but in general it is sub-optimal.

If $\bar{\theta}_1$ is an estimate of θ_1 derived from past observations in an EB sampling scheme, substitution in (5.2.8) leads to an estimate $\bar{\xi}_G$ of the Bayes cut-off. We now have, approximately

$$E_n W(\bar{\xi}_G) = W(\xi_G) + (1/2)\,\mathrm{var}\,(\bar{\xi}_G) W''(\xi_G). \tag{5.2.9}$$

Note that $W''(\xi_G) > 0$. Following the methods given in section 3.4 an estimate of $\mathrm{var}\,(\bar{\xi}_G)$ can be calculated. This enables us to make an assessment of the relative goodness of the Bayes cut-off rule in practice.

5.2.4 *Nuisance parameters*

Suppose that the distribution of X, $F(x|\lambda, \omega)$ depends on the parameter of primary interest, λ, and an unknown nuisance parameter, ω. A typical example is the $N(\lambda, \omega)$ distribution, where the variance ω is the nuisance parameter. Two cases are worth distinguishing:

1. the parameter ω is fixed, i.e. its marginal prior distribution is degenerate at ω,
2. ω has a non-degenerate prior distribution.

In case 1 an estimate of ω will usually be obtainable if m or some of the m_i values are greater than one. Then the EB rule is constructed as before with ω replaced by its estimate. Case 2 can be complicated, and the simplest approach seems to be through assumption of a parametric form of prior distribution for ω. Further simplification can be effected by using the non-optimal strategy of restricting decision rules to be cut-off rules based explicitly only on the natural estimate of λ. As usual, all calculations are much simplified if low-dimensional sufficient statistics exist. We give two examples illustrating this discussion.

Example 5.2.1 Let $X \overset{d}{=} N(\lambda, \theta)$, $P(\Lambda = \lambda_j) = \theta_j$, $j = 1, 2$, $P(\Omega = \omega) = 1$, i.e. we have case 1 above. The natural estimate of ω is the within-groups mean square

$$\hat{\omega} = \frac{1}{(M-n)} \sum_{i=1}^{n} \sum_{j=1}^{m_i} (x_{ij} - x_{i\cdot}),$$

where x_{ij} is the jth observation at the ith past component, $x_{i\cdot}$ is the mean of the ith group of observations and $M = m_1 + m_2 + \cdots + m_n$. Estimating θ_1 according to (5.2.8) the EB rule is

$$\text{choose } \lambda_1 \text{ if } \bar{x} \leqslant \bar{\xi}_G, \qquad \lambda_2 \text{ if } \bar{x} > \bar{\xi}_G$$

where $\bar{x}$ is the mean of the current m observations, and

$$\xi_G = \begin{cases} \dfrac{\lambda_1 + \lambda_2}{2} + \dfrac{\hat{\omega}/m}{(\lambda_2 - \lambda_1)} \ln\left[\dfrac{\bar{x}_\cdot - \lambda_1}{\lambda_2 - x_\cdot}\right], & \lambda_1 < \bar{x} < \lambda_2 \\ -\infty, & \bar{x}_\cdot \geqslant \lambda_2 \\ +\infty, & \bar{x}_\cdot \leqslant \lambda_1. \end{cases} \tag{5.2.10}$$

In order to compare the performance of this rule with that of the rule given by (5.2.5) consider the following special case: m_i are independent realizations of $M \overset{d}{=} U(1, 9)$, $m = 5$, $\lambda_1 = -1/\sqrt{5}$, $\lambda_2 = +1/\sqrt{5}$, $\omega = 1$. Here we have

$$E(\Lambda) = (\theta_2 - \theta_1)/\sqrt{5}, \qquad \text{var}(\Lambda) = \{1 - (\theta_2 - \theta_1)^2\}/5,$$
$$E(\bar{x}_\cdot) = (-\theta_1 + \theta_2)/\sqrt{5}, \qquad \text{var}(\bar{x}_\cdot) = \{\omega E(1/M) + \text{var}(\Lambda)\}/n.$$

Taking $n = 100$, $\theta_1 = 0{\cdot}8$, $\theta_2 = 0{\cdot}2$ and using the approximation (5.2.9) we have $\xi_G = 0{\cdot}6931/\sqrt{5}$, $W(\xi_G) = 0{\cdot}112$, and

$$W''(\xi_G) = 25\{\theta_1(\xi_G\sqrt{5}+1)\phi(\xi_G\sqrt{5}+1) - \theta_2(\xi_G\sqrt{5}-1)\phi(\xi_G\sqrt{5}-1)\}$$
$$= 25(0{\cdot}15225).$$

If we take $\omega = 1$ as known we should get a result close to the entry at $n = 100$, $\theta_1 = 0{\cdot}8$ for $E_n W(\bar{\xi}_G)$ in Table 5.1. In fact (5.2.9) gives

$$E_n W(\xi_G) \simeq 0{\cdot}0041 + W(\xi_G) = 0{\cdot}1163$$

and the table entry is 0·116.

Taking account of the variability in the estimate of ω gives only a slight increase in the expected loss; $E_n W(\xi_G) \simeq 0{\cdot}1168$.

Example 5.2.2 Suppose that $X = N(\lambda, \omega)$ and that the conditional p.d.f. of $W|\Lambda = \lambda_j$ is

$$h(\omega|\lambda_j) = \{(A_j/2)^{\nu/2}/\Gamma(\nu/2)\} w^{-(1+\nu)/2} \exp(-A_j/2\omega), \qquad j = 1, 2.$$

Let $\bar{x}$ be the mean of the m current observations and $Q = \sum_{i=1}^{n} (\bar{x}_i - x)^2$; these are sufficient statistics for λ and ω. Suppose that a cut-off rule is adopted, i.e. choose λ_1 if $\bar{x} \leqslant \xi$, λ_2 otherwise. Then the expected loss is

$$W(\xi) = \theta_1 \int_{x>\xi} \int f(\bar{x}|\lambda_1, \omega) h(\omega|\lambda_1) d\bar{x}\, dw$$
$$+ \theta_2 \int_{x<\xi} \int f(\bar{x}|\lambda_2, \omega) h(\omega|\lambda_2) d\bar{x}\, dw$$

where

$$f(x|\lambda_j, \omega) = (2\pi\omega)^{-m/2} \exp[-\{m(\bar{x} - \lambda_j)^2 + Q\}/(2\omega)], \qquad j = 1, 2.$$

Performing the appropriate integrations, the Bayes cut-off is seen to be the solution of the following equation in ξ:

$$\frac{\theta_1 A_1^{\nu/2}}{\{m(\xi - \lambda_1)^2 + A_1\}^{(m+\nu)/2-1}} = \frac{\theta_2 A_2^{\nu/2}}{\{m(\xi - \lambda_2)^2 + A_2\}^{(m+\nu)/2-1}}$$

which can be reduced to a quadratic in ξ.

In order to construct an empirical version of this rule, estimates of $\theta_1, \theta_2 = 1 - \theta_1, \nu, A_1, A_2$ are needed, and can be obtained by the method of maximum likelihood, or the method of moments, or otherwise. Using the method of moments, let $\bar{x}_i$, $\hat{\omega}_i$ be the estimates of λ and ω at component i. Then, using the symbol $\rightarrow$ to mean 'is an estimate of', we have:

1. $\bar{x}_1 + \cdots + \bar{x}_n \rightarrow n\theta_1\lambda_1 + n\theta_2\lambda_2$
2. $\bar{x}_1^2 + \cdots + \bar{x}_n^2 \rightarrow n(\theta_1\lambda_1^2 + \theta_2\lambda_2^2)$
   $+ ((\theta_1 A_1 + \theta_2 A_2)/(\nu - 2))(1/m_1 + \cdots + 1/m_n)$
3. $\bar{x}_1\hat{\omega}_1 + \cdots + \bar{x}_n\hat{\omega}_n \rightarrow n(\lambda_1 A_1 \theta_1 + \lambda_2 A_2 \theta_2)/(\nu - 2)$
4. $\hat{\omega}_1 + \cdots + \hat{\omega}_n \rightarrow n(\theta_1 A_1 + \theta_2 A_2)/(\nu - 2)$.

The reason for writing $\rightarrow$ instead of $=$ is that solutions of the equations thus obtained may not exist. We shall refer to relations 1, etc., as equation 1, etc. when $\rightarrow$ is replaced by $=$.

Suppose that a non-trivial solution to equation 1 exists, i.e. a solution $\bar{\theta}_1$ such that $0 < \bar{\theta}_1 < 1$. Then taking a trial value $v^{(0)}$ of v, the linear equations 2 and 3 can be solved for $A_1^{(0)}$ and $A_2^{(0)}$. If both of these are greater than 0 equation 4 can be used to calculate a new trial v, this process being repeated until all equations are satisfied, or until the magnitude of the difference between the left and right sides of equation 4 is minimized.

5.3 $k \geqslant 3$ simple hypotheses

5.3.1 $m_i = m = 1$

Let $\lambda_1 < \lambda_2 < \cdots < \lambda_k$ be given possible values of the single parameter Λ, and $\theta_1, \theta_2, \ldots, \theta_k$ the prior probabilities of these parameter values. So, the prior distribution function of Λ has steps of height θ_j at λ_j, $j = 1, 2, \ldots, k$. The Bayes rule is derived by essentially the same reasoning as applied in the case $k = 2$. It states that λ_1 or $\lambda_2, \ldots, \lambda_k$ is selected according as $\theta_1 f(x|\lambda_1)$, $\theta_2 f(x|\lambda_2), \ldots, \theta_k f(x|\lambda_k)$ is greatest when x is observed.

As far as EB implementation is concerned we have here the problem of estimating the finite mixing distribution described above, i.e. the prior distribution of Λ. Methods of estimating $\theta_1, \theta_2, \ldots, \theta_k$ have been discussed in Chapter 2. We shall consider just one example in some detail.

Example 5.3.1 Assume that the data distribution of X for given Λ is $N(\lambda, 1)$, and that $k = 3$. In the EB setting we have to estimate θ_1, θ_2, and $\theta_3 = 1 - \theta_1 - \theta_2$. Using the method of moments we note that

$$\begin{aligned} E(X_G) &= \lambda_1\theta_1 + \lambda_2\theta_2 + \lambda_3(1 - \theta_1 - \theta_2) \\ E(X_G) &= 1 + \lambda_1^2\theta_1 + \lambda_2^2\theta_2 + \lambda_3^2(1 - \theta_1 - \theta_2) \end{aligned} \qquad (5.3.1)$$

Denoting the sample first and second moments calculated from past observations by m_1' and m_2', (5.3.1) suggests that we take as estimates of θ_1 and θ_2

$$\begin{aligned} \bar{\theta}_1 &= \\ \bar{\theta}_2 &= \end{aligned} \begin{bmatrix} (\lambda_1 - \lambda_3) & (\lambda_2 - \lambda_3) \\ (\lambda_1^2 - \lambda_3^2) & (\lambda_2^2 - \lambda_3^2) \end{bmatrix}^{-1} \begin{bmatrix} m_1' - \lambda_3 \\ m_2' - 1 - \lambda_3^2 \end{bmatrix} \qquad (5.3.2)$$

whenever $\theta_1, \theta_2 \geqslant 0$, and $\theta_1 + \theta_2 \leqslant 1$. When (θ_1, θ_2) falls outside these boundaries a point within or on the boundary must be chosen according to some additional criterion. One possibility is to select the boundary point closest to $(\bar{\theta}_1, \bar{\theta}_2)$. Since $m'_r \to E(X^r_G)$, (P), $r = 1, 2$, as $n \to \infty$, it follows that such an EB rule is a.o. (P), and all $W(\cdot)$ being $\leqslant 1$, a.o. (E) is implied.

The normal distribution has monotone likelihood ratio, hence the acceptance regions A_1, A_2, A_3 are the intersections of certain intervals. Let ξ be the solution of

$$f(x|\lambda_r)/f(x|\lambda_s) = \theta_s/\theta_r, \qquad r < s; \qquad r = 1, 2; \qquad s = 2, 3.$$

Then

$$A_1 = [-\infty, \xi_{12G}] \cap [-\infty, \xi_{13G}]$$
$$A_2 = (\xi_{12G}, +\infty] \cap [-\infty, \xi_{23G}]$$
$$A_3 = (\xi_{13G}, +\infty] \cap (\xi_{23G}, +\infty].$$

Note that it is possible for A_2 to be empty. When $\xi_{23G} > \xi_{12G}$ we have

$$W(\delta_G) = \theta_1 \int_{\xi_{1\cdot G}}^{+\infty} w(x - \lambda_1)dx + \theta_2 \int_{-\infty}^{\xi_{12G}} w(x - \lambda_2)dx$$
$$+ \theta_2 \int_{\xi_{23G}}^{+\infty} w(x - \lambda_2)dx + \theta_3 \int_{-\infty}^{\xi_{\cdot 3G}} w(x - \Lambda_3)dx,$$

where $\xi_{1\cdot G} = \min(\xi_{12G}, \xi_{13G})$, $\xi_{\cdot 3G} = \max(\xi_{13G}, \xi_{23G})$, and $w(u)$ denotes the standard normal density. Obvious modifications are required in (5.3.4) when $\xi_{23G} > \xi_{12G}$.

The EB rule is obtained by replacing θ_r, θ_s in (4.3.3) by $\check{\theta}_r$, $\check{\theta}_s$, and $W(\check{\delta}_G)$ is defined similarly to $W(\delta_G)$. Using (4.3.4), it is easy to find $W(\check{\delta}_G)$ for any given $\check{\delta}_G$. Hence, if the joint distribution of θ_1 and θ_2 can be determined, it is in principle also easy to calculate $E_n W(\check{\delta}_G)$, for any given set of θ's. In fact, such calculations are tedious, and Table 5.2 gives results of a very limited study of the performance of δ_G. The non-Bayes rule, T, to which the table refers, is defined by $A_1 = [-\infty, (\lambda_1 + \lambda_2)/2]$, $A_2 = ((\lambda_1 + \lambda_2)/2, (\lambda_2 + \lambda_3)/2]$, and $A_3 = ((\lambda_2 + \lambda_3)/2, +\infty]$.

The results of Table 5.2 are substantially in agreement with those of Table 5.1.

No new principle is involved when $k > 3$ and we shall, therefore, not discuss this more general case in detail. As k increases, estimation of G becomes more troublesome, and we refer to earlier comment on this

Table 5.2 *Three simple hypotheses* $H_1:\lambda = -1$, H_2: $\lambda = 0$, $H_3:\lambda = +1$ *with* $N(\lambda, 1)$ *kernel distribution;* $W(\cdot)$ *is given for* 10 *EB rules each based on* $n = 100$, *with* $\theta_1 = 0{\cdot}1$, $\theta_2 = 0{\cdot}1$, $\theta_3 = 0{\cdot}8$

W(Bayes)		0·168
$W(T)$		0·339
W(EB): sample	1	0·169
	2	0·199
	3	0·172
	4	0·312
	5	0·190
	6	0·399
	7	0·172
	8	0·168
	9	0·168
	10	0·179
Mean W(EB)		0·213, s.e. = 0·025

problem (Chapter 2). Robbins (1964) has dealt with the case of k hypotheses, and Deely and Kruse (1968), amongst others, have treated the same problem. The closely related compound decision problem of choosing between k simple hypotheses has been studied by Samuel (1965), but as far as the author is aware, no studies of the behaviour of EB rules for $k > 3$ and finite n have been reported.

5.3.2 *Variable* $m_i \geqslant 1$, $m > 1$

In the notation of section 5.2.2, the Bayes rule chooses $\lambda_1, \lambda_2, \ldots, \lambda_k$ according as $\theta_1 f(x|\lambda_1)$, $\theta_2 f(x|\lambda_2), \ldots, \theta_k f(x|\lambda_k)$ is greatest. Estimation of $\theta_1, \theta_2, \ldots, \theta_k$ in order to construct empirical versions of the Bayes rule is somewhat more complicated here than in the previous section, but follows the methods suggested in section 5.2.2; see also section 3.8. Qualitatively, results similar to those illustrated in Example 5.2.1 can be expected for the performance of EB rules relative to Bayes rules. Details have not been worked out. Here, as elsewhere, a more interesting question is the assessment of the performance of the EB rule for a given set of previous data. A possible approach to this problem is by using a bootstrap procedure.

Consider for the present discussion the case $k = 3$ as in section 5.2.2. Suppose that estimates θ_1, θ_2 are obtained from the given set of past observations. Now W(Bayes) is a function of θ_1, θ_2 and we shall write it as $W_B(\theta_1, \theta_2)$. We can now estimate W(Bayes) by $W_B(\hat{\theta}_1, \hat{\theta}_2)$. Then treating $(\hat{\theta}_1, \hat{\theta}_2)$ as if it is the true (θ_1, θ_2), we can obtain an estimate of $E_n W$(EB), which is also a function of θ_1, θ_2; write it as $E_n W_{EB}(\theta_1, \theta_2)$. In theory we estimate it by $E_n W_{EB}(\hat{\theta}_1, \hat{\theta}_2)$. The actual evaluation of $E_n W_{EB}(\hat{\theta}_1, \hat{\theta}_2)$, or of $E_n W_{EB}(\theta_1, \theta_2)$, may be feasible only by simulation.

The steps in this first stage, then, are:

1. Find an estimate $(\hat{\theta}_1, \hat{\theta}_2)$ of (θ_1, θ_2).
2. Generate sample sizes $m'_1, m'_2, \ldots, m'_n$ according to the scheme that produced the observed $m_1, m_2, \ldots, m_n$. If the mechanism is known it can be used, otherwise use the bootstrap approach of sampling from the empirical distribution of M.
3. Generate a $\lambda_1, \lambda_2, \lambda_3$ sequence according to the probabilities $\hat{\theta}_1, \hat{\theta}_2, \hat{\theta}_3 = 1 - \hat{\theta}_1 - \hat{\theta}_2$.
4. Calculate estimates (θ'_1, θ'_2) – actually they are estimates of $(\hat{\theta}_1, \hat{\theta}_2)$ – and find the corresponding W(EB), i.e. $W_{EB}(\hat{\theta}_1, \hat{\theta}_2)$. Note that $W_{EB}(\hat{\theta}_1, \hat{\theta}_2) = W_B(\hat{\theta}_1, \hat{\theta}_2) +$ a positive quantity.
5. Repeat steps 2–4 a number of times and calculate the mean of the $W_{EB}(\hat{\theta}_1, \hat{\theta}_2)$ values; it is an estimate of $E_n W_{EB}(\hat{\theta}_1, \hat{\theta}_2)$.

The next stage in these calculations is to consider the variability of $W_B(\hat{\theta}_1, \hat{\theta}_2)$ as an estimate of $W_B(\hat{\theta}_1, \hat{\theta}_2)$. This is essentially also a bootstrap operation; the sequence of $W_B(\hat{\hat{\theta}}_1, \hat{\hat{\theta}}_2)$ values which can be obtained from repetitions of step 4 gives an estimate of the sampling distribution of $W_B(\theta_1, \theta_2)$. However, to obtain a sampling distribution of $E_n W_{EB}(\theta_1, \theta_2)$ the entire sequence of steps 1–4 has to be repeated, a straightforward procedure in principle but involving many steps.

5.4 Two composite hypotheses

5.4.1 *Introduction*

Any non-simple hypothesis will be referred to as a composite hypothesis. In this section we deal with two composite hypotheses of the form: $H_1 : \lambda < \lambda_0$, $H_2 : \lambda \geqslant \lambda_0$. In specifying these hypotheses it is clearly implied that $G(\lambda)$ will not be restricted to the class of finite step-functions with jumps at given points $\lambda_1, \lambda_2, \ldots, \lambda_k$. It may be a continuous distribution, or it may be discrete with an infinite number of jumps. The possibility of it being a finite step-function is not excluded, but the values of $\lambda_1, \lambda_2, \ldots, \lambda_k$ will not usually be assumed known.

Two composite hypotheses of the type $H_1: \lambda < \lambda_0$, $H_2: \lambda \geqslant \lambda_1$, with $\lambda_1 > \lambda_0$, are often considered in classical theory of testing hypotheses. The prior distribution implied by these alternatives would have the appearance of the two 'tails' of a distribution, the central portion having been removed. Such a prior distribution is felt to be rather unrealistic in the EB context, and will not be considered. In practice we may expect to encounter problems in which the real interest centres on the question of whether λ is $< \lambda_0$ or $\geqslant \lambda_1$, but the possibility of a λ in the interval $[\lambda_0, \lambda_1)$ is not excluded. A suitable formulation of the problem would be in terms of three composite hypotheses, H_1, H_2 and $H_3: \lambda_0 < \lambda \leqslant \lambda_1$, H_3 representing an 'indifference' state.

Although we shall devote attention mainly to the 0–1 loss function, the loss system described in section 1.5 is important in having special significance for the exponential family of distributions. The reason is that EB rules can be developed without explicit estimation of G, under this system for such distributions. For this reason it has played an important part in the development of the EB approach to testing of hypotheses.

5.4.2 *The loss system* $|\lambda - \lambda_0|$

Details of the Bayes solution for this case are given in section 1.5. Essentially the solution is contained in (1.5.2), and it states that H_1 or H_2 should be chosen according as the posterior mean of Λ given x is $<$ or $\geqslant \lambda_0$.

When we consider the exponential family of distributions, (1.5.2) can be put in the form

$$C(x)p_G(x+1) - \lambda_0 p_G(x) \lesseqgtr 0. \qquad (5.4.1)$$

In certain cases $\delta_G(x) = C(x)p_G(x+1)/p_G(x)$ is monotonic in x, and then the acceptance regions A_1 and A_2 for H_1 and H_2 are $A_1 = \{x: x < x_0\}$ and $A_2 = \{x: x \geqslant x_0\}$, where x_0 is such that $\delta_G(x_0) < \lambda_0$, $\delta_G(x_0 + 1) \geqslant \lambda_0$.

We consider the Poisson case in detail. First we observe that $\delta_G(x)$ is non-decreasing for all $G(\lambda)$; this is discussed in section 1.9. Therefore the Bayes decision rule can be formulated simply in terms of the point of dichotomy, x_0.

Now, referring to our discussion of the Poisson case in section 1.9, a natural formulation of an EB rule is in terms of the EB point of dichotomy $\hat{x}_0$, given by

$$\delta_n(\hat{x}_0) < \lambda_0, \qquad \delta_n(\hat{x}_0 + 1) \geqslant \lambda_0, \tag{5.4.2}$$

where $\delta_n(x)$ is the EB point estimator defined in section 1.9. Since $\delta_n(x)$ is not, in general, a 'smooth' function of x, the inequalities (5.4.2) can be satisfied by more than one value $\hat{x}_0$. However, for any given set of past observations, $W(\delta_n)$ can be found by substituting $\hat{A}_1$ and $\hat{A}_2$ for A_1 and A_2 in (1.5.1).

It is easy to see that since $\delta_n(x) \to \delta_G(x)$, (P), as $n \to \infty$, $\hat{A}_1 \to A_1$, $\hat{A}_2 \to A_2$, and by arguments similar to those of section 3.2, δ_n is a.o. As in most of this work, exact results for finite n can be easily obtained in principle, but in reality only approximate results can be found with a moderate amount of trouble.

Previous results for δ_n have shown that it is rather poor compared with the 'best' non-Bayes procedure unless n is quite large. A similar tendency may be expected here but no numerical results are available at present.

5.4.3 *The 0–1 loss structure: parametric G families*

The Bayes solution for this case is essentially given by (1.5.1). Since calculation of the posterior median of Λ is required, so that the result depends on

$$\int_{\lambda_0}^{\infty} f(x|\lambda) dG(\lambda),$$

it is generally not possible to find the solution in terms of the mixed p.d.f., as in section 5.4.2. In order to obtain an EB rule, we shall have to find an explicit estimate of G, or an estimate of an approximate G. Thus we have to consider the use of such methods as were developed for smooth EB point estimation. Of these, the case where G is known to belong to a certain parametric family, is the most straightforward. Here, and elsewhere, the procedures for estimating G are the same as those outlined for the problem of point estimation. Details follow for two examples, the normal and binomial each with conjugate prior $G(\lambda)$.

1. $f(x|\lambda) = N(\lambda, 1)$, $dG = N(\mu, \sigma^2)$. Since the posterior distribution

of Λ given x is normal with

$$E(\Lambda|x)=\frac{x+\mu/\sigma^2}{1+1/\sigma^2}=\text{median}(\Lambda|x),$$

the Bayes may be stated as follows,

$$\begin{aligned}&\text{accept } H_1 \quad \text{if } x<\xi_G\\ &\text{accept } H_2 \quad \text{if } x\geqslant \xi_G,\end{aligned}$$

where

$$\xi_G=\lambda_0(1+1/\sigma^2)-\mu/\sigma^2. \tag{5.4.3}$$

See also Example 1.3.1.

The Bayes risk, $W(\xi_G)$, is

$$W(\xi_G)=\int_{-\infty}^{\xi_G}\int_{\lambda_0}^{\infty} f(x|\lambda)dG(\lambda)dx+\int_{\xi_G}^{\infty}\int_{-\infty}^{\lambda_0} f(x|\lambda)dG(\lambda)dx, \tag{5.4.4}$$

and since the joint distribution of Λ and X_G is normal, $W(\xi_G)$ can be found from the tables of the bivariate normal distribution.

Estimation of μ and σ^2 has been discussed in section 3.7.3. The EB rule is developed by replacing $G(\lambda)$ in the formulation of the Bayes rule by the estimated, i.e. empirical, $G(\lambda)$. Consequently we obtain an EB 'cut-off', $\check{\xi}_G$, on replacing μ and σ^2 in (5.4.3) by $\check{\mu}$ and $\check{\sigma}^2$. For the EB rule we have $W(\check{\xi}_G)$, given by (5.4.4) with ξ_G replaced by $\check{\xi}_G$. Since $(\check{\mu},\check{\sigma}^2)\to(\mu,\sigma^2)$, (P), this rule is a.o. because all losses are bounded by 0 and 1. Some details of the performance of this EB rule for small n are given in section 5.4.5.

2. $p(x|\lambda)=\text{Bin}(N,\lambda)$, $dG(\lambda)=\lambda^{p-1}(1-\lambda)^{q-1}\,d\lambda/B(p,q)$. Estimation of p and q can be performed by the method explained in section 3.9.2. The posterior p.d.f. of Λ is

$$dG(\lambda|x)=\frac{\lambda^{p+x-1}(1-\lambda)^{N+q-x-1}}{B(p+x,N+q-x)}d\lambda,$$

and the posterior median $\lambda_{0\cdot5|x}$ is found from

$$\int_0^{\lambda_{0\cdot5|x}} dG(\lambda|x)=0{\cdot}5.$$

Then H_1 or H_2 is accepted according as $\lambda_{0\cdot5|x}$ is $<\lambda_0$ or $\geqslant\lambda_0$. The EB rule is defined similarly, with the estimates $(\check{p},\check{q})$ replacing (p,q) in the formula for $G(\lambda|x)$. Again, the EB rule is clearly a.o.; its performance for finite n is examined briefly in section 5.4.5.

5.4.4 *The 0–1 loss structure: approximating G*

The idea of approximating an unknown G by a step-function G_k has been explored at some length in connection with the problem of point estimation. Since we now want to estimate the median of the posterior distribution, it seems more appropriate to approximate G by a continuous G_k^* of the form defined in section 2.4.3. It may be observed that G_k corresponds to approximation of G by a discrete distribution, this being analogous to the procedure used in calculating means from grouped data. Approximation by G_k^* is tantamount to dG being approximated by a histogram.

The arguments in section 2.7 can easily be re-framed to justify approximation of G by G_k^* through the maximization of $\int \log\{f_{G_k^*}(x)\}\,dF_G(x)$. In EB applications this implies estimate of λ by maximizing the approximate likelihood

$$L = \sum_{i=1}^{n} \log\{f_{G_k^*}(x_i)\} \text{ w.r.t. variation in } \lambda.$$

Since $L/n \to \int \log\{g_{G_k^*}(x)\}dF_G(x)$, (P), as $n \to \infty$, the EB rule based on the estimated G_k^* is a.o. as $k \to \infty$.

Successul application of this process in practice depends partly on the goodness of the approximate Bayes rule for small values of k. This aspect of the approximate procedure is examined for normal

Table 5.3 *Normal-normal case, decision between two hypotheses $H_1{:}\lambda \leqslant \lambda_0$, $H_2{:}\lambda > \lambda_0$. Values of $W(\xi_G)$ and $W(\xi_k^*)$ are given for various σ^2, k, λ_0 when G is approximated by the continuous distribution G_k^*; $dF(x|\lambda) = N(\lambda, 1)$, $dG(\lambda) = N(0, \sigma^2)$*

		$\sigma^2 = 0{\cdot}1$			$\sigma^2 = 0{\cdot}5$	
$G(\lambda_0)$	$W(\xi_G)$	$W(\xi_3^*)$	$W(\xi_5^*)$	$W(\xi_G)$	$W(\xi_3^*)$	$W(\xi_5^*)$
0·5	0·402	0·402	0·402	0·304	0·304	0·304
0·6	0·372	0·375	0·379	0·289	0·292	0·301
0·7	0·296	0·301	0·297	0·248	0·256	0·255
0·8	0·199	0·200	0·199	0·183	0·190	0·183
0·9	0·099	0·100	0·100	0·097	0·097	0·099

Table 5.4 *Decision between* $H_1: \lambda \leqslant \lambda_0, H_2: \lambda > \lambda_0$ *in the case of a* $N(\lambda, 1)$ *kernel with* $G(\lambda) = 1 - e^{-\lambda/\sigma}$, *with* $\sigma^2 = 0{\cdot}1$. *Approximation of G by* G_3^*

$G(\lambda_0)$	*W(Bayes)*	*W(approx. Bayes)*	$W(T)$
0·5	0·42	0·44	0·42
0·7	0·27	0·28	0·40
0·9	0·09	0·09	0·33

$F(x|\lambda)$ in the examples which follow. The performance of smooth EB rules, based on the G_k^* approximation, is studied in section 5.4.5.

1. $f(x|\lambda) = N(\lambda)$, $dG(\lambda) = N(\mu, \sigma^2)$. Table 5.3 gives details of the approximation of G by G_k^*, summarizing values of $W(\xi_G)$, $W(\xi_k^*)$, for various σ^2, k, λ_0, where ξ_k^* denotes the approximate Bayes rule obtained on replacing G by G_k^*. In terms of $W(\cdot)$ the approximation is clearly excellent.

2. $G(\lambda) = 1 - e^{-\lambda/A}$. There is no essential difference between this example and (1) above. It only illustrates the performance of the approximate Bayes procedure when G is J-shaped, representing a useful practical extreme. Details are given in Table 5.4. Again we see that the approximate procedure is very good, as judged by $W(\cdot)$.

5.4.5 *Performance of EB rules for finite n*

Although general results regarding asymptotic optimality can be formulated, it does not appear possible to obtain similarly general results for small n. Rather laborious calculations are required in studies of particular cases for small n. The examples which follow indicate that the EB rules can be satisfactory, and in some circumstances substantially better than 'conventional' rules, for quite small n.

1. $f(x|\lambda) = N(\lambda, 1)$, $dG(\lambda) = N(\mu, \sigma^2)$. When G is known to be normal, estimates of μ and σ^2 can be based on the mean and sample variance of past observations, $\bar{x}$ and s^2. For a given $\bar{x}$ and s^2, defining a certain $\check{\xi}_G$, $W(\check{\xi}_G)$ can be found from tables of the bivariate normal distribution. Alternatively direct numerical integration of the joint normal distribution can be used. Employing the latter method, as well as numerical integration over the joint sampling distribution of $\bar{x}$ and s^2, the values of $E_n W(\check{\xi}_G)$ given in Table 5.5 were found.

Table 5.5 *Values of* $E_n\{W(\check{\xi}_G)\}$, *where* $\check{\xi}_G$ *is an empirical Bayes cut-off based, on the knowledge that* $G(\lambda)$ *is a* $N(\mu, \sigma^2)$ *distribution,* $F(x|\lambda) = N(\lambda, 1)$

	$G(\lambda_0)$	$E_{10}\{W(\check{\xi}_G)\}$	$E_{20}\{W(\check{\xi}_G)\}$	$W(T)$	$W(\xi_G)$
$\sigma^2 = 0{\cdot}1$	0·5	0·475	0·469	0·402	0·402
	0·6	0·459	0·449	0·399	0·372
	0·7	0·409	0·383	0·390	0·296
	0·8	0·313	0·281	0·372	0·199
	0·9	0·198	0·147	0·338	0·099
$\sigma^2 = 0{\cdot}5$	0·5	0·406	0·373	0·304	0·304
	0·6	0·384	0·353	0·299	0·289
	0·7	0·314	0·285	0·284	0·248
	0·8	0·221	0·200	0·256	0·183
	0·9	0·110	0·104	0·205	0·097

The distribution of the 'smooth' EB $W(\hat{\xi}_k^*)$ is most easily studied by Monte Carlo methods, as in similar point estimation problems. Table 5.6 contains estimates of $E_n W(\hat{\xi}_k^*)$ for σ^2, k, λ_0 as in Table 5.5. In the preparation of this table, maximization of L was carried out subject to the constraint $\lambda_{j+1} - \lambda_j > \varepsilon$, $\varepsilon > 0$, to avoid computational difficulties. Table 5.7 gives information on $P[W(\check{\xi}_k^*) > W(T)]$, in the form of frequencies of the event $[W(\check{\xi}_k^*) > W(T)]$ occurring in a certain number of Monte Carlo trials.

Certain expected trends are evident in these tables. For $G(\lambda_0)$ close to 0·5, $E_n W(\check{\xi}_k^*)$ is greater than $W(T)$, but it becomes substantially

Table 5.6 *Values of* $E_n\{W(\check{\xi}_5^*)\}$, *where* $\check{\xi}_5^*$ *is the smooth EB cut-off in the case* $F(x|\lambda) = N(\lambda, 1)$, $G(\lambda) = N(0, \sigma^2)$

	$G(\lambda_0)$	$n = 10$	$n = 20$	$n = 50$
$\sigma^2 = 0{\cdot}1$	0·5	0·478 ± 0·003	0·473 ± 0·003	0·475 ± 0·004
	0·6	0·475 ± 0·007	0·453 ± 0·006	0·455 ± 0·012
	0·7	0·440 ± 0·012	0·398 ± 0·011	0·368 ± 0·018
	0·8	0·381 ± 0·017	0·299 ± 0·013	0·259 ± 0·021
	0·9	0·272 ± 0·020	0·182 ± 0·014	0·130 ± 0·012
$\sigma^2 = 0{\cdot}5$	0·5	0·404 ± 0·011	0·403 ± 0·010	0·359 ± 0·007
	0·6	0·406 ± 0·016	0·387 ± 0·010	0·340 ± 0·008
	0·7	0·376 ± 0·023	0·341 ± 0·016	0·284 ± 0·006
	0·8	0·267 ± 0·022	0·210 ± 0·006	0·205 ± 0·006
	0·9	0·132 ± 0·011	0·118 ± 0·008	0·112 ± 0·004

Table 5.7 *Frequency distributions of* $W(\check{\xi}_5^*)$ *in two classes:* (1) $W(\xi_G) \leqslant W(\check{\xi}_5^*) < W(T)$; (2) $W(T) \leqslant W(\check{\xi}_5^*) \leqslant 1{\cdot}0$

		$n=10$			$n=20$			$n=50$		
	$G(\lambda_0)$	(1)	(2)	*Total*	(1)	(2)	*Total*	(1)	(2)	*Total*
$\sigma^2 = 0{\cdot}1$	0·5	0	200	200	0	200	200	0	50	50
	0·6	31	169	200	47	153	200	5	45	50
	0·7	119	81	200	138	62	200	39	11	50
	0·8	130	70	200	163	37	200	44	6	50
	0·9	151	49	200	181	19	200	48	2	50
$\sigma^2 = 0{\cdot}5$	0·5	0	50	50	0	50	50	0	50	50
	0·6	12	38	50	5	45	50	15	35	50
	0·7	18	32	50	15	35	50	32	18	50
	0·8	41	9	50	47	3	50	47	3	50
	0·9	45	5	50	48	2	50	49	1	50

smaller than $W(T)$ as $G(\lambda_0)$ increases. As before, the relative gain in using a Bayes approach diminishes as σ^2 increases. Previous results, showing that the smooth EB approach can be satisfactory for small n, are confirmed.

2. $f(x|\lambda) = N(\lambda, 1)$, $G(\lambda) = 1 - e^{-\lambda/A}$. When $G(\lambda)$ is known to belong to the exponential family of distributions, the EB rule may be based on the estimate

$$\check{A} = \max(0, \bar{x})$$

of A, where $\bar{x}$ is the sample mean of the past observations. In this

Table 5.8 *Binomial kernel distribution*

$$P(X = x) = \binom{10}{x} \lambda^x (1-\lambda)^{10-x},$$

Beta prior distribution with parameters $p = 10$, $q = 9$, *testing* $H_1\colon \lambda \leqslant \lambda_0$ *against* $H_2\colon \lambda > \lambda_0$, $n = 50$, *'parametric G' EB approach*

Values of $E_{50}W(\cdot)$

$G(\lambda_0)$	$W(T)$	$E_{50}W(\check{\xi}_G)$	$W(\xi_G)$
0·9	0·229	$0{\cdot}102 \pm 0{\cdot}001$	0·100
0·7	0·306	$0{\cdot}253 \pm 0{\cdot}005$	0·245
0·5	0·295	$0{\cdot}324 \pm 0{\cdot}014$	0·295

case an explicit formula for W_G cannot be found, and ξ_G as well as $W(\check{\xi}_G)$ must be found by numerical integration. The same applies to ξ_k^*, $W(\xi_k^*)$ and $E_n W(\xi_k^*)$. These results need little comment; they indicate that a rather skew G can also be approximated satisfactorily by G_k^*, with small k.

3. $p(x|\lambda) = \text{Bin}(N, \lambda)$, $dG(\lambda) = \lambda^{p-1}(1-\lambda)^{q-1}/B(p,q)d\lambda$. The EB rule, $\check{\xi}_G$, based on moment estimates $\hat{p}$, $\hat{q}$ of p, q, when G is known to be member of the beta family of distributions, has been described in section 5.4.3. Data on the performance of $\check{\xi}_G$ in the case $p = 10$, $q = 9$ are given in Table 5.8. For comparison $W(T)$ is shown, where $T = \lambda_0$, a non-Bayes rule. Previous experience suggests that the performance of a smooth EB rule based on G_k^* would be only slightly worse than the performance of $\check{\xi}_G$.

CHAPTER 6

Bayes and empirical Bayes interval estimation

6.1 Introduction

In this chapter we consider interval estimation of a single parameter and region estimation of a vector of parameters. The Bayesian decision theoretic approach considers the choice of a particular interval or region for an unknown parameter as a decision-making problem. Optimal intervals are sought to minimize an expected loss. The Bayesian inferential approach does not make use of the notion of a loss function. Instead it considers the posterior distribution of an unknown parameter as an inferential statement and may set regions for the parameters under the derived posterior distribution. They may be quantitatively similar to regions obtained by a non-Bayes approach but their interpretation is clearly different.

The empirical Bayes approach aims at obtaining estimates of the Bayes regions which converge to the true Bayes regions when the amount of previous data increases. With respect to previous data sets in the typical EB sampling scheme the EB regions are random. Therefore a further requirement of EB regions may be that the converge probability in the usual relative frequency sense is as good as that of a classical, i.e. non-Bayes, interval. In particular, if a classical confidence interval can be constructed with a given probability level α, a good EB interval should have at least the same level α if it is of the same length. Alternatively it should have shorter length if it is of the same level.

6.2 Intervals for single parameters

Suppose that the single parameter λ of the data d.f. $F(x|\lambda)$ of the observable random variable X is to be estimated by an interval (a, b)

in the parameter space. Let λ be a realization of a r.v. Λ with d.f. $G(\lambda)$. Suppose that $\mathbf{x} = (x_1, \ldots, x_m)$ is a set of independent observations on X. The problem is to obtain (a, b) based on data $\mathbf{x}$ which is optimal in some sense.

6.2.1 *Optimal intervals*

Let A be the set of ordered pairs (a, b) such that

$$A = \{(a, b); -\infty < a, b < \infty, a < b\}.$$

If $G(\lambda)$ is known, the Bayesian decision theory seeks the ordered pair (a, b) from A which minimizes the expected value of a loss $L((a, b), \lambda)$ incurred by using (a, b) when λ is the true value of the unknown parameter. A general class of linear loss functions is defined by

$$L(a, b, \lambda) = \begin{cases} c_L(a-\lambda) + c_0(b-a) & \text{if } a \geqslant \lambda \\ c_0(b-a) & \text{if } a < \lambda < b \\ c_U(\lambda - b) + c_0(b-a) & \text{if } b \leqslant \lambda \end{cases} \tag{6.2.1}$$

where c_0, c_L, c_U are known positive constants. For given data $\mathbf{x}$, a and b are functions of $\mathbf{x}$. The overall expected loss of (a, b) is

$$E_G E_D L(a, b, \Lambda) \tag{6.2.2}$$

where E_G is the expectation with respect to G and E_D is the expectation with respect to the joint distribution

$$\prod_{i=1}^{m} F(x_i | \lambda)$$

of $\mathbf{X}$. If $L(a, b, \lambda)$ is bounded, the minimization of the overall expected loss defined in (6.2.2) is the same as minimization of the posterior expected loss given by

$$E_B L(a, b, \Lambda) = \int L(a, b, \lambda) dB(\lambda | \mathbf{x}, G) \tag{6.2.3}$$

where $B(\lambda | \mathbf{x}, G)$ is the posterior d.f. of Λ given data $\mathbf{x}$ and prior d.f. G.

Winkler (1972) takes (6.2.3) as the starting point of a decision theoretic approach to interval estimation and considers the class of loss functions

$$L(a, b, \lambda) = L_L(a-\lambda) + L_U(\lambda - b) + c_0(b-a) \tag{6.2.4}$$

where c_0 is a known non-negative constant, $L_L(\cdot)$ and $L_U(\cdot)$ are monotone non-decreasing functions with $L_L(x) = L_U(x) = 0$ for all $x \leqslant 0$. For the special case (6.2.4) the posterior expectation (6.2.3) becomes

$$\begin{aligned} E_B L(a, b, \Lambda) = {} & B(a|\mathbf{x}, G) E_B\{L_L(a - \Lambda) | \Lambda \leqslant a\} \\ & + \{1 - B(b|\mathbf{x}, G)\} E_B\{L_U(\Lambda - b) | \Lambda \geqslant b\} \\ & + c_0(b - a). \end{aligned}$$

The existence of a minimum of $E_B L(a, b, \Lambda)$ in this case is established by Winkler (1972) for the special class of functions $L_L(\cdot)$ and $L_U(\cdot)$ that are convex. A method of construction of the optimal interval (a, b) is indicated for functions which are differentiable.

The loss function (6.2.1) is of special interest here. It is a special case of (6.2.4) with convex functions and also provides a justification for the use of quantiles of the posterior d.f. $B(\lambda|\mathbf{x}, G)$ as interval estimates as is the practice in a Bayesian inferential approach.

The optimal decision (a, b) for the special loss function (6.2.1) minimizes the quantity

$$\begin{aligned} E_B L(a, b, \Lambda) &= c_L \int_{-\infty}^{a} (a - \lambda) dB(\lambda|\mathbf{x}, G) \\ &\quad + c_U \int_{b}^{\infty} (\lambda - b) dB(\lambda|\mathbf{x}, G) + c_0(b - a) \\ &= c_L \int_{-\infty}^{a} B(\lambda|\mathbf{x}, G) d\lambda \\ &\quad + c_U \int_{b}^{\infty} B(\lambda|\mathbf{x}, G) d\lambda + c_0(b - a). \end{aligned}$$

Thus (a, b) is given by

$$\left. \begin{aligned} B(a|\mathbf{x}, G) &= c_0/c_L \\ B(b|\mathbf{x}, G) &= 1 - c_0/c_U \end{aligned} \right\}$$

so that a solution with $a < b$ exists if $c_0/c_L + c_0/c_U < 1$. One may choose $c_L = c_U = c$ and set $\alpha = 1 - 2c_0/c$. Then the condition for the existence of a solution is $0 < \alpha < 1$ and the problem is to find $100\alpha\%$ limits for λ under the posterior d.f. $B(\lambda|\mathbf{x}, G)$.

Hence under the special loss function (6.2.1) and certain restrictions on the constants c_0, c_L, c_U, the optimal decision theory approach gives a justification for the use of percentiles in the Bayesian credibility interval approach.

6.2.2 *Bayesian credibility intervals*

In the uniparameter case, the Bayesian inferential approach employs intervals around some mode of the posterior d.f. of Λ. In the simplest case, a one-sided $100\alpha\%$ limit is taken as the lower $100(1-\alpha)\%$ or upper $100(1-\alpha)\%$ quantile of the d.f. $B(\lambda|\mathbf{x}, G)$. To obtain two-sided limits, equal tail area quantiles are often used. Thus to construct an interval of the form $[\hat{\lambda}_L^*, \infty)$ with level α, one seeks $\hat{\lambda}_L^*(\mathbf{x}, G, \alpha)$, a function of $\mathbf{x}$, G and α, such that

$$B(\hat{\lambda}_L^*(\mathbf{x}, G, \alpha)|\mathbf{x}, G) = 1 - \alpha.$$

Similarly, to construct an interval of the form $(-\infty, \hat{\lambda}_U^*]$ with level α, one seeks $\hat{\lambda}_U^*(\mathbf{x}, G, \alpha)$ such that

$$B(\hat{\lambda}_U^*(\mathbf{x}, G, \alpha)|\mathbf{x}, G) = \alpha.$$

Thus,

$$\hat{\lambda}_U^*(\mathbf{x}, G, 1-\alpha) = \hat{\lambda}_L^*(\mathbf{x}, G, \alpha).$$

If two sided limits for Λ are required, we look for a pair $[\hat{\lambda}_L^*(\mathbf{x}, G, (1-\alpha)/2), \hat{\lambda}_U^*(\mathbf{x}, G, (1+\alpha)/2)]$ where α is equally distributed in the two tails of the d.f. B. Hence it suffices to find the lower limit $\hat{\lambda}_L^*(\mathbf{x}, G, \alpha)$ in terms of $\mathbf{x}$, G and α.

An alternative to percentiles is to use a highest posterior density (HPD) region (see e.g. Box and Tiao, 1973). This is given by the set of values of λ such that

$$b(\lambda|\mathbf{x}, G) \geqslant c_\alpha \tag{6.2.5}$$

where $b(\cdot|\mathbf{x}, G)$ is the p.d.f. of B and c_α is a constant determined so that the region has probability content α under the d.f. B. The region defined by (6.2.5) has also a justification in terms of Bayesian decision theory if the loss function is taken as

$$L((a, b), \lambda) = c_0(b-a) - I_{a,b}(\lambda)$$

where $I_{a,b}(\lambda) = 1$ if $a < b < \lambda$ and 0 otherwise (see Joshi, 1969).

As in the study of point estimators certain specializations of G are of interest. First we have the parametric G case which is important since a detailed treatment is possible, giving a better understanding of the problems of interval estimation. The finite approximation to G by

G_k is again useful since this provides a practical and versatile family applicable to many problems where exact specification of G is difficult and a completely unspecified G is not attractive due to identifiability requirements.

Suppose that G belongs to the parametric family of $G(\lambda|\xi)$ where $\xi = (\xi_1, \dots, \xi_q)$ is a vector of parameters. In such cases the Bayesian limits are directly determined in terms of ξ and notations such as $\hat{\lambda}_L^*(\mathbf{x}, \xi, \alpha)$ will be used to emphasize this fact.

Suppose next that G is approximated by a step-function $G_k(\lambda)$ with concentration of probability θ_i at the points w_i $(i = 1, \dots, k)$; we can approximate the posterior d.f. of Λ by

$$B_k(\lambda|\mathbf{x}, \xi, m) = \sum_{w_j \leqslant \lambda} \theta_j \left\{ \prod_{i=1}^{m} f(x_i|w_j) \right\} \Big/ h_k(\mathbf{x}, \xi, m)$$

where

$$h_k(\mathbf{x}, \xi, m) = \sum_{j=1}^{k} \theta_j \left\{ \prod_{i=1}^{m} f(x_i|w_j) \right\}$$

and ξ stands for the set of parameters $(\boldsymbol{\theta}, \mathbf{w})$. The Bayesian credibility limits for λ can be obtained approximately as

1. $\hat{\lambda}_L^*(\mathbf{x}, \xi, \alpha) = \frac{1}{2}\{C_L(\mathbf{x}, \xi, \alpha) + D_L(\mathbf{x}, \xi, \alpha)\}$, where

$$C_L(\mathbf{x}, \xi, \alpha) = \inf_{\lambda}[\lambda: B_k(\lambda|\mathbf{x}, \xi, m) \geqslant 1 - \alpha]$$

$$D_L(\mathbf{x}, \xi, \alpha) = \sup_{\lambda}[\lambda: B_k(\lambda|\mathbf{x}, \xi, m) \leqslant 1 - \alpha];$$

2. $\hat{\lambda}_U^*(\mathbf{x}, \xi, \alpha) = \frac{1}{2}\{C_U(\mathbf{x}, \xi, \alpha) + D_U(\mathbf{x}, \xi, \alpha)\}$, where

$$C_U(\mathbf{x}, \xi, \alpha) = \inf_{\lambda}[\lambda: B_k(\lambda|\mathbf{x}, \xi, m) \geqslant \alpha]$$

$$D_U(\mathbf{x}, \xi, \alpha) = \sup_{\lambda}[\lambda: B_k(\lambda|\mathbf{x}, \xi, m) \leqslant \alpha];$$

and similarly for two-sided limits as

$$\left[\hat{\lambda}_L^*\left(\mathbf{x}, \xi, \frac{1-\alpha}{2}\right), \hat{\lambda}_U^*\left(\mathbf{x}, \xi, \frac{1+\alpha}{2}\right)\right].$$

6.2.3 *Empirical Bayes confidence intervals*

When G is not known, the Bayesian limits $\hat{\lambda}_L^*(\mathbf{x}, G, \alpha)$ or $\hat{\lambda}_U^*(\mathbf{x}, G, \alpha)$ cannot be found. Suppose however that when the current observation vector $\mathbf{x}$ is observed, corresponding to the current realization λ of Λ, there are available past observation vectors $\mathbf{x}_1, \dots, \mathbf{x}_n$ corresponding to the realizations $\lambda_1, \ldots, \lambda_n$ of Λ, where $\mathbf{x}_i = (x_{i1}, \ldots, x_{im_i})^T$. Then the unknown G can be estimated from such data, and hence estimates of the Bayesian limits can be obtained. Let $\hat{G}_m$ be a consistent estimator of T at every point of λ in the parameter space Ω. Then under fairly general conditions it can be shown (Rutherford and Krutchkoff, 1969) that

$$\hat{\lambda}_L^*(\mathbf{x}, \hat{G}_n, \alpha) \xrightarrow{p} \hat{\lambda}_L^*(\mathbf{x}, G, \alpha).$$

Thus it is possible in principle to construct consistent estimates of the Bayes limits.

However, the intervals or regions obtained by estimated quantities such as $\hat{\lambda}_L^*(\mathbf{x}, \hat{G}, \alpha)$ no longer possess a fixed probability content α as was intended for the corresponding Bayesian intervals or regions. In fact, these regions themselves are random regions and the probability content of such a region is a random variable with respect to variations in $\hat{G}_n$.

One may impose the desired level of the probability content as a requirement of these regions. We may then say that these regions are EB regions. Since the probability content of a region based on the estimated Bayesian limits is a random variable, there are two ways of imposing a fixed level for it; one is an expected cover requirement and the other is a percentile cover requirement. These criteria are similar to those which are used in statistical tolerance region theory (see Guttman, 1970).

Let $R(\mathbf{x}, G, \alpha)$ be a general Bayesian credibility region for λ based on data $\mathbf{x}$, prior G and credibility level α. Let $R(\mathbf{x}, \hat{G}_m, \alpha)$ be an estimated Bayesian region obtained by directly replacing G by its estimate $\hat{G}_n$ obtained from an EB scheme. This estimated region can be regarded as a random region in the parameter space as far as the sampling variation of $\hat{G}_n$ is concerned and has to be assessed accordingly. In particular, the coverage probability of $R(\mathbf{x}, \hat{G}_m, \alpha)$ under the posterior d.f. is a random variable given by

$$C[R(\mathbf{x}, \hat{G}_n, \alpha)] = \int_{R(\mathbf{x}, \hat{G}_m, \alpha)} dB(\lambda \mid \mathbf{x}, G, n).$$

Following the established pattern of treatment of non-Bayes statistical tolerance regions, one can concentrate on two main characteristics of $C[R(\mathbf{x}, \hat{G}_n, \alpha)]$. One is its expectation with respect to variation of $\hat{G}_n$. The other is a designated percentile of its distribution. One can thus introduce the following two criteria:

Expected cover criterion: An estimated region $R(\mathbf{x}, \hat{G}_n, \alpha)$ is said to be an EB region of expected cover β if

$$E_n\{C[R(\mathbf{x}, \hat{G}_n, \alpha)]\} = \beta \tag{6.2.6}$$

where $E_n\{\cdot\}$ is the expectation operator with respect to variation of $\hat{G}_n$.

Percentile cover criterion: An estimated region $R(\mathbf{x}, \hat{G}_n, \alpha)$ is said to be an EB region of percentile cover β with level γ if

$$\Pr_n\{C[R(\mathbf{x}, \hat{G}_n, \alpha)] \geqslant \beta\} = \gamma. \tag{6.2.7}$$

Where $\mathrm{P}_n\{\cdot\}$ is the probability statement with respect to variation of $\hat{G}_n$.

6.2.4 *Construction of EB regions; parametric G priors*

We now consider the case where G belongs to a specified parametric family of distributions $G(\lambda|\xi)$ indexed by a set of unknown parameters ξ. For detailed study we shall take the Bayesian region $R(\mathbf{x}, \xi, \alpha)$ to be the interval

$$R(\mathbf{x}, \xi, \alpha) = \{\lambda : \hat{\lambda}^*(\mathbf{x}, \xi, \alpha) < \lambda < \infty\},$$

defined by the lower limit, $\hat{\lambda}^*(\mathbf{x}, \xi, \alpha)$. The subscript L is deleted for notational convenience and G is replaced by the q-vector $\xi = (\xi_1, \ldots, \xi_q)$. The parameter ξ is the only unknown element of G and can be estimated from an EB scheme as discussed in Chapter 2.

Let $\hat{\xi}_n$ be the ML estimator of ξ based on the data of an EB scheme. Then an estimated Bayesian region can be defined as

$$R(\mathbf{x}, \hat{\xi}, \alpha) = \{\lambda : \hat{\lambda}^*(\mathbf{x}, \hat{\xi}_n, \alpha) < \lambda < \infty\}.$$

The probability content of the estimated region $R(\mathbf{x}, \hat{\xi}, \alpha)$ is no longer α under the posterior d.f. $B(\lambda|\mathbf{x}, \xi, m)$. We shall now demonstrate how the criteria (6.2.6) and (6.2.7) can be employed to construct EB regions and hence an EB lower limit. Although the discussion here concentrates only on a lower limit, the case of an upper limit or of two sided limits can be treated in a similar manner.

(a) Expected cover EB limit

The coverage probability of the region $R(\mathbf{x}, \hat{\xi}, \alpha)$ is

$$C[R(\mathbf{x}, \hat{\xi}, \alpha)] = 1 - B[\hat{\lambda}^*(\mathbf{x}, \hat{\xi}_n, \alpha)|\mathbf{x}, \xi, m].$$

The exact value of the expectation or the percentile of $C[R(\mathbf{x}, \hat{\xi}, \alpha)]$ can be obtained in principle if the exact distribution of $\hat{\xi}_n$ is obtained. However, this will not be the case in general. Thus an exact solution to the problem of constructing EB limits, satisfying expected or percentile cover criteria, is not possible except in some special cases. Approximate solutions to the problem can be obtained by using techniques from statistical tolerance region theory. The specific methods employed here are implicit in the original works of Wilks (1941) and Wald (1942); the work of the former author has recently been extended to a general case by Atwood (1984).

Let $\psi(\xi)$ be a $q \times 1$ vector whose ith element is the bias of the ith element, $\hat{\xi}_{ni}$, of $\hat{\xi}_n$. Also let $V(\xi)$ be a $q \times q$ matrix whose (i, j)th element is the covariance of $\hat{\xi}_{ni}$ and $\hat{\xi}_{nj}$. Evaluation of $\psi(\xi)$ and $V(\xi)$ to terms of order n^{-1} can be performed as indicated in Chapter 2.

One can expand the coverage probability $C[R(\mathbf{x}, \hat{\xi}_n, \alpha)]$ in a Taylor series of $(\hat{\xi}_{ni} - \xi_i)$. Retaining terms of $O(n^{-1})$ in the expansion, we get, after taking expectation with respect to variation in $\hat{\xi}_n$,

$$
\begin{aligned}
&E_n\{C[R(\mathbf{x}, \hat{\xi}_n, \alpha)]\} \\
&= \alpha + \sum_{i=1}^{q} \psi_i(\xi) b(\hat{\lambda}^*(\mathbf{x}, \xi, \alpha) | \mathbf{x}, \xi, m) \frac{\partial \hat{\lambda}^*(\mathbf{x}, \xi, \alpha)}{\partial \xi_i} \\
&\quad + \tfrac{1}{2} \sum_{i=1}^{q} \sum_{j=1}^{q} \Gamma_{ij}(\xi) \left[b(\hat{\lambda}^*(\mathbf{x}, \xi, \alpha) | \mathbf{x}, \xi, m) \frac{\partial^2 \hat{\lambda}^*(\mathbf{x}, \xi, \alpha)}{\partial \xi_i \partial \xi_j} \right. \\
&\quad \left. + \left\{ \frac{\partial b(y | \mathbf{x}, \xi, m)}{\partial y} \right\}_{y = \hat{\lambda}^*(\mathbf{x}, \xi, \alpha)} \frac{\partial \hat{\lambda}^*(\mathbf{x}, \xi, \alpha)}{\partial \xi_i} \frac{\partial \hat{\lambda}^*(\mathbf{x}, \xi, \alpha)}{\partial \xi_j} \right].
\end{aligned}
$$

Following the approach of Atwood (1984), we define the quantities:

$\mathbf{B}_{10} = (\partial / \partial \lambda) B(\lambda | \mathbf{x}, \xi, m)$,

$\mathbf{B}_{20} = (\partial^2 / \partial \lambda^2) B(\lambda | \mathbf{x}, \xi, m)$,

$\mathbf{B}_{01}$ as a $q \times 1$ vector whose ith element is $\dfrac{\partial}{\partial \xi_i} B(\lambda | \mathbf{x}, \xi, m)$

and

$\mathbf{B}_{02}$ as a $q \times q$ matrix whose (i, j)th element is $(\partial^2 / \partial \xi_i \partial \xi_j) B(\lambda | \mathbf{x}, \xi, m)$

where all the derivatives are evaluated at $\lambda = \hat{\lambda}^*(\mathbf{x}, \xi, \alpha)$. We can now rewrite the expected value of $C[R(\mathbf{x}, \hat{\xi}_n, \alpha)]$ as

$$\begin{aligned} &E\{C[R(\mathbf{x}, \hat{\xi}_n, \alpha)]\} \\ &\quad = \alpha - \psi(\xi)^{\mathrm{T}}\mathbf{B}_{01} + \mathbf{B}_{10}^{-1}\{\mathbf{B}_{01}^{\mathrm{T}}\Gamma(\xi)\mathbf{B}_{11}\} - \tfrac{1}{2}\operatorname{tr}\{\mathbf{B}_{02}\Gamma(\xi)\} \\ &\quad = \alpha + K(\mathbf{x}, \xi, \alpha), \end{aligned}$$

say, where $\Gamma(\xi) = V(\xi) + \psi(\psi)^{\mathrm{T}}$.

To construct an EB lower limit with an expected cover β, we need to seek α for which $E\{C[R(\mathbf{x}, \hat{\xi}_n, \alpha)]\}$ is equal to β. The equation

$$E\{C[R(\mathbf{x}, \hat{\xi}_n, \alpha)]\} = \beta$$

can be solved iteratively for α. Let $\alpha^*(\beta)$ be the solution. Then the required EB region is given by $R(\mathbf{x}, \hat{\xi}_n, \alpha^*(\beta))$ and the corresponding EB lower limit by $\hat{\lambda}^*(\mathbf{x}, \hat{\xi}_n, \alpha^*(\beta))$.

An approximation to $\alpha^*(\beta)$ can be obtained by steps similar to those of Cox (1975) by using a first-order solution,

$$\alpha^*(\beta) \simeq \beta - K(\mathbf{x}, \hat{\xi}_n, \beta).$$

(b) Percentile cover EB region

To construct an EB region of a given percentile cover β with level γ, we obtain an approximation to the distribution of the quantity $C[R(\mathbf{x}, \hat{\xi}_n, \alpha)]$ induced by random variation in $\hat{\xi}_n$. For this purpose, we need to obtain an approximation to the variance of $C[R(\mathbf{x}, \hat{\xi}_n, \alpha)]$. Again, using a Taylor expansion, we have, to the same order of approximation as before,

$$\begin{aligned} &\operatorname{var}\{C[R(\mathbf{x}, \hat{\xi}_n, \alpha)]\} \\ &\quad = \sum_{i=1}^{q}\sum_{j=1}^{q} \operatorname{cov}(\hat{\xi}_{ni}, \hat{\xi}_{nj}) \left\{ \frac{\partial\hat{\lambda}^*(\mathbf{x}, \xi, \alpha)}{\partial\xi_i} \frac{\partial\hat{\lambda}^*(\mathbf{x}, \xi, \alpha)}{\partial\xi_j} \right\} b(\hat{\lambda}^*(\mathbf{x}, \xi, \alpha) \mid \mathbf{x}, \xi, m) \\ &\quad = \{\mathbf{B}_{01}^{\mathrm{T}} V(\xi)\mathbf{B}_{01}\}/\mathbf{B}_{10}^2. \end{aligned}$$

We may now adopt an approach similar to Wald (1942) and use a normal approximation to the distribution of $C[R(\mathbf{x}, \hat{\xi}_n, \alpha)]$. The γ-percentile of this distribution is then approximated by

$$\alpha + K(\mathbf{x}, \xi, \alpha) + z_\gamma[\{\mathbf{B}_{01}^{\mathrm{T}} V(\xi)\mathbf{B}_{01}\}^{1/2}/\mathbf{B}_{10}]$$

where z_γ is the γ-probability point of a standard normal d.f. Thus an approximate EB region whose coverage has a γ-percentile equal to β is given by $R(\mathbf{x}, \hat{\xi}_n, \alpha^*(\beta, \gamma))$ where

$$\alpha^*(\beta, \gamma) \simeq \beta - K(\mathbf{x}, \hat{\xi}_n, \beta) - z_\gamma[\{\mathbf{B}_{01}^{\mathrm{T}} V(\xi)\mathbf{B}_{01}\}^{1/2}/\mathbf{B}_{10}]_{\alpha=\beta, \xi=\hat{\xi}_n}.$$

An alternative approximation can be obtained by following Wilks (1941) and using a beta distribution approximation to the distribution of $C[R(\mathbf{x}, \hat{\xi}_n, \alpha)]$. Writing $E\{C[R(\mathbf{x}, \hat{\xi}_n, \alpha)]\} = \mu_C$ and $\operatorname{var}\{C[R(\mathbf{x}, \hat{\xi}_n, \alpha)]\} = \sigma_C^2$, we can express the parameters of a beta distribution by

$$\begin{aligned} p^* &= \{\mu_C^2(1-\mu_C) - \mu_C\sigma_C^2\}/\sigma_C^2 \\ q^* &= \{\mu_C(1-\mu_C)^2 - (1-\mu_C)\sigma_C^2\}/\sigma_C^2. \end{aligned} \tag{6.2.8}$$

Let $y_\gamma(p^*, q^*)$ be the $(1-\gamma)$ probability point of a beta (p^*, q^*) distribution. Then

$$\Pr_n\{C[R(\mathbf{x}, \hat{\xi}_n, \alpha)] \geqslant y_r(p^*, q^*)\} \simeq \gamma.$$

Thus for the region $R(\mathbf{x}, \hat{\xi}_n, \alpha)$ to be an EB region of a given percentile cover β, the value of α must satisfy the following relation:

$$y_\gamma(p^*, q^*) = \beta.$$

For given β and γ, the required value for α can be obtained iteratively in the above equation. Let $\alpha^*(\beta, \gamma)$ be the resulting solution; then the region $R(\mathbf{x}, \hat{\xi}_n, \alpha^*(\beta, \gamma))$ is an EB region satisfying the percentile cover criterion and the corresponding EB limit is $\hat{\lambda}^*(\mathbf{x}, \hat{\xi}_n, \alpha^*(\beta, \gamma))$.

6.2.5 *An example with normal F and normal G*

Consider the special case given in Example 1.3.1. We have $F(x|\lambda)$ as a $N(\lambda, \sigma^2)$ distribution function with known σ^2, $G(\lambda|\xi)$ as a $N(\mu_G, \sigma_G^2)$ distribution function with known σ^2, $G(\lambda|\xi)$ as a $N(\mu_G, \sigma_G^2)$ distribution function with known σ_G^2. The current data set is $\mathbf{x} = (x_1, \ldots, x_m)$, being m observations on the r.v. X with d.f. $F(x|\lambda)$. The posterior d.f., $B(\lambda|\mathbf{x}, \xi)$, of Λ is $N(\mu^*, \sigma^{*2})$ with

$$\begin{aligned} \mu^* &= (\sigma_G^2 + d^2)^{-1}(\sigma_G^2\bar{x} + d^2\mu_G) \\ \sigma^{*2} &= (\sigma_G^2 + d^2)^{-1}\sigma_G^2 d^2 \end{aligned}$$

where $\bar{x}$ is the sample mean and $d^2 = \sigma^2/m$. The Bayesian lower credibility limit of level α is

$$\hat{\lambda}^*(\mathbf{x}, \mu_G, \alpha) = (\sigma_G^2 + d^2)^{-1}(\sigma_G^2\bar{x} + d^2\mu_G) + z_{1-\alpha}\sigma_G d(\sigma_G^2 + d^2)^{-1/2}$$

where $z_{1-\alpha}$ is the value such that $\Phi(z_{1-\alpha}) = 1 - \alpha$.

Suppose that μ_G is unknown and an EB scheme with unequal component sample sizes is available as in section 3.8. Let $\bar{x}_1, \ldots, \bar{x}_n$ be the sample means from the previous stages of the EB scheme. Then the ML estimator of μ_G is given by

$$\hat{\mu}_G = \sum_{i=1}^{n} m_i \bar{x}_i \bigg/ \sum_{i=1}^{n} m_i$$

with the sampling variance

$$V(\hat{\mu}_G) = \sum_{i=1}^{n} m_i(m_i\sigma_G^2 + \sigma^2) \bigg/ \left(\sum_{i=1}^{n} m_i\right)^2.$$

The estimated Bayesian lower limit is $\hat{\lambda}^*(\mathbf{x}, \hat{\mu}_G, \alpha)$ and for the region $R(\mathbf{x}, \hat{\mu}_G, \alpha) = [\hat{\lambda}^*(\mathbf{x}, \hat{\mu}_G, \alpha), \infty)$ it induces a coverage probability under the posterior d.f. of

$$C[R(\mathbf{x}, \hat{\mu}_G, \alpha)] = 1 - \Phi\{(\sigma_G^2 + d^2)^{-1/2} d\sigma_G^{-1}(\hat{\mu}_G - \mu_G) + z_{1-\alpha}\}.$$

This problem is a very special case where exact solutions to both types of EB limits can be obtained. We shall also derive approximate solutions by employing the general procedure outlined in section 6.2.1.

(a) Expected cover EB limit

First we note that $\hat{\mu}_G$ is distributed normally with mean μ_G and variance $V(\hat{\mu}_G)$ given above. Next we note that the coverage probability can be rewritten as

$$C[R(\mathbf{x}, \hat{\xi}_n, \alpha)] = 1 - \Phi(\rho Z + z_{1-\alpha})$$

where

$$\rho = (\sigma_G^2 + d^2)^{-1/2} d\sigma_G^{-1} \left\{\sum_{i=1}^{n} m_i(m_i\sigma_G^2 + \sigma^2)\right\}^{1/2} \bigg/ \left(\sum_{i}^{n} m_i\right)$$

and Z is a standard normal $N(0, 1)$ r.v. Thus the exact value of the expected coverage probability is

$$E_n C[R(\mathbf{x}, \hat{\xi}_n, \alpha)] = 1 - \Phi[z_{1-\alpha}(1 + \rho^2)^{-1/2}].$$

For the estimated Bayes lower limit $\hat{\lambda}^*(\mathbf{x}, \hat{\xi}_n, \alpha)$ to be an exact expected cover EB lower limit with level β, we must have

$$z_{1-\alpha}(1 + \rho^2)^{-1/2} = z_{1-\beta}$$

or

$$\alpha = 1 - \Phi\{z_{1-\beta}(1 + \rho^2)^{1/2}\}.$$

Table 6.1 *Value of α to give expected converage β.*

β	α			
	$n=10$		$n=30$	
	Approximate	*Exact*	*Approximate*	*Exact*
0·50	0·500	0·500	0·500	0·500
0·55	0·551	0·551	0·550	0·550
0·60	0·602	0·602	0·600	0·600
0·65	0·653	0·653	0·650	0·650
0·70	0·704	0·703	0·700	0·700
0·75	0·754	0·754	0·750	0·750
0·80	0·805	0·805	0·801	0·800
0·85	0·855	0·855	0·851	0·850
0·90	0·904	0·904	0·901	0·901
0·95	0·953	0·953	0·950	0·950

Thus the required EB lower limit is given by the expression for $\hat{\lambda}^*(\mathbf{x}, \hat{\xi}_n, \alpha)$ with $z_{1-\alpha}$ replaced by $z_{1-\beta}(1+\rho^2)^{1/2}$.

Next we apply the approximate technique of section 6.2.1. To terms of $O(M^{-1})$, where $M=\sum_{i=1}^n m_i$, we have

$$E_n C[R(\mathbf{x}, \hat{\xi}_n, \alpha)] = \alpha + \mathrm{var}(\hat{\xi}_n)\{(\sigma_G^2+d^2)^{-1} d^2 \sigma_G^{-2}\} z_{1-\alpha}\phi(z_{1-\alpha})/2$$

where $\phi(\cdot)$ is the p.d.f. of the $N(0, 1)$ distribution. Thus for the expected coverage probability to be β, α must satisfy the equation

$$\alpha + \rho^2 z_{1-\alpha}\phi(z_{1-\alpha})/2 = \beta.$$

For values of $\beta = 0{\cdot}50$, $(0{\cdot}05)$, $0{\cdot}95$, $n=10$, $m_1 = m_2 = \cdots = m_n = 1$ and $\sigma^2 = \sigma_G^2 = 1$, corresponding values of α are computed from the exact formula as well as the approximate formula. These are given in Table 6.1.

(b) Percentile cover EB limit

The event

$$C[R(\mathbf{x}, \hat{\mu}_G, \alpha)] \geqslant \beta$$

is equivalent to the event

$$(\sigma_G^2+d^2)^{-1/2} d\sigma_G^{-1}(\hat{\mu}_G - \mu_G) + z_{1-\alpha} \leqslant z_{1-\beta},$$

i.e.

$$Z \leqslant (z_{1-\beta} - z_{1-\alpha})/\rho$$

where Z is defined above, a $N(0, 1)$ r.v. Thus

$$\Pr\{C[R(\mathbf{x}, \hat{\boldsymbol{\mu}}_G, \alpha)] \geqslant \beta\} = \Phi\{(z_{1-\beta} - z_{1-\alpha})/\rho^2\}.$$

Hence for the estimated limit $\hat{\lambda}^*(\mathbf{x}, \hat{\boldsymbol{\mu}}_G, \alpha)$ to be an exact EB limit with a percentile cover β at level γ, we need to have

$$(z_{1-\beta} - z_{1-\alpha})/\rho = z_\gamma,$$

i.e. $z_{1-\alpha}$ is replaced by $z_{1-\beta} - \rho z_\gamma$ or α by $1 - \Phi(z_{1-\beta} - \rho z_\gamma)$.

Next we apply the approximate technique of section 6.2.1. We can obtain the variance of $C[R(\mathbf{x}, \hat{\xi}_n, \alpha)]$ to terms of $O(M^{-1})$ as

$$\operatorname{var}\{C[R(\mathbf{x}, \hat{\boldsymbol{\mu}}_G, \alpha)]\} = \rho^2\phi^2(z_{1-\alpha}).$$

Table 6.2 *Values of α to give percentile coverage specified by β and γ*

β	γ	α			
		$n = 10$		$n = 30$	
		Approximate	*Exact*	*Approximate*	*Exact*
0·5	0·5	0·500	0·500	0·500	0·500
0·5	0·6	0·521	0·520	0·507	0·507
0·5	0·7	0·543	0·542	0·514	0·514
0·5	0·8	0·568	0·567	0·522	0·522
0·5	0·9	0·603	0·601	0·534	0·534
0·6	0·5	0·600	0·600	0·600	0·600
0·6	0·6	0·620	0·619	0·607	0·606
0·6	0·7	0·641	0·640	0·613	0·613
0·6	0·8	0·664	0·663	0·621	0·621
0·6	0·9	0·696	0·695	0·632	0·632
0·7	0·5	0·701	0·700	0·700	0·700
0·7	0·6	0·718	0·717	0·706	0·706
0·7	0·7	0·736	0·735	0·712	0·712
0·7	0·8	0·756	0·756	0·719	0·719
0·7	0·9	0·783	0·782	0·729	0·729
0·8	0·5	0·801	0·800	0·800	0·800
0·8	0·6	0·815	0·814	0·805	0·805
0·8	0·7	0·829	0·828	0·810	0·810
0·8	0·8	0·844	0·844	0·815	0·815
0·8	0·9	0·863	0·864	0·823	0·823
0·9	0·5	0·901	0·900	0·900	0·900
0·9	0·6	0·909	0·909	0·903	0·903
0·9	0·7	0·918	0·917	0·906	0·906
0·9	0·8	0·926	0·926	0·910	0·910
0·9	0·9	0·937	0·938	0·914	0·914

We can now proceed to calculate the quantities p^* and q^* of (6.2.8) using

$$\mu_c = \alpha + \rho^2 z_{1-\alpha}\phi(z_{1-\alpha})/2$$
$$\sigma_c^2 = \rho^2\phi^2(z_{1-\alpha}).$$

Let $y_\gamma(p^*, q^*)$ be the $(1-\gamma)$ probability point of a beta distribution with parameters p^* and q^*. Then α must satisfy the equation

$$y_\gamma(p^*, q^*) = \beta.$$

For values of $\beta, \gamma = 0{\cdot}5, (0{\cdot}05), 0{\cdot}95$ and $\sigma^2 = \sigma_G^2 = 1, m_1 = \cdots = m_n = 1$ the approximate value of α obtained from the above equation is compared with the corresponding exact value of α in Table 6.2.

6.3 The multiparameter case: region estimators

Suppose now that the data d.f. $F(x|\boldsymbol{\lambda})$ depends on a p-vector parameter $\boldsymbol{\lambda} = (\lambda_1, \ldots, \lambda_p)$. An estimate of $\boldsymbol{\lambda}$ by a p-dimensional region, R, a subset of the parameter space, is sought so that some optimality criterion is achieved. Let $\boldsymbol{\lambda}$ be a realization of a random vector $\boldsymbol{\Lambda}$ with d.f. $G(\boldsymbol{\lambda})$ and $\mathbf{x} = (x_1, \ldots, x_n)$ be a set of independent observations on r.v. X with d.f. F.

6.3.1 *Optimal region estimators*

As in the single-parameter case, the Bayesian decision theoretic approach is to construct an optimal region based on a chosen loss function. A multiparameter analogue of the loss function (6.2.1) is given by

$$L(R, \boldsymbol{\lambda}) = c_0 \operatorname{Vol}(R) - I(R)$$

where $\operatorname{Vol}(R)$ is the volume of a region R and $I(\cdot)$ is the indicator function which has value zero if $\boldsymbol{\lambda} \in R$ and value one otherwise. Joshi (1969) has shown that the optimal region which minimizes the expected loss, $E_G E_D L(R, \boldsymbol{\Lambda})$, is given by the HPD region

$$R^*(\mathbf{x}, G, \alpha) = \{\boldsymbol{\lambda}: b(\boldsymbol{\lambda}|\mathbf{x}, G) \geqslant K_\alpha\} \quad (6.3.1)$$

where K_α is a constant determined so that the region has probability content α under the posterior d.f. B.

The region defined in (6.3.1) is not necessarily a multidimensional rectangle shape, when $p > 1$. On the other hand, a Bayesian credibility region can take any chosen shape as long as its coverage probability is the given specified value α. A multidimensional rectangular region is given by

$$R^+(\mathbf{x}, G, \alpha) = \{\boldsymbol{\lambda}: \hat{\lambda}_{iL}^+(\mathbf{x}, G, \alpha) < \lambda_i < \hat{\lambda}_{iU}^+(\mathbf{x}, G, \alpha); i = 1, 2, \ldots, p\}. \tag{6.3.2}$$

This region is a multidimensional analogue of the two-sided limits and the quantities $(\hat{\lambda}_{iL}^+, \hat{\lambda}_{iU}^+)$ are determined so that $R^+(\mathbf{x}, G, \alpha)$ has credibility level α. One-sided analogues are obtained when the limits for any λ_i are taken to be negative or positive infinity.

6.3.2 *Emperical Bayes confidence regions; general priors*

We now consider EB confidence regions analogous to those developed for the single-parameter case discussed earlier. The unknown G can in principle be estimated nonparametrically by the ML technique of Lindsay (1983a, b) as summarized in Chapter 2. Once a consistent estimator $\hat{G}_n$ of G is available, consistent estimators of the limits $\hat{\lambda}_{iL}^+$ and $\hat{\lambda}_{iU}^+$ can be obtained. When G is replaced by an estimate $\hat{G}_n$, the region $R(\mathbf{x}, \hat{G}_n, \alpha)$ induces a random coverage probability

$$C[R(\mathbf{x}, \hat{G}_n, \alpha)] = \int_{R(\mathbf{x}, \hat{G}_n, \alpha)} dB(\boldsymbol{\lambda} | \mathbf{x}, G, m)$$

which again needs to be adjusted in terms of α to achieve a desired expected coverage or percentile coverage. The criteria described by (6.2.6) and (6.2.7) are directly applicable in constructing EB confidence regions. In the next section, we demonstrate the application of these criteria to estimated Bayesian regions of the form (6.3.2) when G belongs to a parametric family $G(\boldsymbol{\lambda} | \boldsymbol{\xi})$.

When an EB scheme is available, estimation of the unknown parameter $\boldsymbol{\xi}$ follows the procedures given in section 2.9. It should also be noted that although the discussion in the next section is for univariate multiparameter d.f. $F(x | \boldsymbol{\lambda})$, the calculations are essentially the same for the multivariate multiparameter case.

6.3.3 *Construction of EB confidence regions: parametric priors*

We shall consider the special case of the Bayesian region given by $R^+(\mathbf{x}, \boldsymbol{\xi}, \alpha)$ in (6.3.2) where the upper limits are taken to be infinity for each λ_i. The estimated Bayesian region is thus

$$R^+(\mathbf{x}, \hat{\xi}_n, \alpha) = \{\boldsymbol{\lambda}: \hat{\lambda}_i^+(\mathbf{x}, \hat{\xi}_n, \alpha) < \lambda_i < \infty, i = 1, \ldots, p\} \quad (6.3.3)$$

where the subscript L of the lower limits is omitted for convenience. The case of upper limits can be handled in a similar manner.

The probability content of the region (6.3.3) under the posterior d.f. of $\boldsymbol{\Lambda}$ is

$$\begin{aligned} &C[R^+(\mathbf{x}, \hat{\xi}_n, \alpha)] \\ &= \int_{\hat{\lambda}_p^+(\mathbf{x}, \hat{\xi}_n, \alpha)}^{\infty} \cdots \int_{\hat{\lambda}_1^+(\mathbf{x}, \hat{\xi}_n, \alpha)} b(\boldsymbol{\lambda}|\mathbf{x}, \xi, m) d\lambda_1, \ldots, d\lambda_p \quad (6.3.4) \\ &= I(\hat{\xi}_n, \xi, \alpha), \qquad \text{say.} \end{aligned}$$

We now obtain multiparameter analogues of the results in section 6.2.1, using the same criteria to develop EB regions.

(a) Expected cover EB region

To apply the expected cover criterion to adjust the level α of $R^+(\mathbf{x}, \hat{\xi}_n, \alpha)$ we need to obtain the expected value of $C[R^+(\mathbf{x}, \hat{\xi}_n, \alpha)]$, defined in (6.3.1), as $\hat{\xi}_n$ varies. To terms of order $O(n^{-1})$, we have

$$\begin{aligned} E_n C[R^+(\mathbf{x}, \hat{\xi}_n, \alpha)] = \alpha &+ \psi(\xi)^{\mathrm{T}} \{(\partial/\partial\hat{\xi}_n) I(\hat{\xi}_n, \xi, \alpha)\}_{\hat{\xi}_n = \xi} \\ &+ \operatorname{tr}[\Gamma(\xi)\{(\partial^2/\partial\hat{\xi}_n \partial\hat{\xi}_n^{\mathrm{T}}) I(\hat{\xi}_n, \xi, \alpha)\}]_{\hat{\xi}_n = \xi} \end{aligned}$$

where $\psi(\xi)$ and $\Gamma(\xi)$ are, respectively, the bias vector and mean square error matrix of $\hat{\xi}_n$ and further

1. $(\partial/\partial\hat{\xi}_n) I(\hat{\xi}_n, \xi, \alpha)$ is a $q \times 1$ vector whose ith element is $(\partial/\partial\hat{\xi}_i) I(\hat{\xi}_n, \xi, \alpha)$;
2. $(\partial^2/\partial\hat{\xi}_n \partial\hat{\xi}_n^{\mathrm{T}}) I(\hat{\xi}_n, \xi, \alpha)$ is a $q \times q$ matrix whose (i, j)th element is $(\partial^2/\partial\hat{\xi}_i \partial\hat{\xi}_j) I(\hat{\xi}_n, \xi, \alpha)$.

Let

$$\begin{aligned} &I_s(\hat{\xi}_n, \xi, \alpha, \lambda_s) \\ &= \int_{\hat{\lambda}_p^+}^{\infty} \cdots \int_{\hat{\lambda}_{s+1}^+}^{\infty} \int_{\hat{\lambda}_{s-1}^+}^{\infty} \cdots \int_{\hat{\lambda}_1^+}^{\infty} b(\boldsymbol{\lambda}|\mathbf{x}, \xi, m) d\lambda_1 \cdots d\lambda_{s-1} d\lambda_{s+1} \cdots d\lambda_p \end{aligned}$$

and also,

$$I_{st}(\hat{\xi}_n, \xi, \alpha, \lambda_s, \lambda t) = \int_{\hat{\lambda}_p^+}^{\infty} \cdots \int_{\hat{\lambda}_{t+1}^+}^{\infty}, \int_{\hat{\lambda}_{t-1}^+}^{\infty} \cdots \int_{\hat{\lambda}_{s+1}^+}^{\infty} \int_{\hat{\lambda}_{s-1}^+}^{\infty}$$

$$\cdots \int_{\hat{\lambda}_1^+}^{\infty} b(\boldsymbol{\lambda}|\mathbf{x}, \xi, m) d\lambda_1 \cdots d\lambda_{s-1} d\lambda_{s+1}$$
$$\cdots d\lambda_{t-1} d\lambda_{t+1} \cdots d\lambda_p.$$

We then have

$$\{(\partial/\partial\hat{\xi}_i) I(\hat{\xi}_n, \xi, \alpha)\}_{\hat{\xi}_n=\xi} = -\sum_{s=1}^{p} \frac{\partial \hat{\lambda}_s^+(\mathbf{x}, \xi, \alpha)}{\partial \xi_i} I_s(\xi, \xi, \alpha, \hat{\lambda}_s^+(\mathbf{x}, \xi, \alpha)).$$

Also,

$$\{(\partial^2/\partial\hat{\xi}_i\partial\hat{\xi}_j) I(\hat{\xi}_n, \xi, \alpha)\}_{\hat{\xi}_n=\xi}$$
$$= -\sum_{s=1}^{p} \Bigg[\frac{\partial \hat{\lambda}_s^+(\mathbf{x}, \xi, \alpha)}{\partial \xi_i} \Bigg\{ \frac{\partial \hat{\lambda}_s^+(\mathbf{x}, \xi, \alpha)}{\partial \xi_j} \left(\frac{\partial}{\partial y} I_s(\xi, \xi, \alpha, y) \right)_{y=\hat{\lambda}_s^+}$$
$$- \sum_{t \neq s}^{p} \frac{\partial \hat{\lambda}_t^+(\mathbf{x}, \xi, \alpha)}{\partial \xi_j} I_{st}(\xi, \xi, \alpha, \hat{\lambda}_s^+(\mathbf{x}, \xi, \alpha), \hat{\lambda}_t^+(\mathbf{x}, \xi, \alpha)) \Bigg\}$$
$$+ I_s(\xi, \xi, \alpha, \hat{\lambda}_s^+(\mathbf{x}, \xi, \alpha))(\partial^2/\partial\xi_i\partial\xi_j)\hat{\lambda}_s^+(\mathbf{x}, \xi, \alpha) \Bigg].$$

We need to evaluate first- and second-order derivatives of $\hat{\lambda}_s^+(\mathbf{x}, \xi, \alpha)$. For this purpose, we introduce the following notations:

$$\bar{\mathbf{B}}_s(\xi, \xi, \alpha, y) = \int_{\hat{\lambda}_p^+}^{\infty} \cdots \int_{\hat{\lambda}_{s+1}^+}^{\infty} \int_{y}^{\infty} \int_{\hat{\lambda}_{s-1}^+}^{\infty} \cdots \int_{\hat{\lambda}_1^+}^{\infty} b(\boldsymbol{\lambda}|\mathbf{x}, \xi, n) d\lambda_1 \cdots d\lambda_p$$
$$\bar{\mathbf{B}}_{s10} = (\partial/\partial y)\bar{B}_s(\xi, \xi, \alpha, y)$$
$$\bar{\mathbf{B}}_{s20} = (\partial^2/\partial y^2)\bar{B}_s(\xi, \xi, \alpha, y).$$

Also let $\bar{\mathbf{B}}_{s01}$ be a $q \times 1$ vector whose ith element is

$$(\partial/\partial\xi_i)\bar{\mathbf{B}}_s(\xi, \xi, \alpha, y),$$

and $\bar{\mathbf{B}}_{s11}$ be a $q \times 1$ vector whose ith element is

$$(\partial^2/\partial\xi_i\partial y)\bar{\mathbf{B}}_s(\xi, \xi, \alpha, y),$$

and $\bar{\mathbf{B}}_{s02}$ be a $q \times q$ matrix whose (i, j)th element is

$$(\partial^2/\partial\xi_i\partial\xi_j)\bar{\mathbf{B}}_s(\xi, \xi, \alpha, y).$$

All the derivative given for $\bar{\mathbf{B}}_s, \bar{\mathbf{B}}_{s10}, \bar{\mathbf{B}}_{s20}, \bar{\mathbf{B}}_{s11}, \bar{\mathbf{B}}_{s02}$ are evaluated at $y = \hat{\lambda}_s^+(\mathbf{x}, \xi, \alpha)$.

Next, differentiating both sides of the equation

$$I(\xi, \xi, \alpha) = \alpha \tag{6.3.5}$$

with respect to ξ_i we get a set of equations which can be rewritten in matrix form as

$$(\partial/\partial \boldsymbol{\xi})\hat{\lambda}_s^+(\mathbf{x}, \xi, \alpha) = (\bar{\mathbf{B}}_{s10})^{-1}\bar{\mathbf{B}}_{s01}.$$

Differentiating both sides of (6.3.5) twice with respect to ξ_i first and then ξ_j and rearranging the resulting equations in matrix form, we get

$$\begin{aligned}\bar{\mathbf{B}}_{s10}&\{(\partial^2/\partial \boldsymbol{\xi}\partial \boldsymbol{\xi}^{\mathrm{T}})\hat{\lambda}_s^+(\mathbf{x}, \xi, \alpha)\}\\ &= \bar{\mathbf{B}}_{s02} - (\bar{\mathbf{B}}_{s11}\bar{\mathbf{B}}_{s01}^{\mathrm{T}}/\bar{\mathbf{B}}_{s10} + \bar{\mathbf{B}}_{s01}\bar{\mathbf{B}}_{s11}^{\mathrm{T}}/\bar{\mathbf{B}}_{s10})\\ &\quad - (\bar{\mathbf{B}}_{s10})^{-2}\bar{\mathbf{B}}_{s20}\bar{\mathbf{B}}_{s01}\bar{\mathbf{B}}_{s01}^{\mathrm{T}}.\end{aligned}$$

Thus we have

$$\begin{aligned}E_n\{&C[R^+(\mathbf{x}, \hat{\xi}_n, \alpha)]\}\\ &= \alpha - \sum_{s=1}^{p} \psi(\boldsymbol{\xi})^{\mathrm{T}}\bar{\mathbf{B}}_{s01}\\ &\quad + \tfrac{1}{2}\sum_{s=1}^{p}\sum_{t\neq s}^{p} I_{st}(\xi, \xi, \alpha, \hat{\lambda}_s^+, \hat{\lambda}_t^+)(\bar{\mathbf{B}}_{s10})^{-2}\bar{\mathbf{B}}_{s01}^{\mathrm{T}}\Gamma(\boldsymbol{\xi})\bar{\mathbf{B}}_{t01}\\ &\quad - \tfrac{1}{2}\sum_{s=1}^{p} \operatorname{tr}\Gamma(\boldsymbol{\xi})\bar{\mathbf{B}}_{s02}\\ &\quad + \tfrac{1}{2}\sum_{s=1}^{p} (\bar{\mathbf{B}}_{s10})^{-1}\bar{\mathbf{B}}_{s01}^{\mathrm{T}}\Gamma(\boldsymbol{\xi})\bar{\mathbf{B}}_{s11}\\ &= \alpha + \phi(\boldsymbol{\xi}, \alpha, \mathbf{x}), \quad \text{say.}\end{aligned} \qquad (6.3.6)$$

To construct an expected cover region of size β, we need to find the value of α satisfying

$$\alpha + \phi(\boldsymbol{\xi}, \alpha, \mathbf{x}) = \beta.$$

This equation can be solved iteratively. An approximate solution is

$$\alpha \simeq \beta - \phi(\hat{\boldsymbol{\xi}}_n, \beta, \mathbf{x}).$$

(b) Percentile cover EB region

The above method of obtaining an approximate expression for the expected value of the posterior coverage probability $C[R^+(\mathbf{x}, \xi, \alpha)]$ can be extended to obtain a similar approximation to its variance as

$\hat{\xi}_n$ varies. To $O(n^{-1})$, we have

$$\begin{aligned}\operatorname{var}\{C[R^+(\mathbf{x},\hat{\xi}_n,\alpha)]\} &= \sum_{i=1}^{q}\sum_{j=1}^{q}\operatorname{cov}(\hat{\xi}_i,\hat{\xi}_j)\{(\partial/\partial\hat{\xi}_i)I(\hat{\xi}_n,\xi,\alpha)\}_{\hat{\xi}_n=\xi}\\ &\quad\times\{(\partial/\partial\hat{\xi}_j)I(\hat{\xi}_n,\xi,\alpha)\}_{\hat{\xi}_n=\xi}\\ &= \sum_{s=1}^{p}\bar{\mathbf{B}}_{s01}^{\mathrm{T}}V(\xi)\bar{\mathbf{B}}_{s01}.\end{aligned}$$

To construct a percentile cover EB region, we need to use an approximation to the distribution of $C[R(\mathbf{x},\xi,\alpha)]$. In particular a beta distribution approximation can be made along the lines of section 6.2.1, using equation (6.2.8). While a solution following such steps is straightforward in principle the actual implementation is numerically complicated.

6.4 Bayes statistical tolerance regions

In this section and in section 6.5 we discuss statistical prediction from the Bayes and EB points of view. The emphasis is on setting a tolerance limit or region for a single future observation or a set of future observations, instead of on point prediction. Thus the problem considered here could be regarded as a branch of the general problem of confidence region estimation. Indeed the techniques developed in sections 6.1–6.3 for Bayes and EB confidence region estimation can be readily adapted to obtain analogous results for the present problem.

In non-Bayes tolerance region theory, the established convention is to distinguish between two types of tolerance regions, namely, expected cover and percentile cover regions; see Guttman (1970). A similar distinction is made in Bayesian tolerance region theory. In the following, we summarize Bayes prediction theory, taking as our starting point the assumption of a particular prior distribution of the parameters. This will pave way for the development of EB analogues in section 6.5.

Let X be a continuous r.v. with d.f. $F(x|\boldsymbol{\lambda})$, and a p.d.f. $f(x|\boldsymbol{\lambda})$ depending on the unknown parameter vector $\boldsymbol{\lambda}$. Suppose also that the range, $\mathscr{X}$, of X is independent of $\boldsymbol{\lambda}$. Let $\mathbf{X}=(X_1,\ldots,X_m)$ be a set of m independent copies of X whose realization is $\mathbf{x}=(x_1,\ldots,x_m)$. Let $\boldsymbol{\lambda}$ be a realization of a vector r.v. $\boldsymbol{\Lambda}$ with d.f. $G(\boldsymbol{\lambda})$, concentrated on a parameter space $\mathscr{L}$. Then the posterior d.f. of λ given $\mathbf{X}=x$ is

$$dB(\lambda|\mathbf{x};G,m)=\{h(\mathbf{x};G,m)\}^{-1}\left\{\prod_{i=1}^{m}f(x_i|\lambda)\right\}dG(\lambda) \qquad (6.4.1)$$

where

$$h(\mathbf{x},G,m)=\int_{\mathscr{L}}\left\{\prod_{i=1}^{m}f(x_i|\lambda)\right\}dG(\lambda).$$

When a set of sufficient statistics $\mathbf{T}(\mathbf{X})$ with p.d.f. $p_m(\mathbf{t}|\lambda)$ exists the expression (6.4.1) reduces to

$$dB(\lambda|\mathbf{t};\mathbf{T},m)=\{p_{G,n}(\mathbf{t})\}^{-1}p_m(\mathbf{t}|\lambda)dG(\lambda)$$

where

$$p_{G,n}(\mathbf{t})=\int_{\mathscr{L}}p_m(\mathbf{t}|\lambda)dG(\lambda).$$

Let $\mathscr{X}^{(m)}$ be the space of observations $\mathbf{x}$ and let $\mathscr{G}$ be the event space of $\mathscr{X}$. Let $A^*(\cdot)$ be a statistic with domain $\mathscr{X}^{(m)}$ and range $\mathscr{G}$ such that for each $\mathbf{x}$, it provides a subset $A^*(\mathbf{x})$ of $\mathscr{X}$. Thus under F, $A^*(\mathbf{x})$ induces the probability content

$$C[A^*(\mathbf{x}),\lambda]=\int_{A^*(\mathbf{x})}dF(y|\lambda). \qquad (6.4.2)$$

The problem is to obtain $A^*(\mathbf{x})$ of 'a given shape' such that either of the following criteria is satisfied:

Expected cover criterion: If the coverage probability defined in (6.4.2) is such that

$$E_B\{C[A^*(\mathbf{x}),\mathbf{\Lambda}]\}=\beta \qquad (6.4.3)$$

then $A^*(\mathbf{x})$ is said to have an expected cover of size β.

Percentile cover criterion: If the coverage probability defined in (6.4.2) is such that

$$\Pr_B\{C[A^*(\mathbf{x}),\mathbf{\Lambda}]\geqslant\beta\}=\gamma \qquad (6.4.4)$$

then $A^*(\mathbf{x})$ is said to have a percentile cover of size β with level γ.

By the phrase 'a given shape', we mean, for example, that $A^*(\mathbf{x})$ may be an interval $[u_1(\mathbf{x}),u_2(\mathbf{x})]$ where one of the u_1's can be infinite. For simplicity we consider only those regions $A^*(\mathbf{x})$ of the form $[u(\mathbf{x}),\infty)$ to obtain a lower limit for a future observation on X. An upper tolerance limit or a pair of two-sided limits may be obtained in a similar way. The problem of determining $A^*(\mathbf{x})$ of the form $[u(\mathbf{x}), \infty]$ reduces to that of determining the quantity $u^*(\mathbf{x}, \beta, G)$ or $u^*(\mathbf{x}, \beta, \gamma, G)$, say, which is the solution for $u(\mathbf{x})$ in (6.4.3) or (6.4.4) when $A^*(\mathbf{x})$ is expressed in terms of $u(\mathbf{x})$.

6.4.1 *Expected cover regions*

The problem of determining $u^*(\mathbf{x}, \beta, G)$ which satisfies the expected cover criterion is readily solved since by Fubini's theorem. We can write

$$E_B C[A^*(\mathbf{x}, \boldsymbol{\Lambda})] = \int_{A^*(\mathbf{x})} E_B f(y \mid \boldsymbol{\lambda}) dy,$$

i.e. $u^*(\mathbf{x}, \beta, G)$ is the solution of the equation

$$\int_{u^*(\mathbf{x}, \beta, G)}^{\infty} p(y \mid \mathbf{x}, G, m) dy = \beta,$$

where

$$p(y \mid \mathbf{x}, G, m) = \int_{\mathscr{L}} f(y \mid \boldsymbol{\lambda}) dB(\boldsymbol{\lambda} \mid \mathbf{x}, G, m) \tag{6.4.5}$$

is the predictive d.f. or a r.v. Y representing a future observation y on $\mathbf{X}$. To obtain (6.4.5) the knowledge of functional form of $h(\mathbf{x}; G, m)$ in (6.4.1) is enough. For we have

$$p(y \mid \mathbf{x}, G, m) = \{h(\mathbf{x}, G, m)\}^{-1} \int_{\mathscr{L}} f(y \mid \boldsymbol{\lambda}) \left\{ \prod_{i=1}^{m} f(x_i \mid \boldsymbol{\lambda}) \right\} dG(\boldsymbol{\lambda})$$

$$= \{h(\mathbf{x}; G, m)\}^{-1} h((y, \mathbf{x}); G, m+1) \tag{6.4.6}$$

where $(y, \mathbf{x})$ is treated as a 'pooled-sample' of the current observation $\mathbf{x}$ and the future observation y and $h(\cdot; G, m+1)$ is evaluated exactly the same as $h(\cdot; G, m)$ with m replaced by $m+1$.

Further reduction of (6.4.6) into simpler forms can be achieved when there exists a set of sufficient statistics for $\boldsymbol{\lambda}$. Let $T(\mathbf{X})$ and $T(Y, \mathbf{X})$ be the sufficient statistics for $\boldsymbol{\lambda}$ based on $\mathbf{X}$ and the pooled sample $(Y, \mathbf{X})$ respectively. First we note that (6.4.6) reduces to

$$p(y \mid \mathbf{x}, G, m) = p(y \mid T(\mathbf{x}), G)$$

$$= \{p_{G,m}(\mathbf{t})\}^{-1} \int f(y \mid \boldsymbol{\lambda}) p_m(\mathbf{t} \mid \boldsymbol{\lambda}) dG(\boldsymbol{\lambda})$$

where $\mathbf{t}$ is a realization of $T(\mathbf{X})$ when $\mathbf{X}=\mathbf{x}$. Next, we use the identity

$$f(y|\boldsymbol{\lambda})p_m(\mathbf{t}|\boldsymbol{\lambda}) = r(y|T(y,\mathbf{x}))p_{m+1}(T(y,\mathbf{x})|\boldsymbol{\lambda}),$$

where $r(y|T(y,\mathbf{x}))$ is the conditional p.d.f. of Y given $T(Y,\mathbf{X}) = T(y,\mathbf{x})$ and is independent of $\boldsymbol{\lambda}$. Hence we get

$$p(y|\mathbf{x},G,m) = \{p_{G,m}(\mathbf{t})\}^{-1} r(y|T(y,\mathbf{x}))p_{G,m+1}(T(y,\mathbf{x})).$$

Thus to evaluate the predictive density in this case, it is required to compute only the unconditional p.d.f. of $T(\mathbf{X})$ and the conditional p.d.f. of Y given $T(Y,\mathbf{X})$. Evaluation of these quantities is straightforward once G is known.

6.4.2 *Percentile cover region*

Next consider the problem of determining $u^*(\mathbf{x},\beta,\gamma,G)$ satisfying the criterion (6.4.4) which requires a given size β for the $100(1-\gamma)\%$ percentile of the coverage $C[A^*(\mathbf{x}),\boldsymbol{\Lambda}]$ under the posterior d.f. $B(\boldsymbol{\lambda}|\mathbf{x},G,m)$. Let the set of values of $\boldsymbol{\lambda}$ for which the coverage probability defined in (6.4.2) exceeds β be

$$\Delta[\boldsymbol{\lambda},u(\mathbf{x}),\beta] = \{\boldsymbol{\lambda}: C[A^*(\mathbf{x}),\boldsymbol{\lambda}] \geqslant \beta\}.$$

Then $u^*(\mathbf{x},\beta,r,G)$, satisfying (6.4.4), is the solution of the equation

$$\int_{\Delta[\boldsymbol{\lambda},u^*(\mathbf{x},\beta,\gamma,G),\beta]} dB(\boldsymbol{\lambda}|\mathbf{x},G,m) = \gamma. \qquad (6.4.7)$$

Further simplification of (6.4.7) could be achieved if we confine $f(\cdot|\boldsymbol{\lambda})$ to more special forms like location–scale types as illustrated in the following section.

6.4.3 *Location and scale family d.f.* $F(\cdot|\boldsymbol{\lambda})$

Let the p.d.f. $F(\cdot|\boldsymbol{\lambda})$ be of the form $F\{(x-\lambda_1)/\lambda_2\}$ where $F(\cdot)$ is of known form and $\boldsymbol{\lambda}=(\lambda_1,\lambda_2)$. Then

$$C[A^*(\mathbf{x}),\boldsymbol{\lambda}] = 1 - F\{(u(\mathbf{x})-\lambda_1)/\lambda_2\}$$

so that

$$\Delta[\boldsymbol{\lambda},u(\mathbf{x}),\beta] = \{(\lambda_1,\lambda_2); \lambda_1 + \lambda_2\eta_{1-\beta} \geqslant u(\mathbf{x})\}, \qquad (6.4.8)$$

where $\eta_{1-\beta}$ is such that $F(\eta_{1-\beta}) = 1-\beta$ and is known since $F(\cdot)$ is known. Three separate cases are investigated in detail:

1. *λ_2 known:* in this case the unknown parameter is λ_1 and we denote the $(1-\gamma)$ probability point of the posterior d.f. $B^{(1)}(\lambda_1|\mathbf{x},G,m)$ of Λ_1 by $w^{(1)}(\mathbf{x},1-\gamma,G)$. Then (6.4.8) and (6.4.7) give
$$u^*(\mathbf{x},\beta,\gamma,G)=\lambda_2\eta_{1-\beta}+w^{(1)}(\mathbf{x},1-\gamma,G). \qquad (6.4.9)$$
2. *λ_1 known:* denote the $(1-\gamma)$ probability point of the posterior d.f. $B^{(2)}(\lambda_2|\mathbf{x},G,m)$ of the unknown parameter Λ_2 by $w^{(2)}(\mathbf{x},1-\gamma,G)$. Then from (6.4.8) and (6.4.7) we get the relation
$$u^*(\mathbf{x},\beta,\gamma,G)=\lambda_1+\eta_{1-\beta}w^{(2)}(\mathbf{x},1-\gamma,G). \qquad (6.4.10)$$
3. *Both λ_1 and λ_2 unknown:* the required $u^*(\mathbf{x},\beta,\gamma,G)$ for this case is explicitly obtainable if the exact posterior d.f. of $\Lambda_1+\eta_{1-\beta}\Lambda_2$ is known. Let $w^*(\mathbf{x},1-\gamma,G)$ be the $(1-\gamma)$ probability point of this posterior d.f. Then (6.4.8) and (6.4.7) give the result
$$u^*(\mathbf{x},\beta,\gamma,G)=w^*(\mathbf{x},1-\gamma,G). \qquad (6.4.11)$$
If an explicit form of the posterior d.f. of $\Lambda_1+\eta_{1-\beta}\Lambda_2$ is not available, $u^*(\mathbf{x},\beta,\gamma,G)$ needs to be obtained by solving the equation
$$\int_0^\infty\int_{u^*(\mathbf{x},\beta,\gamma,G)-\lambda_2\eta_{1-\beta}} b(\boldsymbol{\lambda}|\mathbf{x},G,m)\,d\lambda_1\,d\lambda_2=\gamma.$$

(a) A general result for the single-parameter case

Let the d.f. F be indexed by a single parameter λ and also let $d_c(\lambda)$ be the c probability point of F. The d.f. F is assumed to belong to a family such that $d_c(\lambda)$ decreases as λ increases; this family of d.f.s has been considered by Aitchison (1964). In this case, we have
$$\Delta[\lambda,u^*(\mathbf{x}),\beta]=\{\lambda:u^*(\mathbf{x})\leqslant d_{1-\beta}(\lambda)\}.$$
Let $w(\mathbf{x},\gamma,G)$ be the γ probability point of the posterior d.f. of Λ. Then the following subset of the parameter space has probability γ under the posterior d.f. of Λ:
$$\{\lambda:w(\mathbf{x},\gamma,G)\geqslant\lambda\}.$$
However, this set is equivalent to the set given below:
$$\{\lambda:d_{1-\beta}[w(\mathbf{x},\gamma,G)]\leqslant d_{1-\beta}(\lambda)\}.$$
Hence the quantity $u^*(\mathbf{x},\beta,\gamma,G)$ which satisfies (6.4.7) must be given by the relation:
$$u^*(\mathbf{x},\beta,\gamma,G)=d_{1-\beta}[w(\mathbf{x},\gamma,G)].$$

6.5 EB tolerance regions

Suppose that G is not known, but that results from the EB sampling scheme described previously are available. Thus we have past observation vectors $\mathbf{x}_1, \ldots, \mathbf{x}_n$ corresponding to realizations $\boldsymbol{\lambda}_1, \ldots, \boldsymbol{\lambda}_n$ of $\boldsymbol{\Lambda}$ where $\mathbf{x}_i = (x_{i1}, \ldots, x_{im_i})$, in addition to the current observation vector $\mathbf{x}$ corresponding to the current realization $\boldsymbol{\lambda}$. Then G can be estimated from previous data; let $\hat{G}_n$ be a consistent estimator of G at every point of $\boldsymbol{\lambda}$ in the parameter space $\mathscr{L}$. Then EB analogues of Bayesian tolerance regions can be developed in general terms by substituting $\hat{G}_n$ for G in the procedures of section 6.4. As discussed in the case of EB confidence regions, the estimated Bayesian tolerance regions obtained in this way no longer induce the required expected coverage or percentile coverage under the posterior d.f. of $\boldsymbol{\Lambda}$. Since $\hat{G}_n$ is a random entity due to variation of $\mathbf{x}_1, \ldots, \mathbf{x}_n$, the estimated Bayesian tolerance regions based on $\hat{G}_n$ are also random and their properties with respect to the posterior d.f. need to be assessed accordingly.

Let a Bayesian expected cover region satisfying (6.4.3) be denoted by $A^*(\mathbf{x}, \beta, G)$ and a Bayesian percentile cover region satisfying (6.4.4) by $A^*(\mathbf{x}, \beta, \gamma, G)$. These regions can be estimated by $A^*(\mathbf{x}, \beta, \hat{G}_n)$ and $A^*(\mathbf{x}, \beta, \gamma, \hat{G}_n)$ respectively. For the region $A^*(\mathbf{x}, \beta, \hat{G}_n)$, the coverage probability under the predictive density function is

$$C[A^*(\mathbf{x}, \beta, \hat{G}_n), G] = \int_{A^*(\mathbf{x}, \beta, \hat{G}_n)} dP(y|\mathbf{x}, G), \qquad (6.5.1)$$

where $P(y|\mathbf{x}, G)$ is the d.f. of the predictive density $p(y|\mathbf{x}, G)$. We can then assess the sampling properties of the coverage probability defined in (6.5.1) as $\hat{G}_n$ varies. In particular, we could again employ the expected cover and percentile cover criteria in terms of this random variation.

Expected cover criterion: An estimated Bayesian expected cover tolerance region $A^*(\mathbf{x}, \alpha, \hat{G}_n)$ of size α is said to satisfy an expected cover criterion of size β if α satisfies the relation:

$$E_n C[A^*(\mathbf{x}, \alpha, \hat{G}_n), G] = \beta,$$

where $E_n(\cdot)$ is the expectation with respect to the variation of $\hat{G}_n$.

Percentile cover criterion: An estimated Bayesian expected cover tolerance region $A^*(\mathbf{x}, \alpha, \hat{G}_n)$ of size α is said to satisfy a percentile cover criterion with size β at level γ if α satisfies the relation:

$$\Pr_n\{C[A^*(\mathbf{x}, \alpha, \hat{G}_n), G] \geqslant \beta\} \geqslant \gamma$$

Practical construction of $A^*(\mathbf{x}, \alpha, \hat{G}_n)$ is not straightforward for a general unspecified G. For parametric priors a method of construction is discussed in section 6.5.1.

A similar type of treatment can be given to the estimated Bayesian percentile cover region $A^*(\mathbf{x}, \beta, \gamma, \hat{G}_n)$. By replacing γ by α we can consider the coverage of $A^*(\mathbf{x}, \beta, \alpha, \hat{G}_n)$ under the d.f. $F(\cdot|\boldsymbol{\lambda})$:

$$C[A^*(\mathbf{x}, \beta, \alpha, \hat{G}_n), \boldsymbol{\lambda}] = \int_{A^*(\mathbf{x}, \beta, \alpha, \hat{G}_n)} dF(y|\boldsymbol{\lambda}).$$

The region in $\mathscr{L}$ defined by

$$\Delta[\boldsymbol{\lambda}, A^*(\mathbf{x}, \beta, \alpha, \hat{G}_n)] = \{\boldsymbol{\lambda}: C[A^*(\mathbf{x}, \beta, \alpha, \hat{G}_n), \boldsymbol{\lambda}] \geqslant \beta\},$$

has a probability content under the posterior d.f. as given by

$$C[\Delta[\boldsymbol{\lambda}, A^*(\mathbf{x}, \beta, \alpha, \hat{G}_n)], G] = \int_{\Delta[\boldsymbol{\lambda}, A^*(\mathbf{x}, \beta, \alpha, \hat{G}_n)]} dB(\boldsymbol{\lambda}|\mathbf{x}, G, n). \tag{6.5.2}$$

The coverage probability defined in (6.5.2) is a random quantity due to variation of $\hat{G}_n$. Thus an adjustment of α in the expression of $A^*(\mathbf{x}, \beta, \alpha, \hat{G}_n)$ can be made in terms of the expected value or in terms of the percentile value of this coverage probability. Further development of EB analogues of percentile cover Bayesian tolerance regions also requires specialization of the form of d.f. $F(\cdot|\boldsymbol{\lambda})$.

6.5.1 *Construction of EB tolerance regions: parametric G priors*

The general formulation outlined in the previous section can be specialized to the case of a parametric G prior distribution denoted by $G(\boldsymbol{\lambda}|\xi)$. Here G is known up to a set of parameters ξ so that we will denote the expected cover and percentile cover Bayesian tolerance regions by $A^*(\mathbf{x}, \beta, \xi)$ and $A^*(\mathbf{x}, \beta, \gamma, \xi)$ respectively to emphasize their dependence on ξ. The unknown parameter vector ξ can be estimated by the ML estimator $\hat{\xi}_n$ based on an EB scheme as discussed in section 2.12. Estimated Bayesian tolerance regions can then be obtained by replacing ξ by $\hat{\xi}_n$.

(a) EB analogues of expected cover Bayesian tolerance region
The estimated Bayesian expected cover region $A^*(\mathbf{x}, \beta, \hat{\xi})$ can be assessed by looking at the sampling properties of the coverage probability (6.5.1) under the predictive density function. Consider the special form $A^*(\mathbf{x}, \beta, \xi) = [u^*(\mathbf{x}, \beta, \xi), \infty)$. With α in place of β, the coverage probability defined in (6.5.1) becomes

$$C[A^*(\mathbf{x}, \hat{\xi}, \alpha), \xi] = 1 - P(u^*(\mathbf{x}, \alpha, \hat{\xi}) | \mathbf{x}, \xi, m).$$

The expected value of this coverage probability with respect to variation of $\hat{\xi}$ can be obtained exactly in principle if the exact distribution of $\hat{\xi}$ is known. This not being the case in general, we proceed to obtain an approximate expression for it. The technical results here are now very similar to those of section 6.2.4 with $P(y|\mathbf{x}, \xi, m)$ in place of $B(\lambda|\mathbf{x}, \xi, m)$.

We can define the following derivatives which are all evaluated at $y = u^*(\mathbf{x}, \alpha, \hat{\xi})$:

$$\mathbf{P}_{10} = (\partial/\partial y)P(y|\mathbf{x}, \xi, m)$$
$$\mathbf{P}_{20} = (\partial^2/\partial y^2)P(y|\mathbf{x}, \xi, m)$$
$$\mathbf{P}_{01} = (\partial/\partial \xi)P(y|\mathbf{x}, \xi, m)$$
$$\mathbf{P}_{02} = (\partial^2/\partial \xi \partial \xi^{\mathrm{T}})P(y|\mathbf{x}, \xi, m)$$

Then to terms of $O(n^{-1})$, we have

$$E_n\{C[A^*(\mathbf{x}, \alpha, \hat{\xi}), \xi]\} = \alpha - \psi(\xi)^{\mathrm{T}}\mathbf{P}_{01} + \mathbf{P}_{10}^{-1}\{\mathbf{P}_{01}^{\mathrm{T}}\Gamma(\xi)\mathbf{P}_{11}\} - \tfrac{1}{2}\operatorname{tr}\{\mathbf{P}_{02}\Gamma(\xi)\} \quad (6.5.3)$$

where $\psi(\xi)$ and $\Gamma(\xi)$ are, as before, the bias vector and the mean square error matrix of the ML estimator $\hat{\xi}$. We may also note here, for computational purpose, that the required derivatives may also be evaluated from the posterior d.f.:

$$\mathbf{P}_{10} = E_B\{f(y|\mathbf{\Lambda})\}$$

$$\mathbf{P}_{20} = E_B\left\{\frac{\partial \ln f(y|\mathbf{\Lambda})}{\partial y}\right\}$$

$$\mathbf{P}_{10} = E_B\left\{f(y|\mathbf{\Lambda})\frac{\partial}{\partial \xi}\ln g(\mathbf{\Lambda}|\xi)\right\}$$

$$\mathbf{P}_{20} = E_B\left\{f(y|\mathbf{\Lambda})\left(\frac{\partial^2}{\partial \xi \partial \xi^{\mathrm{T}}}\right)\ln g(\mathbf{\Lambda}|\xi)\right\}.$$

Next, consider the construction of an EB analogue of a Bayesian expected cover region which satisfies the percentile cover criterion. The expected value of the coverage probability $C[A^*(\mathbf{x}, \alpha, \hat{\xi}), \xi]$ is given by (6.5.3). Its variance can be obtained to the same order of approximation as

$$\text{var}_n\{C[A^*(\mathbf{x}, \alpha, \hat{\xi}), \xi]\} = \mathbf{P}_{01}^{\mathrm{T}} V(\xi)\mathbf{P}_{01}/\mathbf{P}_{10}^2.$$

We may now follow an approach of Wald (1942) and apply a normal approximation or follow an approach of Wilks (1941) and apply a beta distribution approximation to the distribution of $C[A^*(\mathbf{x}, \alpha, \hat{\xi})\xi]$. The results are virtually along the same lines as in section 6.2.4 and will not be repeated here.

(b) EB analogues of percentile cover Bayesian tolerance regions
Consider the estimated Bayesian percentile cover region $A^*(\mathbf{x}, \beta, \gamma, \hat{\xi})$, where

$$A^*(\mathbf{x}, \beta, \gamma, \hat{\xi}) = [u^*(\mathbf{x}, \beta, \gamma, \hat{\xi}), \infty).$$

The coverage probability of this region under the d.f. $F(\cdot|\boldsymbol{\lambda})$ is given by

$$C[A^*(\mathbf{x}, \beta, \gamma, \hat{\xi}), \boldsymbol{\lambda}] = 1 - F(u^*(\mathbf{x}, \beta, \gamma, \hat{\xi})|\boldsymbol{\lambda}).$$

Consider the region in the parameter space defined by

$$\Delta[\boldsymbol{\lambda}, u^*(\mathbf{x}, \beta, \gamma, \hat{\xi}), \beta] = \{\boldsymbol{\lambda}: C[A^*(\mathbf{x}, \beta, \gamma, \hat{\xi}), \boldsymbol{\lambda}] \geqslant \beta\}. \quad (6.5.4)$$

The posterior probability of the region defined in (6.5.4) is no longer γ. We can treat $A^*(\mathbf{x}, \beta, \alpha, \hat{\xi})$ as an adjustable region with respect to α and consider the posterior coverage probability defined in (6.5.4) with α in place of γ; this is given by

$$C[\Delta[\boldsymbol{\lambda}, u^*(\mathbf{x}, \beta, \alpha, \hat{\xi}), \beta], \xi] = \int_{\Delta[\boldsymbol{\lambda}, u^*(\mathbf{x}, \beta, \alpha, \hat{\xi}), \beta]} dB(\boldsymbol{\lambda}|\mathbf{x}, \xi, m).$$

6.5.2 *Location and scale family F*

When $F(y|\boldsymbol{\lambda})$ is of the form $F((y - \lambda_1)/\lambda_2)$, we have (6.5.4) with α in place of γ as follows:

$$\Delta[\boldsymbol{\lambda}, u^*(\mathbf{x}, \beta, \alpha, \hat{\xi}), \beta] = \{\boldsymbol{\lambda}: \lambda_1 + \eta_{1-\beta}\lambda_2 \geqslant u^*(\mathbf{x}, \beta, \alpha, \hat{\xi})\}. \quad (6.5.5)$$

1. λ_2 *known:* using the expression (6.4.9), (6.5.5) becomes

$$\Delta[\boldsymbol{\lambda}, u^*(\mathbf{x}, \beta, \alpha, \hat{\xi}), \beta] = \{\lambda_1: \lambda_1 \geqslant w^{(1)}(\mathbf{x}, 1 - \alpha, \hat{\xi})\}.$$

In this case, adjusting α in terms of a preassigned value of γ while taking in account the random variation of $\hat{\xi}$ is the same as obtaining EB confidence limit for λ_1. Thus the methods discussed in section 6.2.4 are directly applicable here.

2. λ_1 *known*: using the expression (6.4.10), (6.5.5) becomes

$$\Delta[\boldsymbol{\lambda}, u^*(\mathbf{x}, \beta, \alpha, \hat{\xi}), \beta] = \{\lambda_2 : \lambda_2 \geqslant w^{(2)}(\mathbf{x}, 1-\alpha, \hat{\xi})\}.$$

In this case again, the problem of adjusting α reduces to that of obtaining a lower EB confidence limit for λ_2. Thus one can again directly apply the methods of section 6.2.4.

3. *Both* λ_1 *and* λ_2 *unknown:* let $\boldsymbol{\Lambda}^*$ be a new r.v. given by $\boldsymbol{\Lambda}_1 + \eta_{1-\beta}\boldsymbol{\Lambda}_2$. Using the expression (6.4.11), (6.5.5) becomes

$$\Delta[\boldsymbol{\lambda}, u^*(\mathbf{x}, \beta, \alpha, \hat{\xi}), \beta] = \{\lambda^* : \lambda^* \geqslant w^*(\mathbf{x}, 1-\alpha, \hat{\xi})\}.$$

Again the problem of adjusting α is reduced to that of an EB confidence lower limit for λ^* if an explicit expression for $w^*(\mathbf{x}, 1-\alpha, \hat{\xi})$ is available. Otherwise approximate solutions need to be sought.

(a) A special single-parameter case

Consider the data d.f. $F(\cdot|\lambda)$ with the special property described in section 6.4.2. For this case, (6.5.4) gives

$$\Delta[\lambda, u^*(\mathbf{x}, \beta, \alpha, \hat{\xi}), \beta] = \{\lambda : u^*(\mathbf{x}, \beta, \alpha, \hat{\xi}) \leqslant d_{1-\beta}(\lambda)\}.$$

Using the monotonicity of $d_c(\cdot)$ again, this set is equivalent to

$$\{\lambda : d_{1-\beta}[w(\mathbf{x}, \alpha, \hat{\xi})] \leqslant d_{1-\beta}(\lambda)\}$$

which is again equivalent to

$$\{\lambda : w(\mathbf{x}, \alpha, \hat{\xi}) \geqslant \lambda\}.$$

Hence the problem of adjusting α again reduces to that of finding an EB upper confidence limit for λ with level γ. The methods of section 6.2.4 are again directly applicable.

6.6 Other approaches to EB interval and region estimation

Empirical Bayes interval estimation is a fairly recent development. In the preceding sections we have provided an EB approach by exploiting connections with the classical problem of tolerance region estimation. This is in the spirit of Cox (1975) and Lwin and Maritz (1976). Technically and conceptually it differs from classical tolerance theory in that the expected cover and the percentile cover are computed conditionally on the current observed data. In Lwin and Maritz (1976) such an approach is explicitly defined, but the actual calculations for specific examples are made in terms of the sampling variation of all the data in the EB scheme, This produces an unconditional probability statement for the EB interval or region. Cox's approach is also unconditional in this sense.

The earliest development of an EB interval estimate seems to be due to Deely and Zimmer (1969) although the first use of the term 'EB interval estimate' appears to be by Cox (1975). In both of these works the prediction interval (i.e. expected cover tolerance interval) approach is used, with particular attention to normally distributed data. Again, these approaches are based on unconditional probability statements. Lord and Cressie (1975) considered prediction limits for λ based on the Λ on X regression in the joint distribution of (X, Λ). A prediction interval for a future observation on Λ is constructed as though Λ is an observable r.v. with observations $\lambda_1, \lambda_2, \ldots, \lambda_k$ in an EB scheme. The uncertainty in not knowing the parameters in the joint distribution is allowed for in the confidence statement. Lord and Cressie apply this method to the case when the distribution of X, conditional on λ, is binomial.

Deely and Lindley (1981) provide a systematic basis for interval estimation in the full Bayes EB approach. The relation (1.14.8) can be used to obtain posterior limits for the current unknown parameter since the posterior distribution is fully determined once the third stage prior distribution function $P(\phi)$ is specified. The approach hinges on the specification of $P(\phi)$. It is also worth noting that the advocates of this approach are not primarily concerned with the usual sampling properties of the intervals.

Morris (1983a, b) also gives an EB interval estimation technique. Although the sampling properties of the intervals are of main interest, the method is similar to the Bayes EB method. Laird and Louis (1987) present a general bootstrap approach to the problem of EB interval estimation, comparing their results with those of Morris.

In the following sections we give brief summaries of Morris and bootstrap intervals, confining discussion to the structure of the FB approach given in section 1.14.2.

6.6.1 *Morris EB interval estimates*

The posterior joint distribution given in (1.14.8) is recast in the form

$$B(\lambda_1, \lambda_2, \ldots, \lambda_k | \mathbf{X}) = \int \prod_{i=1}^{k} B(\lambda_i | x_i) dP(\phi | \mathbf{X}). \qquad (6.6.1)$$

In this equation $B(\lambda_i | x_i, \phi)$ is the posterior d.f. of λ_i given the ith stage data x_i and the prior d.f. $G(\lambda | \phi)$. Also, $P(\phi | \mathbf{x}_1, \mathbf{x}_2, \ldots, \mathbf{x}_k)$ is the posterior distribution function of ϕ derived from the hyper prior d.f. $P(\Phi)$ given the data $(\mathbf{x}_1, \mathbf{x}_2, \ldots, \mathbf{x}_k)$. The evaluation of $P(\phi | \mathbf{x}_1, \mathbf{x}_2, \ldots, \mathbf{x}_k)$ is often performed with $P(\phi)$ replaced by a 'flat' prior. The relation (6.6.1) is used by Morris to obtain approximations for the posterior mean and variance of a specified λ_j.

From (6.6.1) the posterior marginal d.f. of a particular Λ_j is

$$B(\lambda_j | \mathbf{x}_1, \mathbf{x}_2, \ldots, \mathbf{x}_k) = \int B(\lambda_j | x_j, \phi) dP(\phi | \mathbf{x}_1, \mathbf{x}_2, \ldots, \mathbf{x}_k). \qquad (6.6.2)$$

From (6.6.2) it is also useful to express the posterior mean and variance of Λ_j as

$$E(\Lambda_j | \mathbf{x}_1, \mathbf{x}_2, \ldots, \mathbf{x}_k) = E_P E(\Lambda_j | x_j, \phi), \qquad (6.6.3)$$

and

$$\operatorname{var}(\Lambda_j | \mathbf{x}_1, \mathbf{x}_2, \ldots, \mathbf{x}_k) = E_P \operatorname{var}(\Lambda_j | x_j, \phi) + \operatorname{var}_P E(\Lambda_j | x_j, \phi). \qquad (6.6.4)$$

In these expressions the subscript refers to the posterior d.f. $P(\phi | \mathbf{x}_1, \mathbf{x}_2, \ldots, \mathbf{x}_k)$.

Example 6.6.1 We consider the case where $F(x | \lambda)$ is $N(\lambda, \sigma^2)$ with known σ^2, and G is $N(\xi, \tau^2)$, and $m_1 = m_2 = \cdots = m_k = 1$. Morris (1983a) gives explicit approximate expressions for the mean and variance in (6.6.3) and (6.6.4) by employing Taylor series expansions for $E(\Lambda_j | x_j, \phi)$ and $\operatorname{var}(\Lambda_j | x_j, \phi)$ at $\phi = E(\phi | \mathbf{X})$. In these calculations the flat priors for ξ and τ^2 are uniform on $(-\infty, +\infty)$ and $(0, \infty)$ respectively. The following approximations are given:

$$E(\Lambda_j | \mathbf{x}_1, \mathbf{x}_2, \ldots, \mathbf{x}_k) = \hat{c}\bar{x} + (1 - \hat{c}) x_j,$$

$$\operatorname{var}(\Lambda_j | \mathbf{x}_1, \mathbf{x}_2, \ldots, \mathbf{x}_k) = (1 - \hat{c})\sigma^2 + \hat{c}\sigma^2/k + 2(x_j - \bar{x})^2 \hat{c}^2/(k - 3),$$

where

$$\hat{c} = \frac{(k-3)\sigma^2}{\sum_{i=1}^{k}(x_i - \bar{x})^2}.$$

The posterior distribution of λ_j can be approximated by a normal distribution with mean and variance given above. Hence the EB two-sided limits for λ of size α are given by

$$\hat{\lambda}_L, \hat{\lambda}_U = E(\Lambda | x_1, x_2, \ldots, x_k) \pm z_{(1-\alpha)/2}[\text{var}(\Lambda | x_1, x_2, \ldots, x_k)]^{1/2}. \quad (6.6.5)$$

It is claimed that

$$\Pr\{\hat{\lambda}_L < \lambda_j < \hat{\lambda}_U\} \geqslant \alpha,$$

where the probability is calculated according to the joint distribution of (X, Λ), i.e. the probability is unconditional.

Morris (1983b) considers extensions of this example to the case where the data distribution of the ith component is $N(\lambda_i, \sigma_i^2)$. Also, the prior mean of Λ_i is taken to change deterministically according to

$$E_G(\Lambda_i) = \mathbf{a}_i^{\mathrm{T}} \xi,$$

where $\mathbf{a}_i^{\mathrm{T}}$ is a vector of covariates and ξ is a vector of unknown parameters.

6.2.2 *Bootstrap EB intervals*

Suppose that N bootstrap samples are generated from a relevant model, providing N estimates, $\hat{\phi}_1, \hat{\phi}_2, \ldots, \hat{\phi}_k$, of the parameter ϕ of the prior G can be obtained. Methods of generating bootstrap samples are given below. The posterior distribution of λ_j given by (6.6.2) can then be estimated from the bootstrap samples as

$$\hat{B}(\lambda_j | \mathbf{x}_1, \mathbf{x}_2, \ldots, \mathbf{x}_k) = \sum_{i=1}^{N} B(\lambda_j | \mathbf{x}_j, \hat{\phi}_i)/N, \quad (6.6.6)$$

The estimated distribution in (6.6.6) can be used to obtain 'confidence' intervals for λ_j. For a 100α% two-sided interval we need to find $\hat{\lambda}_L$ and

$\hat{\lambda}_U$ such that

$$\frac{1-\alpha}{2} = \int_{-\infty}^{\hat{\lambda}_L} d\hat{B}(\lambda_j|\mathbf{x}_1, \mathbf{x}_2, \ldots, \mathbf{x}_k)$$

$$= \int_{\hat{\lambda}_U}^{+\infty} d\hat{B}(\lambda_j|\mathbf{x}_1, \mathbf{x}_2, \ldots, \mathbf{x}_k).$$

Laird and Louis (1987) suggest three types of models for producing bootstrap samples, appropriate when there is just one observation at each component:

1. Construct the empirical distribution function H_k of the observations $x_1, x_2, \ldots, x_k$. Then generate N samples of size k from this distribution. Each of these samples is used to calculate an estimate of ϕ, thus creating the sequence $\hat{\phi}_1, \hat{\phi}_2, \ldots, \hat{\phi}_k$.
2. Using the data of the EB scheme the prior G is estimated by the nonparametric ML method described in section 2.10. Let the estimate of G be $\hat{G}_k$. Then perform the following operations N times (i.e. for $i = 1, 2, \ldots, N$): generate a sample of λ values $(\lambda_{1i}^*, \lambda_{2i}^*, \ldots, \lambda_{ki}^*)$ using this $\hat{G}_k$. Then generate an observation x_{ji}^* from $F(x|\lambda_{ji}^*)$ for $j = 1, 2, \ldots, k$. This creates a set of x observations which can be used to calculate an estimate $\hat{\phi}_i$ of ϕ. In this model the form of F is assumed known while G is not specified.
3. This model is like 2 above except that a parametric form of $G(\lambda|\phi)$ of G is assumed. Initially the parameter ϕ is estimated by a standard method such as maximum likelihood, yielding $\hat{\phi}$. Then the distribution $G(\lambda|\hat{\phi})$ is used exactly like $\hat{G}_k$ in 2.

CHAPTER 7

Alternatives to empirical Bayes

7.1 Introduction

One view of the development of the EB approach is that it is an attractive compromise between the classical non-Bayes and the full Bayes approaches to statistical inference. These represent extremes in that the former uses no prior information whereas the latter requires complete specification of a prior distribution. The EB approach uses previous data to get an estimate of the prior distribution. The previous data and current data are linked in the form of a two-stage sampling scheme by a common prior distribution G of the unknown parameters; see section 1.8.

The EB method is actually only one of several methods of more effective utilization of data from such a two-stage sampling scheme. Established competitors of the EB method are: compound decision theory, the full Bayesian multiparameter approach and a modified likelihood approach. These methods treat the EB scheme as a multivariate case where $k = n + 1$ variables are observed, each having a distribution with its own unknown parameter. The joint distribution of the k variables X_i, $i = 1, 2, \ldots, k$, is the product of k individual distributions owing to the independence of the X_i's.

Compound decision theory assumes no prior distribution but uses a compound loss structure, i.e. the loss in designing a decision rule is typically taken to be the sum of losses incurred in making decisions about the k parameters. The full Bayesian approach does not use compound loss but requires specification of a hyper-prior distribution of the k parameters.

7.2 The multiparameter full Bayesian approach

This technique was introduced briefly by Lindley (1962) as a Bayesian alternative to the compound decision theoretic approach in a discussion of a paper presented by Stein (1962). A more systematic exposition was given by Lindley (1971) for the special case when the individual component distributions $F(x_i|\lambda_i)$ are $N(\lambda_i, \sigma^2)$, and there are $m_i \geqslant 1$ observations on X_i in the EB sampling scheme. In section 1.14 this approach is introduced for the case $m_i = 1$. The hyper-prior distribution used in this case is a mixture of $N(\mu, \tau^2)$ distributions with respect to a diffuse prior distribution of μ. Both σ^2 and τ^2 are assumed known. This construction of a hyper prior is based on the concept of exhangeability due to de Finetti (1964). The full Bayesian developments up to and including that in Lindley (1971) can be deduced as special cases of a general theory developed by Lindley and Smith (1972); a summary is given in the next section.

7.2.1 *The general Bayesian linear model*

The following assumptions characterize the classes of data and prior distributions of the general Bayesian linear model of Lindley and Smith (1972). Let $\mathbf{Y}, \boldsymbol{\theta}_1, \boldsymbol{\theta}_2, \boldsymbol{\theta}_3$ be $N \times 1, p_1 \times 1, p_2 \times 1, p_3 \times 1$ vectors. Let $\mathbf{A}_1$, $\mathbf{A}_2$, $\mathbf{A}_3$, $\mathbf{C}_1$, $\mathbf{C}_2$, $\mathbf{C}_3$ be $N \times p_1$, $p_1 \times p_2$, $p_2 \times p_3$, $p_1 \times p_1$, $p_2 \times p_2$, $p_3 \times p_3$ matrices, $\mathbf{C}_1$, $\mathbf{C}_2$, $\mathbf{C}_3$ being positive definite. Also assume that the conditional distributions of $\mathbf{Y}$ given $\boldsymbol{\theta}_1$, $\boldsymbol{\theta}_1$ given $\boldsymbol{\theta}_2$, $\boldsymbol{\theta}_2$ given $\boldsymbol{\theta}_3$ are as follows:

$$
\begin{aligned}
\mathbf{Y}|\boldsymbol{\theta}_1 &\text{ is } N(\mathbf{A}_1\boldsymbol{\theta}_1, \mathbf{C}_1)\\
\boldsymbol{\theta}_1|\boldsymbol{\theta}_2 &\text{ is } N(\mathbf{A}_2\boldsymbol{\theta}_2, \mathbf{C}_2)\\
\boldsymbol{\theta}_2|\boldsymbol{\theta}_3 &\text{ is } N(\mathbf{A}_3\boldsymbol{\theta}_3, \mathbf{C}_3)
\end{aligned}
\tag{7.2.1}
$$

The variables $\boldsymbol{\theta}_1$ and $\boldsymbol{\theta}_2$ are unobservables, while $\boldsymbol{\theta}_3$ is assumed to be given. The matrices $\mathbf{A}_i$ are 'design' matrices and are also assumed known. The matrices $\mathbf{C}_i$ are in general unknown parameters, but for the first stage of the development of the model they also are assumed known. The assumptions that $\boldsymbol{\theta}_1|\boldsymbol{\theta}_2$ and $\boldsymbol{\theta}_2|\boldsymbol{\theta}_3$ have distributions are regarded as representations of exchangeability between elements of $\boldsymbol{\theta}_1$.

The main result deduced from the above specifications is that the conditional distribution of $\boldsymbol{\theta}_1$ given $\mathbf{y}$ and $\boldsymbol{\theta}_3$ is $N(\mathbf{Qq}, \mathbf{Q})$, where

$$
\begin{aligned}
\mathbf{q} &= \mathbf{A}_1^{\mathrm{T}}\mathbf{C}_2^{-1}\mathbf{y} + (\mathbf{C}_2 + \mathbf{A}_2\mathbf{C}_3\mathbf{A}_2^{\mathrm{T}})^{-1}\mathbf{A}_2\mathbf{A}_3\boldsymbol{\theta}_3\\
\mathbf{Q}^{-1} &= \mathbf{A}_1^{\mathrm{T}}\mathbf{C}_1^{-1}\mathbf{A}_1 + (\mathbf{C}_2 + \mathbf{A}_2\mathbf{C}_3\mathbf{A}_2^{\mathrm{T}})^{-1}
\end{aligned}
\tag{7.2.2}
$$

The full Bayesian approach regards this posterior distribution as the final result, summarizing all data and prior information. Its mean or its mode, these two being identical in the case considered, is regarded as a reasonable point estimate of $\boldsymbol{\theta}_1$.

When $\mathbf{C}_3^{-1} = 0$ the posterior mean $\mathbf{Qq}$ becomes $\mathbf{Q}_0\mathbf{q}_0$ with

$$\begin{aligned} \mathbf{q}_0 &= \mathbf{A}_1^{\mathrm{T}}\mathbf{C}_1^{-1}\mathbf{y} \\ \mathbf{Q}_0^{-1} &= \mathbf{A}_1^{\mathrm{T}}\mathbf{C}_1^{-1}\mathbf{A}_1 + \mathbf{C}_2^{-1} - \mathbf{C}_2\mathbf{A}_2(\mathbf{A}_2^{\mathrm{T}}\mathbf{C}_2^{-1}\mathbf{A}_2)^{-1}\mathbf{A}_2^{\mathrm{T}}\mathbf{C}_2^{-1}. \end{aligned} \tag{7.2.3}$$

Alternatively we may write the posterior mean as

$$E(\boldsymbol{\theta}_1 \mid \mathbf{y}, \boldsymbol{\theta}_3) = (\mathbf{A}_1^{\mathrm{T}}\mathbf{C}_1^{-1}\mathbf{A}_2 + \mathbf{C}_2^{-1})^{-1}(\mathbf{A}_1^{\mathrm{T}}\mathbf{C}_1^{-1}\mathbf{A}_1\hat{\boldsymbol{\theta}}_1 + \mathbf{C}_2^{-1}\mathbf{A}_2\hat{\boldsymbol{\theta}}_2) \tag{7.2.4}$$

where

$$\begin{aligned} \hat{\boldsymbol{\theta}}_1 &= (\mathbf{A}_1^{\mathrm{T}}\mathbf{C}_1^{-1}\mathbf{A}_1)^{-1}\mathbf{A}_1^{\mathrm{T}}\mathbf{C}_1^{-1} \\ \hat{\boldsymbol{\theta}}_2 &= \{(\mathbf{A}_1\mathbf{A}_2)^{\mathrm{T}}(\mathbf{C}_1 + \mathbf{A}_1\mathbf{C}_2\mathbf{A}_1)^{-1} \\ &\quad \times (\mathbf{A}_1\mathbf{A}_2)\}^{-1}(\mathbf{A}_1\mathbf{A}_2)^{\mathrm{T}}(\mathbf{C}_1 + \mathbf{A}_1\mathbf{C}_2\mathbf{A}_1)^{-1}\mathbf{y}. \end{aligned} \tag{7.2.5}$$

Some remarks are in order. One can interpret $\mathbf{C}_3^{-1} = 0$ as reflecting prior ignorance at the third stage, and when this holds we note that the posterior mean is independent of $\boldsymbol{\theta}_3$. The posterior mean, $E(\boldsymbol{\theta}_1 \mid \mathbf{y}, \boldsymbol{\theta}_3)$, is then completely determined by the design matrices and the observed data. This is useful for application in a two-stage scheme where the second stage has a proper probabilistic structure which cannot be ignored. This is precisely the nature of an EB scheme, so that the full Bayesian approach is an alternative to the EB methods discussed in previous chapters.

Example 7.2.1 The one-way ANOVA model. We now give an application of the above general result to the case considered in section 1.14. Let the vectors and matrices of the general linear Bayesian model be specialized as follows:

$$\mathbf{Y}^{\mathrm{T}} = (x_{11}, \ldots, x_{1m_1}, x_{21}, \ldots, x_{2m_2}, x_{k1}, \ldots, x_{km_k})$$

$$\boldsymbol{\theta}_1^{\mathrm{T}} = (\lambda_1, \ldots, \lambda_k), \quad \theta_2 = \mu.$$

$$\mathbf{A}_1 = \begin{bmatrix} \mathbf{1}_{m_1} & \mathbf{0}_{m_1} & \cdots & \mathbf{0}_{m_1} \\ \mathbf{0}_{m_2} & \mathbf{1}_{m_2} & \cdots & \mathbf{0}_{m_2} \\ \vdots & \vdots & & \vdots \\ \mathbf{0}_{m_k} & \mathbf{0}_{m_k} & & \mathbf{1}_{m_k} \end{bmatrix}.$$

where $\mathbf{1}_r$ is a column vector of r ones, and $\mathbf{0}_r$ is a column vector of r

zeroes. Also,

$$\mathbf{A}_2 = \mathbf{1}_k, \quad \mathbf{C}_1 = \mathbf{I}_N\sigma^2, \quad \mathbf{C}_2 = \mathbf{I}_k\tau^2,$$

where $\mathbf{I}_r$ is an $r \times r$ identity matrix. Then we have

$$\mathbf{q}^{\mathrm{T}} = \sigma^{-2}(m_1\bar{x}_{1\bullet}, \ldots, m_k\bar{x}_{k\bullet}),$$

where $\bar{x}_{i\bullet} = (x_{i1} + \cdots + x_{im})/m_i$ is the mean of the observations at component i, and $\mathbf{Q}_0$ is a $k \times k$ matrix with elements

$$q_{ij}^0 = \begin{cases} \tau^2(1-\omega_i)\left(1 + \sum_{r \neq i} \omega_r\right)\left(\sum_r \omega_r\right)^{-1}, & i = j \\ \tau^2(1-\omega_i)(1-\omega_j)\left(\sum_r \omega_r\right)^{-1}, & \text{otherwise,} \end{cases}$$

with $\omega_i = m_i\sigma^2/(m_i\sigma^2 + \tau^2)$. See also Smith (1973).

The full Bayes estimate of λ_i is found to be

$$E(\Lambda_i|\mathbf{y}) = (m_i\bar{x}_i\tau^2 + \mu^*\sigma^2)/(m_i\tau^2 + \sigma^2) \tag{7.2.6}$$

where $\mu^* = \sum_{i=1}^k \omega_i\bar{x}_i / \sum_{i=1}^k \omega_i$. The posterior variance of Λ_i is q_{ii}^0 given above. When every $m_i = 1$ these results reduce to the corresponding expressions in section 1.14.

Example 7.2.2 Ridge regression. Suppose we specialize the general Bayesian linear model as follows:

$\mathbf{A}_1$ is an $n \times p$ matrix $\mathbf{X}$ of design variables

$\mathbf{A}_2$ is a $p \times p$ identity matrix $\mathbf{I}_p$

$\mathbf{C}_1 = \mathbf{I}_n\sigma^2$

$\mathbf{C}_2 = \mathbf{I}_p\tau^2$

$\boldsymbol{\theta}_1$ is a $p \times 1$ random vector $\boldsymbol{\beta}$

$\boldsymbol{\theta}_2$ is a $p \times 1$ random vector $\mathbf{1}\zeta$, where ζ is a random variable.

Then we have an $n \times 1$ data vector $\mathbf{Y}$ with $N(\mathbf{X}\boldsymbol{\beta}, \mathbf{I}_n\sigma^2)$ distribution, while $\boldsymbol{\beta}$ has a $N(\mathbf{1}\zeta, \mathbf{I}_p\tau^2)$ distribution, and the r.v. ζ represents third-stage prior ignorance. From (7.2.3) we get

$$q_0 = \mathbf{X}^{\mathrm{T}}\mathbf{Y}/\sigma^2$$
$$\mathbf{Q}^{-1} = \{(\mathbf{X}^{\mathrm{T}}\mathbf{X})/\sigma^2 + \mathbf{I}_p/\tau^2 - (p\sigma^2)^{-1}\mathbf{J}_p\},$$

where $\mathbf{J}_p$ is a $p \times p$ matrix all of whose elements have the value 1. Hence we get

$$E(\boldsymbol{\beta}|\mathbf{Y}=y)=\{\mathbf{I}+(\mathbf{X}^{\mathrm{T}}\mathbf{X})^{-1}(\mathbf{I}_p-\mathbf{J}_p/p)\sigma^2/\tau^2\}^{-1}\hat{\boldsymbol{\beta}}, \qquad (7.2.7)$$

where $\hat{\boldsymbol{\beta}}$ is the usual least squares estimator of $\boldsymbol{\beta}$. The result (7.2.7) is the full Bayesian analogue of the ridge regression model proposed in a non-Bayes context by Hoerl and Kennard (1970).

The assumption that the components of $\boldsymbol{\beta}$ are independent r.v.s with common mean and variance is interpreted as a representation of exchangeability within the multiple regression equations. An alternative Bayesian model is to take only a two-stage version of the general linear Bayesian model with the same special forms of the vectors and matrices above but $\zeta = 0$ a constant. This gives the estimate

$$\hat{\boldsymbol{\beta}}^{+}=\{\mathbf{1}+(\mathbf{X}^{\mathrm{T}}\mathbf{X})^{-1}\sigma^2/\tau^2\}^{-1}\hat{\boldsymbol{\beta}}, \qquad (7.2.8)$$

which is of the same form as the ridge regression estimator.

Lindley and Smith (1972) give an example where the assumption of exchangeability may be realistic. It is in an educational testing context where the p regressor variables might be the results of p tests applied to students, and the dependent variable $\mathbf{Y}$ a measure of the students' performance after training. Individual regression coefficients may then be regarded as exchangeable after a rescaling of the regressor variables so that $\mathbf{X}^{\mathrm{T}}\mathbf{X}$ becomes a correlation matrix.

As introduced above, the full Bayes approach deals with simultaneous estimation of all of the elements of $\boldsymbol{\theta}_1$, the 'first-stage' parameters of the general linear model. In the standard EB framework the different elements of $\boldsymbol{\theta}_1$ represent different components of the EB scheme, and usually the last element of $\boldsymbol{\theta}_1$ is of interest as the current parameter. The full Bayes approach, as presented, requires specification of parametric forms of the priors and hyper priors; in particular it seems that it is strongly dependent on normality assumptions. However, a later development by Deely and Lindley (1981) extends it to general non-normal data and prior distributions. This development, which has been termed **Bayes empirical Bayes** considers estimation of only a single parameter, at the $(n+1)$th component of the EB scheme. The transition to the simultaneous estimation of all parameters in the $k=n+1$ components is then straightforward.

7.2.2 *Bayes empirical Bayes*

Consider the EB scheme as in (1.8.1). Following Deely and Lindley (1981) we specify

$f(\mathbf{x}_i|\lambda_i)$, the conditional p.d.f. of $\mathbf{X}_i|\lambda_i$, $i=1,2,\ldots,k$,

$g(\boldsymbol{\lambda}|\boldsymbol{\theta})$, the conditional prior p.d.f. of $\boldsymbol{\Lambda}|\boldsymbol{\theta}$,

$a(\boldsymbol{\theta}|\boldsymbol{\phi})$, the conditional hyper-prior p.d.f. of $\boldsymbol{\theta}|\boldsymbol{\phi}$.

As before, the assumptions of p.d.f.s $g(\boldsymbol{\lambda}|\boldsymbol{\theta})$ and $a(\boldsymbol{\theta}|\boldsymbol{\phi})$ are justified by exchangeability. Putting

$$C(\mathbf{x},\boldsymbol{\lambda},\boldsymbol{\theta},\boldsymbol{\phi}) = \prod_{i=1}^{k} f(\mathbf{x}_i|\lambda_i)\}g(\boldsymbol{\lambda}|\boldsymbol{\theta})a(\boldsymbol{\theta}|\boldsymbol{\phi}),$$

the posterior p.d.f. of $\boldsymbol{\Lambda}$ given $\mathbf{x}_1,\mathbf{x}_2,\ldots,\mathbf{x}_k$ and $\boldsymbol{\phi}$ is

$$dB(\boldsymbol{\lambda}|\mathbf{x}_1,\mathbf{x}_2,\ldots,\mathbf{x}_k;\boldsymbol{\phi}) = \left\{\int\!\!\int C(\mathbf{x},\boldsymbol{\lambda},\boldsymbol{\theta},\boldsymbol{\phi})d\boldsymbol{\theta}d\boldsymbol{\lambda}\right\}^{-1}\int C(\mathbf{x},\boldsymbol{\lambda},\boldsymbol{\theta},\boldsymbol{\phi})d\boldsymbol{\theta}.$$

The integrals appearing in the above expression cannot readily be evaluated when the distributions involved are not normal. Deely and Lindley (1981), following Lindley (1961), used approximations in terms of the derivatives of $a(\boldsymbol{\theta}|\boldsymbol{\phi})$ and of the 'marginal' p.d.f.

$$h(\mathbf{x}|\boldsymbol{\theta}) = h(\mathbf{x}_1,\mathbf{x}_2,\ldots,\mathbf{x}_k|\,\boldsymbol{\theta}) = \int\left\{\prod_{i=1}^{k} f(\mathbf{x}_i|\lambda_i)\right\}g(\boldsymbol{\lambda}|\boldsymbol{\theta})d\boldsymbol{\lambda}$$

to derive the result

$$dB(\boldsymbol{\lambda}|\mathbf{x}_1,\mathbf{x}_2,\ldots,\mathbf{x}_k,\boldsymbol{\phi}) \simeq b(\boldsymbol{\lambda}|\mathbf{x}_1,\mathbf{x}_2,\ldots,\mathbf{x}_k,\hat{\boldsymbol{\theta}}),$$

where

$$b(\boldsymbol{\lambda}|\mathbf{x}_1,\mathbf{x}_2,\ldots,\mathbf{x}_k,\boldsymbol{\theta}) = \{h(\mathbf{x}|\boldsymbol{\theta})\}^{-1}\prod_{i=1}^{k} f(\mathbf{x}_i|\boldsymbol{\lambda})g(\boldsymbol{\lambda}|\boldsymbol{\theta}),$$

is the posterior p.d.f. of $\boldsymbol{\lambda}$ given $\mathbf{x}_1,\mathbf{x}_2,\ldots,\mathbf{x}_k$ and $\boldsymbol{\theta}$, and $\hat{\boldsymbol{\theta}}$ is the estimate of $\boldsymbol{\theta}$ obtained by maximizing $h(\mathbf{x}|\boldsymbol{\theta})$ w.r.t. $\boldsymbol{\theta}$.

The result above is interesting for two reasons. First, it is a practically applicable result which requires no knowledge of hyper priors or their parameters. Second, it gives a justification for the standard EB procedures with parametric prior $g(\boldsymbol{\lambda}|\boldsymbol{\theta})$ as an approximation to the full Bayesian solution.

7.3 Likelihood-based approaches

7.3.1 *A modified likelihood approach*

A brief introduction to a 'likelihood type' approach to the EB scheme is given in section 1.14. Recapitulating, the main points are:

1. The pairs $(\mathbf{x}_i, \lambda_i)$, $i = 1, 2, \ldots, k$ of the EB scheme (1.8.1) are regarded as independent realizations of r.v. pairs $(\mathbf{X}_i, \Lambda_i)$ with joint p.d.f. $f(\mathbf{x}_i|\lambda_i)g(\boldsymbol{\lambda}|\boldsymbol{\theta})$.
2. The joint p.d.f. of all pairs is
$$L(\mathbf{x}, \boldsymbol{\lambda}) = \prod_{i=1}^{k} f(x_i|\lambda_i)g(\lambda_i|\boldsymbol{\theta}), \tag{7.3.1}$$
and it is regarded as a 'likelihood function' for the unobservables λ_i, $i = 1, 2, \ldots k$.
3. The marginal p.d.f. of the x_i's, given by
$$h_k(\mathbf{x}|\boldsymbol{\theta}) = \int L(\mathbf{x}, \boldsymbol{\lambda})d\boldsymbol{\lambda},$$
is regarded as a 'likelihood function' for the parameter $\boldsymbol{\theta}$.
4. The function $L(\mathbf{x}, \boldsymbol{\lambda})$ is used to obtain a 'likelihood' estimate of $\boldsymbol{\lambda}$ in terms of $\mathbf{x}$ and $\boldsymbol{\theta}$; let $\hat{\boldsymbol{\lambda}}(\mathbf{x}, \boldsymbol{\theta})$ be this estimate.
5. The 'likelihood' $h_k(\mathbf{x}|\boldsymbol{\theta})$ is used to obtain a $\hat{\boldsymbol{\theta}}$ estimate of $\boldsymbol{\theta}$.
6. The estimate for $\boldsymbol{\lambda}$ is finally given as $\hat{\boldsymbol{\lambda}}(\mathbf{x}, \hat{\boldsymbol{\theta}})$.

The theory behind the above approach at first seems to be somewhat different from that behind the standard EB procedure. In the above form it seems to have appeared first in Nelder (1972) and Finney (1974). As will be shown below, it is in agreement with the standard EB method for the parametric G case. The following example is from Finney (1974).

Example 7.3.1 Suppose that $\mathbf{X}_i$ is the single random variable X_i and that the p.d.f. $f(x_i|\lambda_i)$ is $N(\lambda_i, \sigma^2)$. Suppose also that $g(\lambda_i|\boldsymbol{\theta})$ is $N(\mu, \tau^2)$; note that $\boldsymbol{\theta} = (\mu, \tau^2)^{\mathrm{T}}$. The logarithm of the modified likelihood function $L(\mathbf{x}, \boldsymbol{\lambda})$ is then

$$\text{Constant} - \sum_{i=1}^{k} (x_i - \lambda_i)^2/(2\sigma^2) - \sum_{i=1}^{k} (\lambda_i - \mu)^2/(2\tau^2).$$

Maximization of this function leads to the estimates

$$\hat{\mu} = \bar{x} = \sum_{i=1}^{k} x_i/k$$

$$\hat{\lambda}_i = \bar{x} + \{1 - \tau^2/(\sigma^2 + \tau^2)\}(x_i - \bar{x}).$$

Thus the modified likelihood approach leads to the usual EB estimate. It should be noted that the approach in this example differs slightly from the general outline given above in that the parameter μ of the prior distribution is also estimated using the likelihood function.

Questions have arisen regarding the modified likelihood approach, mainly directed at steps 2 and 4 above. Can one justify the use of $L(\mathbf{x}, \boldsymbol{\lambda})$ to construct estimates of $\boldsymbol{\lambda}$? What are the properties of such estimates? These questions are prompted by the fact that $L(\mathbf{x}, \boldsymbol{\lambda})$ is not a likelihood in the usual sense because $\boldsymbol{\lambda}$ is an unobservable random variable.

A justification can be given in terms of distance measures between distributions. In particular, the Kullback–Leibler criterion can be employed. The distance between the empirical distributions of x_i and λ_i is minimized if $\sum \ln g(\lambda_i|\boldsymbol{\theta})$ is minimized. We can regard $\boldsymbol{\lambda}$ as a 'point' intermediate between the empirical distribution of the x_i's and the theoretical distribution of the λ_i's. We can then regard the total distance between the empirical distribution of the x_i's and the theoretical distribution of the λ_i's as being minimized when we minimize

$$\sum_{i=1}^{k} \ln f(x_i|\lambda_i) + \sum_{i=1}^{k} \ln g(\lambda_i|\boldsymbol{\theta})$$

which is just the logarithm of $L(\mathbf{x}, \boldsymbol{\lambda})$, in (7.3.1).

7.3.2 *Empirical regression estimation*

The joint p.d.f. of observables and unobservables introduced in section 7.3.1 can be refactored as

$$L(\mathbf{x}, \boldsymbol{\lambda}) = b(\boldsymbol{\lambda}|\mathbf{x}, \boldsymbol{\theta})h_k(\mathbf{x}, |\boldsymbol{\phi}),$$

where the entities on the right side are defined as in section 7.2.1. Under regularity conditions, including that the range of $f(\cdot|\lambda)$ be independent of λ_i, maximizing $L(\mathbf{x}, \boldsymbol{\lambda})$ amounts to maximizing the posterior p.d.f. $b(\boldsymbol{\lambda}|\mathbf{x}, \boldsymbol{\theta})$ with respect to $\boldsymbol{\lambda}$. Hence the step 4 of section 7.3.1 is equivalent to obtaining the posterior mode of $\boldsymbol{\lambda}$. One can therefore consider replacing step 4 by a more general procedure such as:

4a. Obtain a summary of $b(\boldsymbol{\lambda}|\mathbf{x}, \boldsymbol{\theta})$ such as the posterior mode, mean or a percentile. Each of these will in general be a function of $\mathbf{x}$ and $\boldsymbol{\theta}$.

Using the posterior mean in 4a leads to an alternative modified likelihood approach which is equivalent to a standard EB approach with quadratic loss function. This alternative modified likelihood approach appeared in the literature of genetic selection in a paper by Fairfield-Smith (1936). Developments along these lines were re-examined and given prominence by Rao (1975), who referred to this approach as **empirical regression estimation** (ERE). The use of the posterior mean in 4a is justified by a regression construction, i.e. regression of the unobservable $\boldsymbol{\lambda}$ on the observations $\mathbf{x}$. Apart from this change, the ERE is a modified likelihood procedure, and follows the steps 1–4 in section 7.3.1. Thus it may also be regarded as a special case of the EB method. Further developments of the ERE approach have been restricted mainly to estimates that are linear in the observations. This has the advantage that only first- and second-order moments need to be specified or estimated in practical applications. Indeed, the early ERE methods can be regarded as forerunners of the linear EB methods described in Chapters 3 and 4. In the following subsections a brief account of early developments in ERE methods is given with special reference to examples that were examined in this area. Material is drawn largely from Rao (1975).

(a) A general linear model for the ERE appraoch

The basic model

Consider the vector random variables $\boldsymbol{\lambda}$ and $\mathbf{y}$, where $\boldsymbol{\lambda} = (\lambda_1, \lambda_2, \ldots, \lambda_k)$, $\mathbf{y} = (y_1, y_2, \ldots, y_k)$ and let $U_i = \lambda_i + e_i$, $i = 1, 2, \ldots, k$. The e_i's have a distribution F such that $F_F(\mathbf{e}) = \mathbf{0}$ and $\text{cov}_F(\mathbf{e}) = \sigma^2\mathbf{V}$, $\mathbf{V}$ being a known $k \times k$ matrix. The λ_i's are the parameters of interest and they are assumed to be generated by a prior distribution G. Also, put $\boldsymbol{\lambda} = \boldsymbol{\theta} + \boldsymbol{\eta}$, where $\boldsymbol{\theta}$ is an unknown parameter and $\boldsymbol{\eta}$ is a random vector with $E_G(\boldsymbol{\eta}) = \mathbf{0}$ and $\text{cov}_G(\boldsymbol{\eta}) = \tau^2\mathbf{W}$, where $\mathbf{W}$ is a known matrix.

We shall be concerned with the regression of Λ (or $\mathbf{\Theta}$) on $\mathbf{Y}$ for fixed values of σ^2, τ^2 and $\boldsymbol{\phi}$ (or $\boldsymbol{\beta}$) and note that

$$E(\mathbf{\Lambda}|\mathbf{Y}=\mathbf{y})=\boldsymbol{\phi}+(\tau^2\mathbf{W}+\sigma^2\mathbf{V})^{-1}\sigma^2\mathbf{V}(\mathbf{y}-\boldsymbol{\phi}).$$

This follows from the results on linear Bayes estimation in section 4.4.

A prediction model for genetic breeding ability

A variant of the basic model is the case when $\boldsymbol{\theta}$ depends linearly on a set of covariates. This model has been used in genetics, and elsewhere, and is specified by the further relation $\boldsymbol{\theta}=\mathbf{Z}\boldsymbol{\beta}$. Here $\mathbf{Z}$ is a $k \times s$ matrix of known elements representing covariates, and $\boldsymbol{\beta}$ is a vector of unknown parameters. See also section 4.7 for a discussion of concomitant variables in the EB framework. The problem in the genetic context is to obtain a predictor for $\boldsymbol{\eta}$ which is regarded as the true animal breeding ability.

The regression of $\boldsymbol{\eta}$ on $\mathbf{Y}$ is given by

$$E(\boldsymbol{\eta}|\mathbf{Y}=\mathbf{y})=(\tau^2\mathbf{W}+\sigma^2\mathbf{V})^{-1}\sigma^2\mathbf{V}(\mathbf{y}-\mathbf{Z}\boldsymbol{\beta}).$$

The Gauss–Markov model with random coefficients

Another variant of the basic model is obtained when $\boldsymbol{\lambda}=\mathbf{Z}\boldsymbol{\theta}$, and $\boldsymbol{\theta}$ is a realization of a random variable $\mathbf{\Theta}$ such that $E(\mathbf{\Theta})=\boldsymbol{\beta}$ and $\operatorname{cov}(\mathbf{\Theta})=\tau^2\mathbf{\Gamma}$. In this formulation $\boldsymbol{\beta}$ is an $s\times 1$ vector of unknown parameters, τ^2 is an unknown parameter and $\mathbf{\Gamma}$ is a known $s\times s$ matrix. The regression of $\mathbf{\Theta}$ on $\mathbf{Y}$ is given by

$$E(\mathbf{\Theta}|\mathbf{Y}=\mathbf{y})=\boldsymbol{\beta}+\{\tau^2\mathbf{\Gamma}+\sigma^2(\mathbf{Z}^{\mathrm{T}}\mathbf{V}^{-1}\mathbf{Z})^{-1}\}^{-1}$$
$$\times\,\sigma^2(\mathbf{Z}^{\mathrm{T}}\mathbf{V}^{-1}\mathbf{Z})^{-1}(\hat{\boldsymbol{\theta}}-\boldsymbol{\beta}),$$

where

$$\hat{\boldsymbol{\theta}}=(\mathbf{Z}^{\mathrm{T}}\mathbf{V}^{-1}\mathbf{Z})^{-1}\mathbf{Z}^{\mathrm{T}}\mathbf{V}^{-1}\mathbf{Y}$$

is the usual weighted least squares estimate of $\boldsymbol{\theta}$. This result follows from the application of the linear Bayes method to $\hat{\boldsymbol{\theta}}$ whose sampling distribution has the moments

$$E(\hat{\boldsymbol{\theta}}|\boldsymbol{\theta})=\boldsymbol{\theta},\qquad \operatorname{cov}(\hat{\boldsymbol{\theta}}|\boldsymbol{\theta})=(\mathbf{Z}^{\mathrm{T}}\mathbf{V}^{-1}\mathbf{Z})^{-1}\sigma^2.$$

(b) Empirical regression estimates

The regression functions obtained in subsection (a) above are in terms of parameters σ^2, τ^2 and $\boldsymbol{\phi}$, or $\boldsymbol{\beta}$. These parameters are generally not known, and in order to estimate them appropriate sampling schemes and data are needed.

Basic model

Consider the EB scheme (1.8.1) where m_i observations are made on X_i at the ith component. Now let $\mathbf{Y}^{\mathrm{T}} = (\bar{X}_1, \bar{X}_2, \ldots, \bar{X}_k)$. Its ith element $\bar{X}_i = (X_{i1} + X_{i2} + \cdots + X_{im_i})/m_i$ can be written as $\bar{X}_i = \lambda_i + e_i$, $i = 1, 2, \ldots, k$. The r.v. $\mathbf{e}$ has the mean and variance structure of the basic model with $\mathbf{V}$ a $k \times k$ diagonal matrix with ith diagonal element $1/m_i$. Since the λ_i's are independent realizations of $\boldsymbol{\Lambda}$ with mean μ_G we can take $\boldsymbol{\theta}$ in the basic model to be the $k \times 1$ vector all of whose elements have the value μ_G. The covariance matrix of $\mathbf{H}$ is $\tau^2 \mathbf{I}_k$. The parameters μ_G and τ^2 can be estimated

$$\bar{\mu}_G = \sum_{i=1}^{k} \bar{x}_i / k$$

$$\bar{\tau}^2 = \sum_{i=1}^{k} \bar{x}_i^2 / k - \bar{\sigma}^2 \left\{ \sum_{i=1}^{k} (1/m_i)/k \right\},$$

where $\bar{\sigma}^2$ is the usual 'within-groups' sample variance. The ERE of λ is now obtained as

$$\bar{\boldsymbol{\lambda}}(\mathrm{ERE}) = \bar{\mu}_G \mathbf{1} + (\bar{\tau}^2 \mathbf{I} + \bar{\sigma}^2 \mathbf{V})^{-1} \bar{\sigma}^2 \mathbf{V}(\mathbf{y} - \bar{\mu}_G \mathbf{1}).$$

An application of this ERE is found in a genetic selection problem. Suppose that $\boldsymbol{\lambda}$ is a vector of genetic variables and that a linear function, $\mathbf{a}^{\mathrm{T}}\boldsymbol{\lambda}$, for given known $\mathbf{a}$ is a genetic value of an individual. Then a good index for selecting and comparing individuals with respect to the genetic value is $\mathbf{a}^{\mathrm{T}}\bar{\boldsymbol{\lambda}}(\mathrm{ERE})$. A more general treatment of this type of genetic selection problem is given by Rao (1977).

Prediction of genetic breeding ability

Consider the model for prediction of genetic breeding ability, $\boldsymbol{\eta} = \boldsymbol{\lambda} - \mathbf{Z}\boldsymbol{\beta}$, as before. We assume that the covariance matrix of $\boldsymbol{\eta}$ is $\tau^2 \mathbf{I}_k$. Using $\bar{x}_i$ in the place of y_i, the regression of $\boldsymbol{\eta}$ on $\mathbf{y}$ can be expressed as

$$E(\boldsymbol{\eta} | \mathbf{Y} = \mathbf{y}) = \{\sigma^2/(\sigma^2 + \tau^2)\}(\mathbf{y} - \mathbf{Z}\boldsymbol{\beta}).$$

Given an EB scheme as in (1.8.1) σ^2 can be estimated as in the previous section. Adjusting for the covariates, $\sigma^2 + \tau^2$ can be estimated by $(\mathbf{y} - \mathbf{Z}\hat{\boldsymbol{\beta}})^{\mathrm{T}}(\mathbf{y} - \mathbf{Z}\hat{\boldsymbol{\beta}})/(k - r)$, where r is the rank of $\mathbf{Z}$, and $\boldsymbol{\beta} = (\mathbf{Z}^{\mathrm{T}}\mathbf{V}^{-1}\mathbf{Z})^{-1}\mathbf{Z}^{\mathrm{T}}\mathbf{V}^{-1}\mathbf{y}$.

The Gauss–Markov model with random coefficients

The parameters $\boldsymbol{\beta}$ and τ^2 are to be estimated when the prior distribution in unknown. Suppose that $\boldsymbol{\beta} = \mathbf{1}\gamma$. Then, based on $\mathbf{y}$, with

$y_i = \bar{x}_i$, moment estimates for γ and τ^2 are given in Rao (1975) as

$$\hat{\gamma} = \mathbf{1}^{\mathrm{T}}(\mathbf{Z}^{\mathrm{T}}\mathbf{V}^{-1}\mathbf{Z})\hat{\boldsymbol{\beta}}/\mathbf{1}^{\mathrm{T}}(\mathbf{Z}^{\mathrm{T}}\mathbf{V}^{-1}\mathbf{Z})\mathbf{1}$$
$$\hat{\tau}^2 = \{(\hat{\beta} - \mathbf{1}\hat{\gamma})^{\mathrm{T}}(\mathbf{Z}^{\mathrm{T}}\mathbf{V}^{-1}\mathbf{Z})^{-1}(\boldsymbol{\beta} - 1\gamma)/(r-3) - \hat{\sigma}^2\}/D,$$

where

$$D = [\mathrm{tr}\,(\mathbf{Z}^{\mathrm{T}}\mathbf{V}^{-1}\mathbf{Z})^{-1}\boldsymbol{\Gamma} - \{\mathbf{1}^{\mathrm{T}}(\mathbf{Z}^{\mathrm{T}}\mathbf{V}^{-1}\mathbf{Z})^{-1}\boldsymbol{\Gamma}(\mathbf{Z}^{\mathrm{T}}\mathbf{V}^{-1}\mathbf{Z})^{-1}\mathbf{1}\}/\{\mathbf{1}^{\mathrm{T}}(\mathbf{Z}^{\mathrm{T}}\mathbf{V}^{-1}\mathbf{Z})^{-1}]/(r-1)\},$$

and $\hat{\sigma}^2$ is the usual estimate of the error variance,

$$\hat{\sigma}^2 = (\mathbf{Y} - \mathbf{Z}\hat{\boldsymbol{\beta}})^{\mathrm{T}}\mathbf{V}^{-1}(\mathbf{Y} - \mathbf{Z}\hat{\boldsymbol{\beta}})/(k - r + 2).$$

It should be noted here that $\hat{\tau}^2$ and $\hat{\sigma}^2$ are not unbiased estimates of τ^2 and σ^2.

The situation above is the same as the exchangeability within the regression coefficients as considered by Lindley and Smith (1972). Thus the ERE for $\boldsymbol{\beta}$ obtained in this way is an analogue if the ridge regression estimate of Hoerl and Kennard (1970), with shrinkage to $\hat{\gamma}$ instead of the origin of the regression coefficients.

When $\boldsymbol{\beta}$ has no structure previous data are needed in the form

$$\mathbf{y}_1, \mathbf{y}_2, \mathbf{y}_n, \mathbf{y},$$

where $\mathbf{y}_i$ is generated by a realization of $\boldsymbol{\theta}_i$ of $\boldsymbol{\Theta}$. This case has been treated by Rao (1975). A related more general multivariate linear case is considered by Efron and Morris (1972). The data structure now is such that one is dealing with different linear models whose parameters are estimated simultaneously.

7.4 Compound estimation and decision theory

This is another alternative to EB techniques, which could be regarded as a stimulus for the creation of the EB approach. Originated by Robbins (1951) before the introduction of his EB approach in Robbins (1955), compound decision theory deals with the same sampling scheme (1.8.1). The difference now is that no *a priori* distribution is assumed as generating the parameter values $\lambda_1, \lambda_2, \ldots, \lambda_k$. In introducing compound decision theory Robbins (1951) considered the problem of decision between two simple hypotheses, assuming normal data distributions. Robbins did not proceed to point estimation, which was taken up by Stein (1955) and James and Stein (1961), who produced the James–Stein estimator.

Much further literature has since been devoted to both compound estimation and decision. Our present purpose is not to give an extensive review of the subject, but only to give an outline of the basic ideas, tracing the connection with EB methods. In particular, the aim is to reinforce the main theme that following the techniques suggested by the EB approach provides one route by which compound estimators or decision rules can be developed.

7.4.1 *Compound estimation*

(a) Component and compound decision problems

In section 1.14 a linear compound estimator for simultaneous estimation of the k parameters λ_i, $i=1,2,\ldots,k$, in the EB scheme was introduced. The key to the construction of a compound estimator is in minimizing a compound risk function rather than a component risk function. The special case dealt with in section 1.14 assumes that the observation x_i at the ith component is generated by a $N(\lambda_i, \sigma^2)$ distribution. Essentially the same technique can be used in a more general setting without assumption of a prior distribution.

Suppose that $X_1, X_2, \ldots, X_k$ are independent r.v.s and that for $i=1,2,\ldots,k$ the distribution function of X_i is $F(x_i|\lambda_i)$, depending on the parameter λ_i. Suppose also that $m_i \geqslant 1$ independent observations, x_{ij}, $j=1,2,\ldots,m_i$, are made on X_i. Let $\mathbf{x}_i$ be the vector whose elements are x_{ij}, $j=1,2,\ldots,k$; if $m_i=1$, $\mathbf{x}_i=x_i$. An estimator is sought for the vector $\boldsymbol{\lambda}=(\lambda_1, \lambda_2, \ldots, \lambda_k)$.

Since the X_i are mutually independent, any two estimation problems concerning λ_i and λ_j, $i \neq j$, can be regarded as unrelated. Hence it may be thought that an estimator of λ_i should depend on $\mathbf{x}_i$ alone and be of the form

$$\hat{\lambda}_i^{(0)} = \delta^{(0)}(\mathbf{x}_i). \tag{7.4.1}$$

However, for the vector $\boldsymbol{\lambda}$ the simple estimator $\hat{\boldsymbol{\lambda}}^{(0)}$ whose elements are $\lambda_i^{(0)}$, $i=1,2,\ldots,k$, is not necessarily optimal in term of compound risk. The compound risk function is

$$R^{(k)}(\hat{\boldsymbol{\lambda}}, \boldsymbol{\lambda}) = E_D L^{(k)}(\hat{\boldsymbol{\lambda}}, \boldsymbol{\lambda}), \tag{7.4.2}$$

where D is the distribution of $\mathbf{X}$ and E_D is expectation w.r.t. D; the compound loss $L^{(k)}(\hat{\boldsymbol{\lambda}}, \boldsymbol{\lambda})$ is the average of the component losses $L_i(\hat{\lambda}_i, \lambda_i)$, $i=1,2,\ldots,k$. The expected value w.r.t. D of the component loss is called the **component risk**. A typical choice for the form of component loss is $L_i(\hat{\lambda}_i, \lambda_i) = (\hat{\lambda}_i - \lambda_i)^2$ so that

$$L^{(k)}(\hat{\boldsymbol{\lambda}}, \boldsymbol{\lambda}) = \sum_{i=1}^{k} (\hat{\lambda}_i - \lambda_i)^2/k. \tag{7.4.3}$$

The best estimator of $\boldsymbol{\lambda}$ may be such that the estimator of λ_i depends not only on $\mathbf{x}_i$ but also on $\mathbf{x}_j$ for $j \neq i$. In general the estimator of λ_i should be of the form

$$\hat{\lambda}_i^+ = \delta_i(\mathbf{x}_1, \mathbf{x}_2, \ldots, \mathbf{x}_k) = \delta_i(\mathbf{x}), \tag{7.4.4}$$

In (7.4.4) it is assumed that the decisions are to be made after all observations on each variable have been made. A special class of rules of the type of (7.4.4) is the symmetric rules satisfying also

$$\delta_i(q\mathbf{x}) = q\delta_i(\mathbf{x}), \qquad q \in Q, \tag{7.4.5}$$

where Q is the set of permutations of the integers $1, 2, \ldots, k$. The rule defined in (7.4.1) belongs to the class of symmetric rules.

Rules of the type $\boldsymbol{\lambda}^{(0)}$ are called **non-compound** so as not to confuse terminology with that of simple EB rules, and those of (7.4.5) are **symmetric** compound rules. A non-compound rule can be obtained by using any conventional estimate of λ_i in the place of $\delta_i^{(0)}(\mathbf{x}_i)$. On the other hand, compound rules are obtained by direct consideration of the compound loss. Special forms of $\hat{\boldsymbol{\lambda}}^+$ have been motivated by the fact that the 'best' symmetric rule is often a non-compound rule being a functional of a symmetric function $g(\boldsymbol{\lambda})$ of $\boldsymbol{\lambda}$. If it is possible to estimate $g(\boldsymbol{\lambda})$ from the entire data set replacing it by the estimate produces a compound estimate. To illustrate, we consider a special class of estimators in the next section.

(b) Optimal linear estimators

Consider estimators of λ_i of the form

$$\hat{\lambda}_i = a_{0i} + \sum_{j=1}^{m_i} a_{ji} X_{ij}. \tag{7.4.6}$$

We shall determine the constants a_{ji} in such a way as to minimise the compound risk

$$R^{(k)}(\hat{\boldsymbol{\lambda}}, \boldsymbol{\lambda}) = E_D k^{-1} \sum_{i=1}^{k} \left\{ \left(a_{0i} + \sum_{j=1}^{m_i} a_{ji} X_{ij} \right) - \lambda_i \right\}^2.$$

Straightforward calculations show that

$$a_{ji} = \sigma_\lambda^2/(m_i\sigma_\lambda^2 + \omega_\lambda), \qquad j = 1, 2, \ldots, m_i,$$
$$a_{0i} = \bar{\lambda}\omega_\lambda/(m_i\sigma_\lambda^2 + \omega_\lambda),$$

where $E(X_i|\lambda_i) = \lambda_i$, $\text{var}(X_i|\lambda_i) = \sigma_i^2, \bar{\lambda} = \sum_{i=1}^k \lambda_i/k$, $\omega_\lambda = \sum_{i=1}^k \sigma^2/k_i$, and $\sigma_\lambda^2 = \sum_{i=1}^k (\lambda_i - \bar{\lambda})^2/k$. Thus the optimal linear estimator is defined by

$$\hat{\lambda}_i^* = \bar{\lambda} + (1 - c_{\lambda i})(\bar{X}_i - \bar{\lambda}), \tag{7.4.7}$$

where

$$\bar{X}_i = \sum_{j=1}^{m_i} X_{ij}/m_i, \qquad c_{\lambda i} = \omega_\lambda/(m_i\sigma_\lambda^2 + \omega_\lambda).$$

We note that $\hat{\lambda}^*$ is a non-compound rule and a function of $\bar{\lambda}$ and the $c_{\lambda i}$ which are symmetric functions of the λ_i. The quantities $\bar{\lambda}$ and $c_{\lambda i}$ are generally unknown in practice, but they can be estimated using the full set of data, as we show below. Replacing these unknowns by their estimates produces a compound estimator, because each $\hat{\lambda}_i^*$ is replaced by a quantity depending not only on the observations in component i but also on all other observations.

(c) Linear compound estimates of means

A special structure for σ_i^2

We now assume that

$$\sigma_i^2 = (A - 1)\lambda_i^2 + B\lambda_i + C, \tag{7.4.8}$$

a relation satisfied by many important data distributions. We also assume that A, B and C are common to all component problems, and that every $m_i \geqslant 1 - A$. Estimates of $\bar{\lambda}$ and $c_{\lambda i}$ can now be obtained by the following steps: let

$$\bar{X}_\bullet = \sum_{i=1}^k m_i\bar{X}_i \Big/ \sum_{i=1}^k m_i,$$

$$S_{XX} = \sum_{i=1}^k m_i(\bar{X}_i - \bar{X}_\bullet)^2/(k-1),$$

$$U_i = m_i(m_i - A + 1)^{-1}[(A-1)\bar{X}_i^2 + B\bar{X}_i + C].$$

Then unbiased estimates of $\bar{\lambda}$, ω and σ^2 are given by

$$\hat{\bar{\lambda}} = \bar{X}.$$

$$\bar{\omega}_\lambda = \sum_{i=1}^{k} U_i/k$$

$$\bar{\sigma}_\lambda^2 = S_{XX} - \bar{\omega}_\lambda k^{-1}\left(\sum_{i=1}^{k} m_i^{-1}\right).$$

These unbiased estimates of ω_λ and σ_λ^2 can be negative and the truncated estimates defined as follows can be used instead:

$$\hat{\sigma}_\lambda^2 = \max(\bar{\sigma}_\lambda^2, 1/k), \qquad \hat{\omega}_\lambda = \max(\bar{\omega}_\lambda, 1/k).$$

The quantity $c_{\lambda i}$ is then estimated by

$$\hat{c}_{\lambda i} = \hat{\omega}_\lambda/(m_i\hat{\sigma}_\lambda^2 + \hat{\omega}_\lambda),$$

while the optimal non-compound estimator is estimated according to

$$\hat{\lambda}_i = \hat{\bar{\lambda}} + (1 - \hat{c}_{\lambda i})(\bar{X}_i - \hat{\bar{\lambda}}). \tag{7.4.9}$$

This is now a symmetric compound estimator and it is an obvious analogue of the James–Stein estimator in a more general setting. The estimates proposed above are modifications of those given by Southward and van Ryzin (1972).

No structure assumed for σ_i^2

We now abandon the assumption that σ_i^2 has the special structure in (7.4.8), but we do require that every $m_i \geqslant 2$. This approach has the advantage that a knowledge of the parametric form of F is not needed. The results are therefore applicable to nonparametric cases where we assume only that the means λ_i and the variances σ_i^2 are finite. With $\hat{\sigma}_i^2$ the usual sample variance of the m_i results in the ith component, the estimates are

$$\hat{\omega}_\lambda = \max\left[\sum_{i=1}^{k} \hat{\sigma}_i^2/k, 1/k\right],$$

$$\hat{\sigma}_\lambda^2 = \max\left[S_{XX} - \left(\sum_{i=1}^{k} m_i^{-1}\right)\hat{\omega}_\lambda, 1/k\right],$$

to be used in (7.4.9).

Example 7.4.1 The problem of linear calibration, discussed also in section 8.4.1, provides an interesting example of the application of compound estimation. Suppose that a relation $y = \alpha + \beta x + e$ holds, where y is a measurement by a 'quick' method, x is a measurement by a 'slow' but accurate method, and e is a random error with zero mean and variance σ^2. A calibration experiment yields measurements $x_1, x_2, \ldots, x_n$ by the slow method, and $y_1, y_2, \ldots, y_n$ by the quick method. The problem is to estimate the unknown current x when a measurement y by the quick method has been obtained.

Consider linear estimators $\phi(Y) = \omega_0 + \omega_1 Y$, where Y is the random variable of which y is a realization. Suppose that an estimator of x is chosen from the class (7.4.1) such that, if applied to the previous y_i's, it would result in the smallest overall mean square error. That is, choose ω_0 and ω_1 so as to minimize

$$\sum_{i=1}^{n} E_D\{\phi(Y_i) - x_i\}^2/n,$$

where D refers to the joint distribution of $\mathbf{Y} = (Y_1, Y_2, \ldots, Y_n)$ for fixed $\mathbf{x} = (x_1, x_2, \ldots, x_n)$. This is a sensible criterion because in practice one would generally be concerned with x values falling in the range of values covered by the calibration experiment. Also, the estimator that is best for the known x's is certainly a reasonable candidate as an estimator for the current x.

According to the criterion above the optimal linear estimator of x is

$$\hat{x}(Y) = \bar{x} + \frac{\beta S_{xx}}{\{n/(n-1)\}\sigma^2 + \beta S_{xx}}(Y - \alpha - \beta x),$$

where $S_{xx} = \sum_{i=1}^{n}(x_i - \bar{x})^2/(n-1)$. Defining S_{YY} and S_{xY} similarly we have

$$E_D(\bar{Y}) = \alpha + \beta\bar{x}$$
$$E_D S_{xY} = \beta S_{xx}$$
$$E_D S_{YY} = \sigma^2 + \beta^2 S_{xx}.$$

From these relations estimates of the parameters α, β, σ^2 can be obtained by the method of moments. Replacing these parameters by their estimates in $\hat{x}(Y)$, and $n/(n-1)$ by 1 yields the 'inverse estimator'

$$\hat{x}_{\mathrm{I}}(Y) = \bar{x} + (S_{xY}/S_{YY})(Y - \bar{Y})$$

proposed by Krutchkoff (1967) as an alternative to the 'classical' estimator

$$\hat{x}_{\mathrm{C}}(Y) = \bar{x} + (S_{xx}/S_{xY})(Y - \bar{Y}).$$

The compound mean square error criterion provides a justification for the 'inverse' estimator. A more detailed discussion of the calibration controversy is given in Lwin and Maritz (1982). See also section 8.4.1 where an EB approach to the same problem is illustrated.

(d) Optimal non-linear estimators

One may not wish to start with a class of estimators such as (7.4.6), especially if λ_i is not the mean of the ith component distribution. If the parametric form of F is known it is still possible to obtain an optimal, generally non-linear, estimator of λ_i which minimizes the compound risk defined in (7.4.2). It is of the form

$$\hat{\lambda}_i^*(\mathrm{NL}) = \frac{\sum_{u=1}^k \lambda_u \{\prod_{j=1}^{m_i} f(x_{ij}|\lambda_u)\}}{\sum_{u=1}^k \{\prod_{j=1}^{m_i} f(x_{ij}|\lambda_u)\}} \tag{7.4.10}$$

The formula above is readily obtained from (1.3.2) by noting that G may be replaced by the empirical c.d.f. of the λ_i so that $W(\delta)$ is formally identical to the compound risk (7.4.1). The estimator in (7.4.10) again is not practically useful since it depends on the λ_i values themselves, but its form does indicate how a non-linear compound estimator could be derived. Let $\hat{\lambda}_i$ be a standard non-compound estimate of λ_i depending only on the observations in the ith component. Replacing λ_i by $\hat{\lambda}_i$ in (7.4.10) produces a compound non-linear estimator. The estimate of every λ_i then becomes a weighted average of the conventional estimates, and so induces a shrinkage to an average, a feature of all compound estimators.

(e) Performance of compound estimators

The applicability of compound estimators has been the source of much controversy since the appearance of the original James–Stein estimator. The justification of the compound estimators is in the compound loss. Thus, although the component problems may be unrelated, the use of only the information at the ith component to construct an estimate of λ_i, does not necessarily produce an optimal result in terms of the compound loss.

A telling result is that, for the original James–Stein estimator, the compound risk is

$$R^{(k)}(\hat{\boldsymbol{\lambda}}^+, \boldsymbol{\lambda}) = \sigma^2[1 - E_D\{(k-3)^2/S\}], \tag{7.4.11}$$

where S is a τ^2 r.v. with $k-1$ degrees of freedom. This $R^{(k)}(\hat{\boldsymbol{\lambda}}^+, \boldsymbol{\lambda})$ is smaller than the corresponding risk of the best non-compound estimator, i.e. the ML estimator. Much work has been done extending the James–Stein estimators to more general situations. The case of the exponential family of data distributions has been treated by Hudson (1978); an outline of this work follows.

Let X_i of the ith component have a p.d.f. belonging to the exponential family (3.4.5). Then the sample information of the ith component is summarized in a sufficient statistic T_i also having a p.d.f. in the exponential family, say,

$$f(t|\lambda_i) = \exp\{\lambda_i t - \psi(\lambda_i)\} k_i(t). \tag{7.4.12}$$

Suppose that $\theta_i = E_F(T_i)$, $i = 1, 2, \ldots, k$, are the parameters of interest to be estimated. Consider a subclass of (7.4.12) such that

$$E_F\{(T_i - \theta_i)l(T_i)\} = E_F\{a(T_i)l'(T_i)\} \tag{7.4.13}$$

for some function $a(T_i)$ and for all continuous functions $l(\cdot)$ such that $E|a(T_i)l'(T_i)| < \infty$. Hudson (1978) showed that if the p.d.f. of T_i is given by

$$f(t|\lambda_i) = \exp\left\{\theta_i \int b(t)dt - \chi(\lambda_i)\right\} b(t) \exp\left\{-\int tb(t)dt\right\},$$

where $b(t) = 1/a(t)$, then it belongs to the subclass defined by (7.4.13).

Let $S = \sum_{i=1}^{k} (\int a(t_i)dt_i)^2$ and define the compound estimator of θ_i as

$$\hat{\theta}_i^+ = T_i - (k-2)\left\{\int a(t_i)dt_i\right\} \Big/ S. \tag{7.4.14}$$

Then the compound risk of $\hat{\boldsymbol{\theta}}^+$ under squared error losses is

$$R^{(k)}(\hat{\boldsymbol{\theta}}, \boldsymbol{\theta}) = R^{(k)}(\mathbf{T}, \boldsymbol{\theta}) - E\{(k-2)^2/S\}, \tag{7.4.15}$$

where $R^{(k)}(\mathbf{T}, \boldsymbol{\theta})$ is the compound risk of $\mathbf{T}$ as an estimator of $\boldsymbol{\theta}$. The significance of result (7.4.15) is that the compound risk of $\hat{\boldsymbol{\theta}}^+$ is always smaller than the compound risk of $\mathbf{T}$, and it is an analogue of the James–Stein estimator for the exponential family of distributions.

(f) Performance of linear compound estimators

We now give an assessment of the linear compound estimator discussed in section 7.4.1(b). In doing so we modify the earlier results in two ways. First we use a generic non-compound estimator T_i in place of $\bar{X}_i$ and assume that $E_F(T_i) = \lambda_i$, $\text{var}_F(T_i) = \tau_i$. Second we take as our starting point the class of linear estimators

$$\hat{\lambda}_i = a_0 + a_1 T_i. \tag{7.4.16}$$

Note that T_i, and therefore $\hat{\lambda}_i$, need not be linear in the original observations. If $T_i = \bar{X}_i$ we are back to the linear case treated in section 7.4.1(b).

Following the arguments in section 7.4.1(b) the optimal linear estimator of λ_i, in terms of T_i, is

$$\bar{\lambda}_i^* = \bar{\lambda} + (1 - d_{\lambda i})(T_i - \bar{\lambda}),$$

where $d_{\lambda i} = \bar{\tau}/(\sigma_{\lambda}^2 + \bar{\tau})$, where $\bar{\lambda} = \sum_{i=1}^k \lambda_i/k$, $\bar{\tau} = \sum_{i=1}^k \tau_i/k$, and, $\sigma_{\lambda}^2 = \sum_{i=1}^k (\lambda_i - \bar{\lambda})^2/k$. Now suppose that $\hat{\tau}_i$ is any consistent estimate of τ_i from the data of the ith component. Then estimates of the elements of $d_{\lambda i}$ can be obtained as follows:

$$\hat{\bar{\lambda}} = \bar{T} = \sum_{i=1}^{k} T_i/k, \qquad \hat{\bar{\tau}} = \sum_{i=1}^{k} \hat{\tau}_i,$$

and

$$\hat{\sigma}_{\lambda}^2 = \max\left\{ \sum_{i=1}^{k} (T_i - \bar{T})^2/k - (k-1)\hat{\bar{\tau}},\ 1/k \right\}.$$

Replacing $\bar{\lambda}$, τ_i, σ_{λ}^2, $\bar{\tau}$ by their estimates in the formulae for $d_{\lambda i}$ and $\hat{\lambda}_i^*$ yields a compound estimator of λ_i.

It can be shown that, when $\min(m_1, m_2, \ldots, m_k)$ is large and $\tau_i = O(1/m_i)$, then $\bar{\lambda}^*$ is preferred to $\mathbf{T}$ in terms of compound risk if the following inequality holds:

$$\bar{\tau} > \sum_{i=1}^{k} (\lambda_i - \bar{\lambda})^2/k. \tag{7.4.17}$$

This is an important result since it gives a guide as to the practical usefulness of compound estimation. It should be noted that, even in terms of compound risk, the James–Stein type estimators do not dominate the non-compound estimators uniformly in general. The uniform dominance in the exponential family seems an exception rather than the rule. The result (7.4.17) reflects the more common situation that the compound estimators dominate the best non-compound estimators only in a subset of the parameter space. That subset is where the spread of the λ_i values is smaller than the average spread of the data about their central values.

This last property is related to the roles played by the prior variance and the data variance in EB methods. In general EB estimates can only be better than conventional ones if the prior variance is smaller than the data variance. In this context where the dispersion of the λ_i values is regarded as being brought about by a random mechanism, it also seems necessary to assume a physical connectedness between the component problems. In the compound decision context it does not seem to make practical sense to use a compound loss criterion unless there is some such connectedness.

Example 7.4.2 This example is given by Hudson (1978). Let X_i be a gamma r.v. with p.d.f. $f(x|\theta_i)$ given by

$$\ln f(x|\theta_i) = (\theta_i - 1)\ln x - x - \ln\Gamma(\theta_i), \qquad i = 1, 2, \ldots, k, \; k \geqslant 3.$$

Then $T_i = \sum_{i=1}^{m_i} X_{ij}$ has a gamma distribution of the same form as $f(x|\cdot)$ with parameter $\lambda_i = m_i\theta_i$. Let

$$\hat{\lambda}_i^+ = T_i - (1/S)(k-2)\ln T_i,$$

where $S = \sum_{i=1}^{k} (\ln X_i)^2$. Then the compound estimator $\hat{\boldsymbol{\lambda}}^+$ dominates the non-compound unbiased estimator $\mathbf{T} = (T_1, T_2, \ldots, T_k)$.

7.4.2 *Compound decisions between hypotheses*

Compound decision (CD) theory originated with Robbins (1951), who dealt in the first instance mainly with the problem of decision between two simple hypotheses. The particular problem treated in detail is that of choosing between $\lambda = +1$ and $\lambda = -1$ in a $N(\lambda, 1)$ distribution, when one observation x_i is made at each component. In two more recent papers, Copas (1969, 1974) gives a critical review of CD theory with special emphasis on its connection with EB methods. The view presented is that, of the two approaches, CD seems to be the more flexible frame of reference in which to examine decision procedures. The following summary of CD theory is based in part on these two papers by Copas.

(a) Decisions between two simple hypotheses

Suppose that the value of λ_i at the ith component is known to be either $\lambda^{(1)}$ or $\lambda^{(2)}$. We have to make a collection of decisions comprising the assigning of every unknown λ_i, which gave rise to observations x_i, to either $\lambda^{(1)}$ or $\lambda^{(2)}$.

As a somewhat oversimplified example, consider each component experiment to be the random drawing, with replacement, of m items from a batch of fixed size. The proportion of defectives in a batch is λ; suppose that we know every component λ to be either $\lambda^{(1)}$ or $\lambda^{(2)}$ and that we are to 'sentence' batches accordingly. Every observed x_i, $i=1,2,...,k$, is a realization of a Bin(m, λ) r.v. with $\lambda=\lambda^{(1)}$ or $\lambda=\lambda^{(2)}$. We emphasize that the sentencing is not done sequentially and that the entire sequence of batches is sentenced simultaneously when all sample results are to hand. Sequential sentencing is a different case and will be treated later.

Since we make the k decisions simultaneously we may regard the problem as one of selecting a parameter point $(\lambda'_1, \lambda'_2, \ldots, \lambda'_k)$ from a space $\mathbf{\Omega}$ comprising the 2^n points $(\lambda^{(1)}, \lambda^{(1)}, \ldots, \lambda^{(1)}), \ldots,$ $(\lambda^{(2)}, \lambda^{(2)}, \ldots, \lambda^{(2)})$. One standard method of doing this is to maximize the likelihood $L(x, \boldsymbol{\lambda})$ of the observations $\mathbf{x}=(x_1, x_2, \ldots, x_k)^{\mathrm{T}}$ w.r.t. $\boldsymbol{\lambda}$ in the space Ω. Since

$$L(\mathbf{x}, \boldsymbol{\lambda}) = \prod_{i=1}^{k} f(x_i|\lambda_i), \tag{7.4.18}$$

it is maximized by maximizing every individual $f(x_i|\lambda_i)$. This is equivalent to comparing $f(x_i|\lambda^{(1)})$ and $f(x_i|\lambda^{(2)})$, i.e. using the likelihood ratio criterion

$$z_i = f(x_i|\lambda^{(1)})/f(x_i|\lambda^{(2)}), \qquad i=1,2,\ldots,k. \tag{7.4.19}$$

This means that the decision rule is non-compound. The rule is also symmetric in the sense that every batch is sentenced according to the same criterion.

One can recast the problem in the decision theoretic framework. For a typical component we partition the sample space into two regions A_1 and A_2 and our rule, $\delta(x_i)$, is to choose $\lambda^{(1)}$ when $x_i \in A_1$ and $\lambda^{(2)}$ otherwise. Here we take the loss to be $L(\delta, \lambda)=0$ if the correct decision is made and $L(\delta, \lambda)=1$ otherwise. For the entire sequence of decisions the total expected loss is then

$$\begin{aligned} E_D \sum_{i=1}^{k} L\{\delta(x_i), \lambda_i\}/k = E_D C(\delta) \\ = (M_i/k) \int_{A_1} f(x|\lambda^{(1)})dx \\ + (M_2/k) \int_{A_2} f(x|\lambda^{(2)})dx, \end{aligned} \tag{7.4.20}$$

where M_1 and M_2 are the numbers of λ_i's having the values $\lambda^{(1)}$ and $\lambda^{(2)}$ respectively. The symbol E_D indicates expectation with respect to the joint distribution of $X_1, X_2, \ldots, X_k$ given $\lambda_1, \lambda_2, \ldots, \lambda_k$.

Let $\theta = M_1/k$. Then (7.4.19) can be rewritten formally as the expression for $W(\delta)$ in section 1.4. Hence the optimal A_1 can be constructed using the argument of section 1.4 and the rule is readily found to be

$$\begin{aligned} \delta_k(x)\colon \quad & \text{accept } \lambda = \lambda^{(1)} \text{ if } x \leqslant \xi_k, \\ & \text{accept } \lambda = \lambda^{(2)} \text{ otherwise,} \end{aligned} \tag{7.4.21}$$

where ξ_k is the solution of the following equation in x:

$$\theta f(x|\lambda^{(2)}) = (1-\theta) f(x|\lambda^{(1)}).$$

Clearly δ_k generally depends on θ, and this means that the likelihood ratio method will not necessarily produce an optimal rule.

In practice θ will generally be unknown, making determination of the optimal rule impossible. However, it is possible to obtain information about θ from the set of observations x_i, $i = 1, 2, \ldots, k$. Suppose that θ is estimated by $\hat{\theta}$ and that θ is replaced by $\hat{\theta}$ in the construction of the optimal decision rule. If $\hat{\theta}$ is close to θ it is not unreasonable to expect that the compound risk resulting from such a rule might be close to the smallest risk. A rule of the kind under discussion will generally be a compound rule. The question of closeness of the risk of the CD rule to the optimal risk will be taken up in section 7.4.2(c).

We conclude this section by looking at an example studied in detail by Robbins (1951). Let $f(x|\lambda)$ be the $N(\lambda, 1)$ density. Then straightforward calculations yield

$$\xi_k = (\lambda^{(1)} + \lambda^{(2)})/2 - (\lambda^{(2)} - \lambda^{(1)}) \ln\{(1-\theta)/\theta\}$$

where it is assumed that $\lambda^{(2)} \geqslant \lambda^{(1)}$. Now let $\bar{X} = (X_1 + X_2 + \cdots + X_k)/k$. Then $E(\bar{X}) = \theta\lambda^{(1)} + (1-\theta)\lambda^{(2)}$, and $\operatorname{var}(\bar{X}) = 1/k$. Hence a reasonable estimate of θ is $\hat{\theta} = (\bar{x} - \lambda^{(2)})/(\lambda^{(1)} - \lambda^{(2)})$, truncated to lie in the interval $[0, 1]$, and $\hat{\xi}_k$ is obtained by replacing θ by $\hat{\theta}$ in the formula defining ξ_k. By analogy to Example 1.4.1 the compound decision rule becomes

$$\begin{aligned} & \text{if } \bar{x} \leqslant \lambda^{(1)} \text{ choose } \lambda^{(1)} \text{ for all } x \\ & \text{if } \bar{x} \geqslant \lambda^{(2)} \text{ choose } \lambda^{(2)} \text{ for all } x \\ & \text{if } \lambda^{(1)} < \bar{x} < \lambda^{(2)} \text{ choose } \lambda^{(1)} \text{ if } x < \hat{\xi}_k \\ & \qquad\qquad\qquad\quad \text{choose } \lambda^{(2)} \text{ if } x > \hat{\xi}_k. \end{aligned} \tag{7.4.22}$$

which coincides with (5.2.10).

Regarding the CD rule embodied in $\hat{\xi}$ for any given $\hat{\xi}$ the total loss is

$$kC(\hat{\xi}) = \hat{e}_1^{(1)} + \cdots + \hat{e}_{M_1}^{(1)} + \hat{e}_1^{(2)} + \cdots + \hat{e}_{M_2}^{(2)},$$

where the $\hat{e}_i^{(j)} = 1$ or 0 as $x_i^{(j)} \geqslant$ or $< \hat{\xi}_k$ $i = 1, 2, \ldots, M_j$, $j = 1, 2$ where $x_i^{(j)}$ are observations generated by $\lambda^{(j)}$. Since

$$\begin{aligned} E_X(\hat{e}_i^{(1)}) &= P[x_i^{(1)} \geqslant \hat{\xi}] \\ E_X(\hat{e}_i^{(2)}) &= P[x_i^{(2)} < \hat{\xi}], \end{aligned} \tag{7.4.23}$$

and the $x_i^{(j)}$, $i = 1, 2, \ldots, M_j$ are identically distributed for each j, we have

$$E_X C(\hat{\xi}_k) = \frac{M_1}{k} P[x_1^{(1)} \geqslant \hat{\xi}_i] + \frac{M_2}{k} P[x_1^{(2)} < \hat{\xi}_k]. \tag{7.4.24}$$

Without writing down a more explicit expression for the r.h.s. of (7.4.24) it is easy to see that, since $\hat{\xi}_k \to \xi_k$, in probability, as $k \to \infty$, $E_X C(\hat{\xi}_k) \to E_X C(\xi_k)$. Robbins (1951) has given an expression for $E_X C(\hat{\xi})$ which can be evaluated by numerical integration. A Monte Carlo method of estimating $E_X C(\hat{\xi}_k)$ is easily developed. For a given set of $\lambda^{(1)}$ and $\lambda^{(2)}$ values (i.e. given M_1 and M_2), a sequence of observations, $x_1, x_2, \ldots, x_k$, is generated by sampling from appropriate normal distributions. Computation of $\hat{\xi}_k$ and $C(\hat{\xi}_k)$ is then straightforward. Repetition of this process, of x-sampling with the same λ's, yields a sequence of $C(\hat{\xi})$ values whose average is an estimate of $E_X C(\hat{\xi})$. Table 7.1 gives the results of such calculations, in the case

Table 7.1 *Compound decisions between H_1: $\lambda = -1$, H_2: $\lambda = +1$ in the $N(\lambda, 1)$ case. Expected proportions of wrong decisions are tabulated for $k = 100$, $k = 10$, when $\theta_1 = P(\Lambda = -1) = 0{\cdot}1$, $0{\cdot}2$, $0{\cdot}3$, $0{\cdot}4$, $0{\cdot}5$ [Robbins (1951)]*

θ_1	$E_X C(\cdot)$ $k = 100$	$k = 10$
0·1	0·076	0·11 ± 0·01
0·2	0·117	0·15 ± 0·01
0·3	0·144	0·22 ± 0·02
0·4	0·159	0·21 ± 0·02
0·5	0·163	0·20 ± 0·02

$\lambda^{(1)} = -1$, $\lambda^{(2)} = +1$; the results for $k = 100$ are those given by Robbins (1951).

The example we have just considered is a rather simple one of the type involving two simple hypotheses, because the normal distribution belongs to the 'monotone likelihood ratio' family. In other cases A_1 and A_2 may be more complicated, but the principles remain the same.

(b) Two composite hypotheses

Introduction

Instead of restricting the values of the parameter to only two possible numbers, $\lambda^{(1)}$ and $\lambda^{(2)}$, we now allow the k parameter values $\lambda_1, \lambda_2, \ldots, \lambda_k$ to be an arbitrary set of numbers. Two composite hypotheses are represented by a partition of the λ-space into ω_1 and ω_2. Our task is to assign every λ_j to $H_1 : \lambda \in \omega_1$ or $H_2 : \lambda \in \omega_2$, on the basis of x_j, and allowing the possibility that every decision may also be influenced by the observations other than x_j alone.

Following the approach of section 7.4.2(a) we examine the consequences of employing a simple rule, $\delta(x)$, by which we select

$$H_1 \quad \text{when } x_j \in A_1$$

and

$$H_2 \quad \text{when } x_j \in A_2$$

for $j = 1, 2, \ldots, k$; A_1 and A_2 remaining fixed. Let $L(\delta(x), \lambda)$ denote the loss on making decision $\delta(x)$ when the parameter value is λ. Then the total loss on making the decisions for the entire set of observations is

$$kC(\delta) = L[\delta(x_1), \lambda_1] + \cdots + L[\delta(x_k), \lambda_k],$$

and the expected average loss in repeated x-sampling with fixed $\lambda_1, \lambda_2, \ldots, \lambda_k$ is

$$E_X C(\delta) = \frac{1}{k} \sum_{j=1}^{k} \int L[\delta(x_j), \lambda_j] dF(x_j | \lambda_j)$$

$$= \iint L[\delta(x), \lambda] dF(x | \lambda) dG_k(\lambda). \tag{7.4.25}$$

In the expression (7.4.25), $G_k(\lambda)$ is a d.f. with jumps of magnitude $1/k$ at the points $\lambda_{(1)}, \ldots, \lambda_{(k)}$, these being the parameter values $\lambda_1, \ldots, \lambda_k$ arranged in order of magnitude.

Our aim is to determine $\delta(x)$ such that $E_X C(\delta)$ is minimized, and clearly we should be able to do this by the established Bayesian techniques if $G_k(\lambda)$ were known. Knowledge of $G_k(\lambda)$ in this case is equivalent to knowledge of M_1 and M_2 in the case of two simple hypotheses, treated in section 7.4.2. In the more general case it is tantamount to being given the set of values $\lambda_1, \ldots, \lambda_k$ without being told with which x_j any given λ_j is associated. In practice such knowledge is usually not forthcoming; we have seen in the simpler version of this problem that M_1 and M_2 are generally unknown. However, it also appeared that M_1 and M_2 could be estimated using the observed x-values, and the question now is whether the same possibility exists in the more general case, and how it may be exploited.

Define

$$e_i = \begin{cases} 1, & \text{when } x_i \leqslant x \\ 0, & \text{otherwise,} \end{cases}$$

where x is an arbitrary x-value. Then

$$P(e_i = 1) = P(x_i \leqslant x) = \int_{-\infty}^{x} dF(x_i | \lambda_i).$$

Putting

$$F_k(x) = \sum_{i=1}^{k} e_i/k$$
$$= (\text{the number of } x\text{-values} \leqslant x)/k,$$

we see that

$$E_X F_k(x) = \frac{1}{k} \sum_{i=1}^{k} \int_{-\infty}^{x} dF(x_i | \lambda_i)$$
$$= \int \int_{-\infty}^{x} dF(x_i | \lambda) dG(\lambda)$$
$$= F_{G_k}(x).$$

Remembering that var $(F_k(x)) \leqslant 1/(4k)$, we see that $F_k(x)$ is, for every x, an unbiased estimate of $F_{G_k}(x)$.

We emphasize that $F_k(x)$ and $G_{G_k}(x)$ and the relationship between them are obtained by a somewhat different argument from that occurring in the relationship between $F_n(x)$ and $F_G(x)$ in the EB problem. Nevertheless, $F_k(x)$, $F_{G_k}(x)$ and $G_k(\lambda)$ have the mathematical properties of d.f.s, and the connection between them is directly comparable with the relationship between (F_n, F_G, G) in the EB case. An immediate consequence is that the procedure developed for estimating G in the EB case can be applied here. There is this difference: the $\lambda_1, \ldots, \lambda_k$ are not necessarily regarded as originating from a certain population of λ's from which they are obtained by random sampling. Thus the 'r-smooth' method would appear to be more appropriate than the 'parametric smooth' method. However, if the CD problem arises in circumstances where the λ's can be regarded as random observations from a population whose d.f. belongs to a certain parametric family, then a member of that family may possibly be used advantageously as an approximation to G. This possibility has not been explored further. In the following section we give the results of a study of the problem of decision between two composite hypotheses.

A particular case using the '0 − 1' *loss structure*

Let the two hypotheses be $H_1: \lambda < \lambda_0$ or $H_2: \lambda \geqslant \lambda_0$, the loss being 1 when the wrong decision is made, and 0 otherwise. The Bayes solution to this problem is outlined in section 1.4. When

$$R(x) = \int_{-\infty}^{\lambda_0} f(x|\lambda)dG_k(\lambda) \bigg/ \int_{\lambda_0}^{\infty} f(x|\lambda)dG_k(\lambda)$$

is monotonic in x, the regions of acceptance for H_1 and H_2 become, $x \in [-\infty, \xi_{G_k})$ and $x \in [\xi_{G_k}, +\infty]$, where ξ_{G_k} is a Bayes 'cut-off' for x. Thus, if $x < \xi_{G_k}$ accept H_1 and if $x \geqslant \xi_{G_k}$ accept H_2. The value of ξ_{G_k} is determined by finding x such that

$$\int_{-\infty}^{\lambda_0} f(x|\lambda)dG_k(\lambda) = \int_{\lambda_0}^{\infty} f(x|\lambda)dG_k(\lambda). \qquad (7.4.26)$$

Such an x can be determined if the l.h.s. and r.h.s. are continuous functions of x, and x takes all values. When X is a discrete r.v., a suitable convention must be adopted.

When G is not known, but is approximated by G_r^* (see section 2.9), we obtain an estimate of G^* by maximizing.

$$(1/k) \sum_{i=1}^{k} \log\{f_{G^*}(x_i)\}.$$

The motivation for this procedure is exactly the same as before, namely minimization of a 'distance' between $F(x)$ and $F_G(x)$; the reader is referred to section 2.9. We denote the estimated G^* by $\hat{G}_r^*$, and on substituting $\hat{G}_r^*$ for G_k in (7.4.25) a 'compound' cut-off, $\hat{\xi}$, is obtained. Thus the CD rule is

$$\text{accept } H_1 \text{ if } x_i < \hat{\xi}, \quad \text{accept } H_2 \text{ if } x_i \geqslant \hat{\xi}, \quad i = 1, 2, \ldots,$$

When $f(x|\lambda)$ is such as not to yield a monotonic $R(x)$, the process of finding the Bayes and CD rules may be somewhat more cumbersome, but it will not be different in principle. The approximation of G by G^* will be carried out in the same way, and H_1 or H_2 is then chosen according to (7.4.25), with G replaced by $\hat{G}^*$.

Example 7.4.3 Let $f(x|\lambda)$ be the $N(\lambda, 1)$ density. We have seen in section 1.5 that $R(x)$ is monotonic in x for this case. Determination of the compound decision rules follows the methods of Chapter 4, and no further details are needed here. Numerical results relating to this example are given in section 7.4.7.

7.4.4 *Risk convergence*

The risk convergence criterion was introduced for compound decisions as an analogue of asymptotic optimality in EB theory; see Robbins (1951), Samuel (1965). For a typical CD rule, δ, the ith stage decision is $\delta_i(x)$ with the corresponding average compound risk $R^{(k)}\{\boldsymbol{\delta}^{(k)}, \boldsymbol{\lambda}\}$ as defined in (7.4.2) with $\boldsymbol{\delta}^{(k)} = (\delta_1, \delta_2, \ldots, \delta_k)$ in the place of $\hat{\boldsymbol{\lambda}}$. Suppose that the parameter space of every component λ_i is Ω. Then the parameter space of $\boldsymbol{\lambda}$ is $\Omega^{(k)}$, the k-fold Cartesian product of Ω. Let G_k be a distribution over $\Omega^{(k)}$ which assigns probability s/k to $\boldsymbol{\lambda} \in \Omega^{(k)}$ when $\lambda_i = \lambda$ for s values of i; $i = 1, 2, \ldots, k$; $s = 0, 1, \ldots, k$. Then for a typical component, the average risk is

$$R(\boldsymbol{\delta}, \boldsymbol{\lambda}) = \int E_F L\{\boldsymbol{\delta}(x), \boldsymbol{\lambda}\}\, dG_k(\lambda), \tag{7.4.27}$$

where F is the d.f. of X for given $\boldsymbol{\lambda}$. Thus the optimal estimator for the component problem, i.e. minimizing $R(\boldsymbol{\delta}, \boldsymbol{\lambda})$, is the 'Bayes' decision δ_k with respect to the 'prior' G_k, and let this minimum average component risk be $R(\delta_k)$.

Now let the parameter space $\Omega^{(k)}$ be partitioned into equivalence classes by means of the empirical d.f. G_k and suppose that $Z(G_k)$ be the equivalence class of all $\boldsymbol{\lambda} \in \Omega^{(k)}$ which have the specified H as their empirical distribution. If δ_k is used at every component problem it incurs the average component risk $R(\delta_k)$ at every $\boldsymbol{\lambda} \in Z(G_k)$. No simple rule $\boldsymbol{\delta}^{(k)}$ exists which satisfies

$$\sup_{\boldsymbol{\delta}^{(k)} \in Z(G_k)} R^{(k)}(\boldsymbol{\delta}^{(k)}, \boldsymbol{\lambda}) \leqslant R(\delta_k)$$

unless δ_i is some version of δ_k in which case equality holds (Samuel, 1965). This means effectively that non-compound rules cannot minimize the average compound risk, thus justifying compound rules. Samuel (1965) also advanced that CD rules which are risk convergent should be sought.

The risk convergence property: A compound decision rule is said to be risk convergent if for every infinite sequence $\boldsymbol{\lambda}_0 = (\boldsymbol{\lambda}_1, \boldsymbol{\lambda}_2, \ldots)$ and every $\varepsilon > 0$ there exists an inter $N(\boldsymbol{\lambda}_0, \varepsilon)$ such that for all $k > N(\boldsymbol{\lambda}_0, \varepsilon)$

$$R^{(k)}(\boldsymbol{\delta}^{(k)}, \boldsymbol{\lambda}^{(k)}) - R(\delta_k) < \varepsilon,$$

where $\boldsymbol{\lambda}^{(k)}$ is the initial k-vector of $\boldsymbol{\lambda}_0$. Samuel (1965) showed that, under fairly general conditions, no non-compound rule exists which possesses the risk convergence property.

There has been much work published on the rates of convergence of risk-convergent rules. The direct relevance of this type of study to practical applications is not obvious, and will not be dealt with in any detail here.

7.4.5 *Decisions between $q \geqslant 2$ hypotheses*

In this section we consider a more general formulation of the compound decision problem. The r.v. X_i has p.d.f. $f(x|\lambda_i)$ depending on the parameter λ_i, $i = 1, 2, \ldots, k$. Each component parameter λ_i belongs to one of the regions $\Omega_1, \Omega_2, \ldots, \Omega_k$ which constitute a partition of the common parameter space Ω of the λ_i. Thus we have a set of q composite hypotheses H_r: $\lambda_i \in \Omega_r$, $r = 1, 2, \ldots, q$ for each λ_i. On the basis of the observed x_i the unknown λ_i must be assigned to one of $\Omega_1, \Omega_2, \ldots, \Omega_k$, i.e. one of $H_1, H_2, \ldots, H_k$ must be chosen.

Let $\delta(x)$ be a non-compound decision rule which partitions the sample space of X into regions $A_1, A_2, \ldots, A_q$ such that H_j is chosen when $x \in A_j$. The loss incurred by using $\delta(x)$ when the parameter value is λ is $L\{\delta(x), \lambda\}$. The expected loss in repeated x-sampling is the average compound risk,

$$R^{(k)}(\boldsymbol{\delta}^{(k)}, \boldsymbol{\lambda}) = k^{-1} E_D \sum_{i=1}^{k} L\{\delta(X_i), \lambda_i\},$$

where, as in section 7.4.2(b), G_k is the empirical d.f. of the $\boldsymbol{\lambda}_i$. To obtain the optimal non-compound CD rule we find the 'Bayes' rule with 'prior' G_k. Formally this is the same as the EB problem with G_k in the place of G.

As in other CD problems, the key to constructing applicable CD rules is a satisfactory method of estimating G_k. Once such an estimate has been found it can be substituted for G_k to derive a CD rule. Risk convergence and rates of convergence have been discussed by van Ryzin (1966a) in this setting, generalizing work on two simple hypotheses by Hannan and Robbins (1955) and Hannan and van Ryzin (1965).

7.4.6 *Compound decision rules; sequential case*

(a) Introduction

Almost every CD rule that has been developed has a sequential counterpart. The work of Samuel (1965) dealing with the non-existence of non-compound rules that are risk convergent also covers the sequential case. One reason for this simultaneous development is the perceived practical potential of the sequential procedures.

Let us imagine that, instead of all k observations being in hand before the decisions regarding the λ_i values have to be made, the results $x_1, x_2, \ldots$ are obtained sequentially. At every stage the decision $\delta(x_i)$ has to be taken immediately. We are still interested in minimizing the collective expected risk over all k decisions. Denote by $\delta_k(x)$ the 'Bayes' rule for minimizing the expected loss when all decisions are made simultaneously. Suppose now that the rule $\delta_{k-1}(x)$ is used for the first $k-1$ decisions, while $\delta_N(x)$ is employed for the Nth one. The total expected loss is then

$$\sum_{j=1}^{k-1} \int L[\delta_{k-1}(x_j), \lambda_j] dF(x_j | \lambda_j) + \int L[\delta_k(x_k), \lambda_k] dF(x_k | \lambda_k)$$

$$\leqslant \sum_{j=1}^{k-1} \int L[\delta_k(x_j), \lambda_j] dF(x_j | \lambda_j) + \int L[\delta_k(x_k), \lambda_N] dF(x_k | \lambda_k)$$

$$= k E_X C(\delta) \leqslant k E_X C(T), \tag{7.4.28}$$

where T is a conventional rule. Repeated application of (7.4.28) shows that if the new Bayes rule is used at every stage, the resulting overall expected loss will not exceed either $E_X C(\delta_k)$ or $E_X C(T)$. Observe that the symbol $C(\cdot)$ has the same meaning as before. Thus, denoting the sequentially adjusted Bayes rule by 'seq. δ_k', equation (7.4.28) states that

$$E_X C(\text{seq. } \delta_k) \leqslant E_X C(\delta_k) \leqslant E_X C(T).$$

In practice, implementation of a sequential Bayes rule may usually be regarded as an impossibility, but the result given by (7.4.28) has significant implications for the CD problem. They may be stated as follows: if $\hat{\delta}_k$ is a non-sequential CD rule such that $\hat{\delta}_k \to \delta_k$, in probability, as $k \to \infty$, then, for k sufficiently large, $E_X L[\hat{\delta}_k(x_j), \lambda_j]$ will be close to $E_X L[\delta_k(x_j), \lambda_j]$. Now suppose we use the rule T for all decisions when $j \leqslant n_0$ and the CD rule $\hat{\delta}_j$ for $j = n_0 + 1, n_0 + 2, \ldots, k$. Then if n_0 and k are large enough, with n_0/k decreasing as N increases, the effect on the overall expected loss of using T for the first n_0 decisions will become negligible as k increases, while the expected loss for the last $k - n_0$ decisions will be close to the 'sequential Bayes' expected loss for these decisions. The combined effect will be to produce an overall expected loss not exceeding $E_X C(T)$.

(b) Two simple hypotheses

We consider the case of two simple hypotheses based on the $N(\lambda, 1)$ kernel distribution with fixed values $\lambda^{(1)}$ and $\lambda^{(2)}$ of λ representing the two hypotheses. Let the sequential parameter values be $\lambda_1^*, \lambda_2^*, \ldots$, where every λ is either $\lambda^{(1)}$ or $\lambda^{(2)}$. In this case the rule δ_k refers to the Bayes cut-off ξ_k, individual losses are $e_i^{(1)}$ and $e_i^{(2)}$ as defined above.

Let m_1 denote the number of $\lambda^{(1)}$ values among the first n values $\lambda_j^*, j = 1, 2, \ldots, n$. For any $\theta_1 = m_1/n$ and arbitrary small $\varepsilon, \varepsilon' > 0$ we know that

$$|\bar{x} - \theta_1 \lambda^{(1)} - \theta_2 \lambda^{(2)}| < \varepsilon \qquad \text{with probability } (1 - \varepsilon'),$$

for $n \geqslant n(\varepsilon, \varepsilon', \theta_1)$. Hence, noting that the expected losses are given by (7.4.23), that the correlation between any x_i and $\bar{x} \to 0$ as $n \to \infty$, and that the losses are bounded by 0 and 1, we have

$$|E_X L[\hat{\delta}_n(x_n), \lambda_n^*] - E_X L[\delta_n(x_n), \lambda_n^*]| < \varepsilon''(\varepsilon, \varepsilon', \theta_1)$$

for $n \geqslant n(\varepsilon, \varepsilon', \theta_1)$, where $\varepsilon''(\varepsilon, \varepsilon', \theta_1) \to 0$ as $\varepsilon, \varepsilon' \to 0$. Putting $n_0 =$

$\max_{\theta_1} n(\varepsilon, \varepsilon', \theta_1)$, we have

$$|E_X L[\hat{\delta}_n(x_n), \lambda_n^*] - E_X L[\delta_n(x_n), \lambda_n^*]| < \varepsilon_0'',$$

uniformly for $n \geq n_0$.

Now let the rule T be used for the first n_0 of our k decisions. Then the average expected loss over all k decisions is at most

$$\sum_{j=1}^{n_0} \{E_X L[T, \lambda_j^*] - E_X L[\delta_j, \lambda_j^*]\}/k + E_X C(\text{seq. } \delta_k)$$
$$+ (k - n_0)\varepsilon_0''/k \leq E_X C(\delta_k), \quad \text{for } k \text{ sufficiently large.}$$

This also implies, of course, that $E_X C(\text{seq. } \delta_k) \leq E_X C(T)$, for k large enough.

Samuel (1963, 1964), has studied this case in detail, emphasizing the practically very important fact that the order in which the λ values occur does not affect the conclusion that the sequential CD rule will be 'better' than a conventional rule for large k. Some numerical results relating to the performance of the sequential CD rule are given by Samuel (1964).

(c) A finite number of simple hypotheses, composite hypotheses

Since the formulation in section 7.4.6(a) is not restricted to two simple hypotheses, it is clear that the ideas of section 7.4.6(b) can be extended to the case of a finite number of simple hypotheses. Essentially the problem revolves about the determination of consistent estimates of the proportions in which the different parameter values occur, i.e. of G_k (Samuel, 1966).

In the case of the composite hypotheses where the parameter values are not restricted to a finite number of given points, the arguments applicable to the case of a finite number of hypotheses can still be used. If the parameter values are restricted to a finite interval an extension of the 'finite' argument, using standard limiting processes, can be used to show that a sequential CD rule will be better than the conventional rule for large k provided a sequential rule can be developed. Since the main question is again the estimation of G_j, $j = n_0 + 1, \ldots, k$ the idea of approximating G_j, developed for EB procedures, can be used (Maritz, 1968). Restriction of the parameter values to a finite interval means, in practice, that the sequential CD rule will be relatively 'good' if the ultimate spread of parameter values is small relative to $\text{var}(x|\lambda)$. The next section gives some numerical data for a case of two composite hypotheses.

7.4.7 *Performance of CD rules in the case $f(x|\lambda)=N(\lambda,1)$*

(a) Non-sequential case

Two series of examples have been prepared. Two finite populations of $\lambda_{(i)}$, $i=1,\ldots,100$, were generated by sampling from $N(0,\sigma^2)$ populations with $\sigma^2=0{\cdot}1$, $0{\cdot}5$. The point λ_0, representing the division between H_1 and H_2, was chosen such that the proportion, $G_k(\lambda_0)$, of $\lambda_{(i)}\leqslant\lambda_0$ was successively $0{\cdot}5, 0{\cdot}6,\ldots,0{\cdot}9$. The optimum (Bayes) cut-off ξ_{G_k} was determined by finding x satisfying

$$\sum_{\lambda_{(j)}}\lambda_{(j)}f(x_j|\lambda_{(j)})\Big/\sum_{\lambda_{(j)}}f(x_j|\lambda_{(j)})=\tfrac{1}{2}.$$

Sets of observations $x_1,\ldots,x_k$ were generated repeatedly by sampling from $N(\lambda_{(i)},1)$ populations, $i=1,\ldots,k$, and, for every set, estimates $\hat{\lambda}_1,\ldots,\hat{\lambda}_5$ were obtained by ML. The actual losses using $\hat{\xi}_5, \xi_G$ and $T=\lambda_0$ were found for every set of observations and the expectation $E_X C(\hat{\xi}_5)$, of the loss in repeated x-sampling, the λ_i remaining fixed, was estimated by averaging observed values of the losses. Table 7.2 gives a summary of the estimated values of

$$E_X C(\hat{\xi}_k),\quad E_X C(\xi_{G_k})\quad\text{and}\quad E_X C(T).$$

For the larger $G_k(\lambda_0)$ we see that $E_X C(\hat{\xi}_5)$ is reasonably close to $E_X C(\xi_{G_k})$, and less than $E_X C(T)$. However, for smaller $G_k(\lambda_0)$, when $E_X C(\xi_G)$ is close to $E_X C(T)$, it may not be advantageous to use the smooth CD approach.

Table 7.2 *Results for non-sequential CD decision cases with* $k=100$ *and values* $\lambda_1,\ldots,\lambda_k$, *obtained by sampling from* $N(0,\sigma^2)$ *populations;* $F(x|\lambda)=N(\lambda,1)$

	$G_k(\lambda_0)$	$E\{C(\xi_5)\}$	$E\{C(\xi_{G_k})\}$	$E\{C(T)\}$
$\sigma^2=0{\cdot}1$	0·5	0·472 ± 0·004	0·417 ± 0·006	0·417 ± 0·005
	0·6	0·454 ± 0·009	0·388 ± 0·004	0·417 ± 0·005
	0·7	0·365 ± 0·015	0·301 ± 0·002	0·406 ± 0·006
	0·8	0·252 ± 0·015	0·200 ± 0·001	0·378 ± 0·006
	0·9	0·116 ± 0·006	0·100 ± 0·000	0·339 ± 0·005
$\sigma^2=0{\cdot}5$	0·5	0·353 ± 0·013	0·310 ± 0·007	0·310 ± 0·007
	0·6	0·314 ± 0·011	0·278 ± 0·008	0·291 ± 0·009
	0·7	0·279 ± 0·011	0·256 ± 0·007	0·287 ± 0·006
	0·8	0·234 ± 0·018	0·186 ± 0·006	0·270 ± 0·010
	0·9	0·105 ± 0·006	0·098 + 0·001	0·214 ± 0·010

(b) Sequential case

Parameter values were obtained by sampling from (i) a $N(0, \sigma^2)$ population and (ii) a population with the exponential d.f.

$$G(\lambda) = 0 \text{ for } \lambda \leqslant 0, \qquad G(\lambda) = 1 - e^{-\lambda B} \text{ for } \lambda > 0.$$

In theory the $\lambda_{(1)}, \ldots, \lambda_{(k)}$ can be a quite arbitrary collection of numbers, but the populations which we have used are thought to represent reasonable practical 'extremes'.

Because of lengthy computations required when recalculating $\hat{\xi}_{r,n}$ for every additional observation, it was decided to adopt a scheme of 'multiple sampling' in our examples. Let us suppose that $k = nm$. Then a 'good' conventional decision rule is used for the first n observations. After obtaining the next n observations $\hat{\xi}_{r,2n}$ is computed and used for the second group of n observations, and so on, until $\hat{\xi}_{r,k}$ is used for the

Table 7.3 *Results for sequential CD cases with* $k = 50$, *values* $\lambda_1, \ldots, \lambda_k$, *obtained by sampling from distributions (i)* $N(0, \sigma^2)$, *(ii)* $G(\lambda) = 1.-\exp(-\lambda/\sigma)$, $\lambda \geqslant 0$; $\lambda_{(1)}, \ldots, \lambda_{(k)}$, *are randomly ordered;* $F(x, \lambda) = N(\lambda, 1)$

	$G_k(\lambda_0)$	$E\{C(\text{seq. } \xi_{5,k})\}$	$E\{C(\xi_{G_k})\}$	$E\{C(T)\}$
		Distribution (i)		
$\sigma^2 = 0{\cdot}1$	0·5	0·454 ± 0·020	0·412 ± 0·014	0·421 ± 0·013
	0·6	0·435 ± 0·027	0·382 ± 0·012	0·404 ± 0·015
	0·7	0·339 ± 0·021	0·297 ± 0·009	0·398 ± 0·016
	0·8	0·238 ± 0·017	0·198 ± 0·002	0·360 ± 0·017
	0·9	0·169 ± 0·013	0·100 ± 0·000	0·323 ± 0·013
$\sigma^2 = 0{\cdot}5$	0·5	0·380 ± 0·014	0·315 ± 0·019	0·315 ± 0·019
	0·6	0·372 ± 0·017	0·302 ± 0·017	0·316 ± 0·017
	0·7	0·336 ± 0·016	0·247 ± 0·010	0·307 ± 0·018
	0·8	0·234 ± 0·016	0·186 ± 0·007	0·281 ± 0·014
	0·9	0·132 ± 0·015	0·100 ± 0·003	0·251 ± 0·009
		Distribution (ii)		
$\sigma^2 = 0{\cdot}1$	0·5	0·454 ± 0·011	0·407 ± 0·013	0·407 ± 0·012
	0·6	0·481 ± 0·019	0·380 ± 0·009	0·400 ± 0·015
	0·7	0·424 ± 0·026	0·315 ± 0·005	0·400 ± 0·012
	0·8	0·279 ± 0·017	0·201 ± 0·001	0·382 ± 0·016
	0·9	0·184 ± 0·021	0·100 ± 0·000	0·349 ± 0·015
$\sigma^2 = 0{\cdot}5$	0·5	0·350 ± 0·017	0·326 ± 0·014	0·329 ± 0·015
	0·6	0·354 ± 0·021	0·287 ± 0·016	0·322 ± 0·017
	0·7	0·294 ± 0·020	0·235 ± 0·014	0·271 ± 0·016
	0·8	0·255 ± 0·022	0·169 ± 0·009	0·260 ± 0·016
	0·9	0·112 ± 0·005	0·088 ± 0·006	0·166 ± 0·011

last n observations. In the light of (7.4.28), this scheme may be expected to give slightly worse results than the sequential scheme in which $\hat{\xi}_{r,n}$ is recalculated with every new observation.

In our examples, populations with $k=50$ were generated as indicated above and we have used $n=10$. In case (i) we used $\sigma^2=0{\cdot}1$ and $\sigma^2=0{\cdot}5$, and in case (ii) $1/B=0{\cdot}1$ and $1/B=0{\cdot}5$. As before, λ_0 was selected to give $G_k(\lambda_0)=0{\cdot}5, 0{\cdot}6, \ldots, 0{\cdot}9$, respectively. The entries in Table 7.3 are Monte Carlo estimates of $E_X W(\text{seq. } \hat{\xi}_{r,k})$ obtained by repeatedly generating new x-values by adding new random normal deviates to the fixed $\lambda_{(j)}$. The actual losses were computed for every new set of observations using, in each case, the new sequence of values of $\hat{\xi}_{r,n}$. The estimates given in Table 7.3 are averages of these losses.

From (7.4.28) we see that the greatest differences between $E_X C(\text{seq. Bayes})$ and $E_X(\xi_{G_k})$ will occur when the G_n for the subpopulations of the $\lambda_{(1)}, \ldots, \lambda_{(k)}$ differ most from G_k. One obvious way in which we can produce such differences is to arrange the $\lambda_{(j)}$ in order of magnitude. The results of Table 7.4 were obtained on this basis, exactly the same $\lambda_{(j)}$ being used as before. Comparison of Tables 7.3 and 7.4 show that the values of $E_X C(\cdot)$ for the non-random sequences of $\lambda_{(j)}$ tend to be lower. The trend is more noticeable for the large values of the variance of the population of $\lambda_{(j)}$, as one may expect. The importance of these results from the point of view of application of smooth sequential CD procedures in problems of acceptance sampling, when there may be systematic variation in quality is obvious.

Table 7.4 *Results for sequential CD cases with.* $k=50$, *values* $\lambda_1, \ldots, \lambda_k$, *obtained by sampling from distributions (i)* $N(0, \sigma^2)$, *(ii)* $G(\lambda)=1-\exp(-\lambda/\sigma)$, $\lambda \geqslant 0$; $\lambda_{(1)}, \ldots, \lambda_{(k)}$ *arranged in increasing order of magnitude;* $F(x|\lambda)=N(\lambda, 1)$

	$G_k(\lambda_0)$	$E\{C(\text{seq. } \hat{\xi}_{5,k})\}$		$G_k(\lambda_0)$	$E\{C(\text{seq. } \hat{\xi}_{5,k})\}$
	Distribution (i)			Distribution (ii)	
$\sigma^2=0{\cdot}1$	0·5	$0{\cdot}336 \pm 0{\cdot}026$	$\sigma^2=0{\cdot}1$	0·5	$0{\cdot}421 \pm 0{\cdot}035$
	0·6	$0{\cdot}319 \pm 0{\cdot}023$		0·6	$0{\cdot}399 \pm 0{\cdot}038$
	0·7	$0{\cdot}338 \pm 0{\cdot}015$		0·7	$0{\cdot}348 \pm 0{\cdot}024$
	0·8	$0{\cdot}268 \pm 0{\cdot}013$		0·8	$0{\cdot}279 \pm 0{\cdot}012$
	0·9	$0{\cdot}179 \pm 0{\cdot}018$		0·9	$0{\cdot}186 \pm 0{\cdot}018$
$\sigma^2=0{\cdot}5$	0·5	$0{\cdot}280 \pm 0{\cdot}027$	$\sigma^2=0{\cdot}5$	0·5	$0{\cdot}310 \pm 0{\cdot}024$
	0·6	$0{\cdot}267 \pm 0{\cdot}027$		0·6	$0{\cdot}271 \pm 0{\cdot}023$
	0·7	$0{\cdot}249 \pm 0{\cdot}015$		0·7	$0{\cdot}217 \pm 0{\cdot}014$
	0·8	$0{\cdot}190 \pm 0{\cdot}015$		0·8	$0{\cdot}138 \pm 0{\cdot}012$
	0·9	$0{\cdot}116 \pm 0{\cdot}009$		0·9	$0{\cdot}100 \pm 0{\cdot}010$

7.5 General discussion

While the different approaches mentioned above make use of different probability models, each of them treats the same data set, i.e. that generated in the EB sampling scheme. They can be placed in a spectrum of inferential techniques for data of this sort. At one extreme there is the purely frequentist approach represented by compound decision theory. It can be regarded as the most flexible, assuming no prior distribution of the parameters. The only requirement is knowledge of component loss functions and the convention of taking their average as a reasonable compound loss for simultaneous decision making. At the other end of the spectrum is the full Bayesian (FB) treatment of the EB scheme. Here a special form of prior distribution is assumed not only for the unobserved λ in the second stage of the EB scheme, but also for the **hyperparameters** of the prior distribution of Λ. Exact FB solutions are generally complicated and seem to be not directly useful for practical applications. However, an approximate general solution is available which obviates the need to specify the hyper-prior distribution explicitly. This approximate solution turns out to be formally the same as the EB solution. A second approximate solution that has been used, especially in the linear model situation, employs a non-informative hyper prior. The usual problems associated with choosing non-informative priors remain.

The EB method and other alternatives to it may be regarded as falling somewhere in the middle of the spectrum, being essentially compromises between the two extremes. The two likelihood based methods, modified maximum likelihood (MML), and empirical regression estimation (ERE), are very similar to each other, the latter being historically important because of its early appearance. These two methods do not employ loss functions and they are not decision theoretic in nature. They can be unified as special cases of a more general method which uses a summary function of the posterior distribution of Λ. They do not assume a distribution for the hyperparameters, and produce results which are virtually the same as the corresponding EB solutions. Indeed, the inferential EB approach which does not use loss functions is practically identical to these likelihood-based methods.

While the CD approach is attractive from the frequentist point of view some statisticians have difficulty in accepting it because detailed studies have hinged on the risk convergence property which is not seen as having a natural practical relevance. The CD rules are found to be formally identical to EB rules, the main difference being in the criteria on which their performance is judged. Using a CD rule for a particular decision implies using in it observations whose distributions do not depend on the parameter of that problem. This latter aspect is unsatisfactory from the viewpoint of likelihood inference and the notion of sufficiency; see Copas (1969).

The crucial notion in the FB approach is exchangeability in the parameter sequence. It has been argued that in many simultaneous decision problems the parameters could be exchangeable in that the prior opinion of any particular parameter is the same as that for any other member of the sequence. Lindley and Smith (1972) exploited the fact that one way of ensuring an exchangeable distribution for $\boldsymbol{\lambda}$ is to take it to be of the form (1.14.9). The distributions $G(\boldsymbol{\lambda}|\boldsymbol{\phi})$ and $P(\boldsymbol{\phi})$ are arbitrarily chosen. Initial developments deal mainly with normal data, prior and hyper-prior distributions, but wider applicability has been demonstrated. The exponential data distribution was treated by Deely and Lindley (1981) who showed also that the usual EB solutions with parametric assumptions for $G(\boldsymbol{\lambda}|\boldsymbol{\phi})$ can be regarded as first-order approximations to the FB solutions. The possibility of second-order solutions has been indicated. So far no comparative study has been made of EB and FB solutions in terms of the average risk that is commonly used in the assessment of EB rules.

Finally we reconsider briefly the circumstances in which one may seriously regard an EB approach as a competitor to more conventional analyses. The simplest case of single observations x_i on normally distributed random variables occurring at realizations λ_i of Λ, $i = 1, 2, \ldots, n$, gives a guide. If $\text{var}(X_i|\lambda_i) = \sigma^2$ and $\text{var}(\Lambda) = \sigma_G^2$, the arguments leading to (7.4.17) indicate that one needs $\sigma^2 > \sigma_G^2$. Clearly, if σ_G^2 is zero, or close to zero, pooling of the data and estimating every λ_i by the grand mean is indicated. On the other hand, if the prior distribution is very diffuse little can be gained by a Bayesian approach. Aside from these considerations there is also the matter of the amount of previous data available for constructing the EB rule. This question has been dealt with in detail at various points. Essentially the EB rules obviously may not be good unless they are reasonably accurate estimates of the Bayes rules.

CHAPTER 8

Applications of EB methods

8.1 Introduction

In this chapter we present summaries of some of the applications of EB methods that have been published. In most branches of statistics there are data sets that are used repeatedly to demonstrate new techniques or modifications of new ones. The reason is that those data sets are considered to be particularly suitable in the sense of being generated by processes closely approximating the models on which the techniques are based. The same is true of EB methods.

Data from non-trivial practical applications seldom follow the somewhat idealized patterns assumed in the development of techniques. Here again, the EB approach is no exception. The EB sampling schemes which have been studied in much detail in previous chapters, and in a large body of other EB literature, are relatively simple, and few actual data sets follow those patterns exactly. This does not, of course, make the study of those simpler schemes useless. The results of those studies, while not necessarily providing answers in particular practical problems, do give valuable guides as to the potential usefulness of the methods, and in this sense, provide qualitative answers.

The examples which have been selected show that practical situations approximating the EB sampling scheme do arise, and they are individually interesting because they illustrate aspects of the theory. Morris (1983b) surveys some important applications of EB methods; some of the examples quoted by Morris are also given in some detail in this chapter. Chapter 2 deals with mixtures, a large topic in its own right, and we could have included many examples on mixtures. By and large, only examples with a clear empirical Bayes connotation have been included, i.e. examples in which decisions about individual components are of interest. Take as an illustration the case of growth curve analysis of which an example is included. If a study is aimed at searching for or quantifying differences between groups of subjects it is not clear that EB methods are relevant. On the other hand, individual estimates may be important, for instance if they are to form part of the basis for clinical advice; the example of Berkey (1982) is a case in point, and others are easily identified.

We have selected several examples showing original data sets in enough detail to make possible the actual calculations associated with EB estimation. They have been selected partly for being of manageable size, for illustrating points of methodology, or for being intrinsically interesting. Some good examples are quoted without data sets, either because they are not readily obtainable or because they are simply too large for inclusion.

In many practical cases there will be concomitant information about the parameter values. Specifically, recall the EB sampling scheme where we have observations $(x_1, x_2, \ldots, x_n)$ when the parameter values are $(\lambda_1, \lambda_2, \ldots, \lambda_n)$. Every x_i is usually thought of as an estimate of the corresponding λ_i. Now it may happen that we also have associated with every x_i an observation c_i on a concomitant variable C. Every c_i is not necessarily an estimate of λ_i but C and Λ may not be independent, so that taking account of the observed c should improve the estimate of λ. However, the emphasis is still on estimating individual λ values, and not on exploring the relationship between Λ and C, for instance, through the regression of Λ on C. The paper by Tsutakawa, Shoop and Marienfeld (1985) pays some attention to concomitant variables in the EB context. Other examples involving concomitant variables are discussed by Raudenbush and Bryk (1985), and Fay and Herriot (1979). Some details of the use of concomitant information are given in section 4.7.

8.2 Examples with normal data distributions

8.2.1 Law school validity studies

Most of the material of this section paraphrases a paper by Rubin (1980) which treats some aspects of data supplied by the Educational Testing Service (ETS) in the USA in yearly reports to law schools participating in a selection procedure. This analysis concentrates on the problem of predicting the first-year grade average in law school (FYA) for applicant students from their Law School Aptitude Test (LSAT) scores and their undergraduate grade point averages

(UGPA). The LSAT is administered and graded by ETS, and its scores range from 200 to 800. UGPA scores range from 1·0 to 4·0. In Rubin's analysis the UGPA scores were multiplied by 200 to give them the same range as the LSAT scores. In what follows UGPA is to be taken as the original score multiplied by 200.

One important objective of the ETS reports is to predict FYA by a linear rule

$$\text{F}\hat{\text{Y}}\text{A} \propto \text{LSAT} + M \times \text{UGPA}$$

and the question for each law school is how to choose the multiplier M in this predictor. One way is for each law school to perform a least squares regression of FYA on LSAT and UGPA and to use the ratio of regression coefficients for M. In Rubin's analysis of 1973 data, in order to predict 1974 FYA, this was the starting point. Data from 83 law schools were used in this study, the results from these schools being treated like the typical sequence of 'past' results in the EB sampling scheme. Every individual result in turn is then treated as the 'current' observation in the terminology of earlier chapters.

The basic data used in the EB analysis comprises two least squares regression coefficients together with an estimated covariance matrix for each law school. As indicated above, the initial estimate of the multiplier M_i for school i is taken to be the ratio r_i of these two coefficients. For the EB analysis Rubin chose to work with $a_i = \arctan(r_i)$ because its distribution is thought to be closer to normal than that of r_i, normality being an underlying assumption in the analysis. By the standard delta method an estimated variance, s_i^2, is calculated for each a_i. These variances differ from school to school. Taking them to be the actual variances we have a situation which is equivalent to the case of a normal data distribution but variable m_i discussed in section 3.9.3. Table 8.1 has been extracted from two tables in Rubin (1980). It shows the least squares estimate of M_i, a_i, s_i^2 for 1973 in columns (2)–(4). Note that, in Rubin's notation, least squares $M_i = 200/r_i$. Column (5) shows the EB estimate of M_i for each school.

Calculation of the EB estimates is according to the following theory. The distribution of a_i for given M_i is $N(\mu_i, s_i^2)$, and the μ_i values are taken to be sampled from a $N(\mu_*, \sigma_*^2)$. Without giving the details here, it should be mentioned that Rubin performed some careful preliminary analyses of the data in order to establish that this model is reasonable. The Bayes estimate of μ_i is $\lambda_i a_i + (1 - \lambda_i)\mu_*$, where $\lambda_i =$

Table 8.1 *Data used in law school validity studies including least squares and EB estimates of multipliers* M_i.
(1) = School ID (2) = Multiplier = M_i *(3) =* $a_i = \arctan (M_i/200)$
(4) = s.d. $(a_i) = s_i$ *(5) = EBE of* M_i

(1)	(2)	(3)	(4)	(5)	(1)	(2)	(3)	(4)	(5)
1	102	1·099	0·148	116	42	76	1·209	0·115	99
2	217	0·745	0·201	146	43	74	1·216	0·110	97
3	211	0·759	0·152	154	44	157	0·906	0·104	143
4	70	1·234	0·159	105	45	96	1·123	0·151	114
5	82	1·184	0·143	107	46	132	0·987	0·119	129
6	149	0·930	0·099	139	47	81	1·188	0·108	100
7	2507	0·080	0·372	151	48	212	0·757	0·176	149
8	119	1·032	0·157	124	49	142	0·954	0·111	134
9	175	0·853	0·172	140	50	114	1·054	0·195	123
10	266	0·645	0·171	161	51	151	0·924	0·182	133
11	179	0·842	0·119	150	52	124	1·016	0·135	125
12	153	0·918	0·138	137	53	123	1·021	0·097	124
13	125	1·012	0·189	126	54	133	0·983	0·114	130
14	127	1·004	0·158	127	55	199	0·787	0·128	156
15	125	1·014	0·211	126	56	158	0·901	0·186	135
16	136	0·974	0·128	131	57	105	1·088	0·173	119
17	136	0·973	0·098	132	58	155	0·912	0·094	143
18	111	1·066	0·173	121	59	163	0·886	0·146	140
19	222	0·733	0·181	150	60	132	0·987	0·174	128
20	82	1·182	0·147	108	61	88	1·158	0·166	113
21	84	1·174	0·108	102	62	151	0·924	0·084	142
22	89	1·152	0·122	108	63	126	1·008	0·138	126
23	89	1·152	0·127	108	64	−24	1·697	0·286	100
24	244	0·687	0·171	157	65	168	0·873	0·259	133
25	152	0·922	0·094	141	66	81	1·187	0·101	99
26	81	1·186	0·077	94	67	132	0·989	0·125	129
27	99	1·111	0·120	112	68	132	0·989	0·132	129
28	104	1·089	0·265	122	69	124	1·016	0·094	125
29	62	1·268	0·194	107	70	202	0·782	0·128	157
30	100	1·107	0·148	116	71	179	0·841	0·197	139
31	158	0·902	0·115	142	72	78	1·198	0·139	105
32	306	0·579	0·076	234	73	198	0·790	0·104	163
33	142	0·955	0·192	130	74	125	1·013	0·088	125
34	115	1·048	0·146	122	75	254	0·666	0·114	182
35	31	1·419	0·124	77	76	152	0·922	0·150	136
36	40	1·375	0·167	94	77	150	0·926	0·105	139
37	94	1·130	0·103	108	78	173	0·858	0·178	139
38	92	1·138	0·146	112	79	140	0·959	0·116	133
39	130	0·995	0·131	128	80	69	1·239	0·161	105
40	68	1·243	0·126	97	81	141	0·956	0·314	128
41	89	1·151	0·104	105	82	72	1·227	0·119	98

$(1 + s_i^2/\sigma_*^2)^{-1}$. The EB estimate is obtained on replacing μ_* and σ_* in these expressions by their estimates as calculated from the data. Rubin estimates μ_* and σ_* by the method of maximum likelihood, i.e. maximizing

$$\prod_{i=1}^{82} [2\pi(s_i^2 + \sigma_*^2)]^{-1/2} \exp[-\tfrac{1}{2}(a_i - \mu_*)^2/(s_i^2 + \sigma_*^2)]$$

w.r.t. μ_* and σ_*. The actual method of calculating the estimates was by using the EM algorithm. See also sections 2.8.2 and 2.9. For the 1973 data shown in Table 8.1 the estimates of μ_* and σ_*^2 are, respectively, 1·008 and 0·0139. For school No. 1 we have $a_1 = 1{\cdot}099$, $s_1 = 0{\cdot}148$, giving the EB estimate of $\mu_1 = 1{\cdot}0433$, and the EB estimate of $M_1 = 200/\tan(1{\cdot}0433) = 116{\cdot}5$·

As Rubin points out, charateristically of EB estimates, the estimates in column (5) do not fluctuate as wildly as those in column (2). Rubin also reports on a validation study of the estimates by looking at the predictions of 1974 performances. It turns out that the EB estimates do perform better according to this important separate criterion.

To conclude this section we remark that the estimation of μ_* and σ_* could have been done by the method of moments, in the manner of the discussions of section 3.9. These estimates are given by

$$(1/82)\sum(a_i - \bar{a})^2/(\sigma_*^2 + s_i^2)^{-1} = 1$$
$$\bar{a} = \sum a_i(\sigma_*^2 + s_i^2)/\sum 1/(\sigma_*^2 + s_i^2).$$

The results calculated from the 1973 data are $\bar{\mu}_* = 1{\cdot}008$, $\bar{\sigma}_*^2 = 0{\cdot}0136$. They are quite close to the ML estimates. Since the method of moments estimate of σ_*^2 is greater than the ML estimate, the corresponding EB estimates are somewhat more variable, but they are still far less variable than the least squares estimates of the multipliers.

8.2.2 *Fitting of growth curves*

Many studies to do with the fitting of growth curves concern human subjects on whom a response variable Y is observed at a succession of age points. If, for example, Y is the height of a child, a plot of Y against age produces a set of points through which one might fit a growth curve. Obviously the associated methodology need not be restricted in application to data of this sort, but the terminology is convenient.

A common type of model in growth curve analysis is that the observed value of the response variable at time t is $y(t) = g(\theta) + e$, where e is taken to be a normally distributed random error, independent at every observation. Berkey (1982) reports a growth study in which the model $g(\boldsymbol{\theta})$ is the Jenss curve, $y(t) = \alpha_0 + \alpha_1 t - \exp(\beta_0 + \beta_1 t)$, thought to be suitable for describing the growth of young children. Here the parameter $\boldsymbol{\theta} = (\alpha_0, \alpha_1, \beta_0, \beta_1)$ is a 4-dimensional vector whose prior distribution is assumed to be $N(\boldsymbol{\mu}_G, \boldsymbol{\Sigma}_G)$. The essentially empirical Bayesian part of Berkey's analysis has to do with estimation of the parameters μ_G and Σ_G from the data on 218 children. On each of these children there were between 11 and 14 (y, t) observations from which least squares estimates of the elements of $\boldsymbol{\theta}$ were calculated, yielding 218 of the 4-dimensional vectors of estimates θ_i, $i = 1, 2, \ldots, 218$. Taking these estimates to be individually unbiased for their respective parameters, the mean vector $\boldsymbol{\mu}_G$ is estimated by the mean of the θ_i values. The matrix $\boldsymbol{\Sigma}_G$ can be estimated by first obtaining the sample covariance matrix $\mathbf{V}$ of the observed least squares estimates. There is a conditional covariance matrix of the least squares estimates for each subject. The average of these matrices is subtracted from $\mathbf{V}$ to give an estimate of $\boldsymbol{\Sigma}_G$. This was the procedure followed in Berkey (1982). In practice the constraint that the estimate $\hat{\boldsymbol{\Sigma}}_G$ should be nonnegative definite may have to be imposed.

For Berkey's data we have

$$\hat{\boldsymbol{\Sigma}}_G = \begin{pmatrix} 24.199 & -1{\cdot}840 & 0{\cdot}620 & 0{\cdot}822 \\ -1{\cdot}840 & 0{\cdot}595 & -0{\cdot}059 & -0{\cdot}092 \\ 0{\cdot}620 & -0{\cdot}059 & 0{\cdot}024 & 0{\cdot}021 \\ 0{\cdot}822 & -0{\cdot}092 & 0{\cdot}021 & 0{\cdot}058 \end{pmatrix}$$

$$\bar{\boldsymbol{\Sigma}} = \begin{pmatrix} 2{\cdot}602 & -0{\cdot}486 & 0{\cdot}065 & 0{\cdot}181 \\ -0{\cdot}486 & 0{\cdot}093 & -0{\cdot}012 & -0{\cdot}032 \\ 0{\cdot}065 & -0{\cdot}012 & 0{\cdot}0025 & 0{\cdot}0021 \\ 0{\cdot}181 & -0{\cdot}032 & 0{\cdot}0021 & 0{\cdot}0126 \end{pmatrix}$$

$$\hat{\boldsymbol{\mu}}_G = (77{\cdot}7785, 6{\cdot}4214, 3{\cdot}2550, -0{\cdot}9919)'.$$

Berkey reports the least squares estimates for six of the children, one of these, child number 083, has the following values:

$$(72{\cdot}746, 8{\cdot}412, 3{\cdot}033, -1{\cdot}216).$$

Unfortunately the matrix $\mathbf{\Sigma}_i$ for this subject is not given, so a matrix close to $\bar{\mathbf{\Sigma}}$ was used in a calculation of the EB estimate for this subject. The matrix actually used was $\mathbf{\Sigma}$ with the diagonal elements replaced by (3·00, 0·11, 0·0030, 0·0160). This substitution was made largely to produce a positive definite $\mathbf{\Sigma}_i$. According to formula (4.4.1), Chapter 4,

$$\text{EB estimate} = (\mathbf{\Sigma}_i^{-1} + \hat{\mathbf{\Sigma}}_G^{-1})^{-1}(\mathbf{\Sigma}_i^{-1}\hat{\boldsymbol{\theta}}_i + \hat{\mathbf{\Sigma}}_G^{-1}\hat{\boldsymbol{\mu}}_G).$$

The result is

$$(73{\cdot}880, 8{\cdot}136, 3{\cdot}072, -1{\cdot}159),$$

which differs somewhat from the result given by Berkey, not only because of the use of $\bar{\mathbf{\Sigma}}$ instead of $\mathbf{\Sigma}_i$ but also because Berkey's method of obtaining an EB estimate is to obtain a posterior mode in the manner of Lindley and Smith (1972).

8.2.3 *Predicting baseball batting averages*

The second column of Table 8.2 gives the batting averages of 18 players over their first $m = 45$ 'at bats' in the 1970 season. Multiplying these averages by 45 gives the integer scores in the third column; call these z_i. These results are extracted from Efron and Morris (1975) according to whom the $Z_i = mY_i$ can be regarded as independent binomial (m, p_i) random variables, $i = 1, 2, \ldots, k$. In order to use previously developed theory connected with the Stein estimator (see for example James and Stein, 1961), Efron and Morris work with

$$X_i = f_m(Y_i) = (m)^{1/2} \arcsin(2Y_i - 1),$$

because the distribution of X_i can be taken as approximately normal with mean $\theta_i = f_m(p_i)$ and unit variance. Developing an EB estimator of $\boldsymbol{\theta}$, the θ_i values are assumed to be independently generated by a $N(\mu, \tau^2)$ distribution. The Bayes estimator for θ_i can be written as

$$\mu + \{1 - (1 + \tau^2)^{-1}\}(X_i - \mu)$$

and an EB version is obtained when estimating μ by $\bar{x}$ and $1/(1 + \tau^2)$ by $(k - 3)/V$, where $V = \sum(x_i - \bar{x})^2$. From the data in Table 8.2 the values of $\bar{x}$ and $(k - 3)/V$ are $-3{\cdot}275$ and 0·791, giving the EB estimate

$$\bar{\delta}^1(x_i) = 0{\cdot}791\bar{x} + 0{\cdot}209x_i = 0{\cdot}209x_i - 2{\cdot}59.$$

For the purpose of this exercise the values in columns (5) and (6) are

Table 8.2 *Column descriptions: (1) Player number (2) Batting average for first 45 at bats* $= y_i$ *(3)* $z_i = 45y_i$ *(4)* $x_i = f_m(y_i)$ *(5) Batting average for the remainder of season* $= p_i$ *(6)* $\theta_i = f_m(p_i)$ *(7)* $\delta^1(x_i)$ *(8) retransformed* $\delta^1(x_i)$ *(9) binomial EBE of* p_i

(1)	(2)	(3)	(4)	(5)	(6)	(7)	(8)	(9)
1	0·400	18	−1·35	0·346	−2·10	−2·87	0·292	0·273
2	0·378	17	−1·66	0·298	−2·79	−2·94	0·288	0·272
3	0·356	16	−1·97	0·276	−3·11	−3·00	0·284	0·270
4	0·333	15	−2·28	0·222	−3·96	−3·07	0·279	0·269
5	0·311	14	−2·60	0·273	−3·17	−3·13	0·275	0·268
6	0·311	14	−2·60	0·270	−3·20	−3·13	0·275	0·268
7	0·289	13	−2·92	0·263	−3·32	−3·20	0·270	0·267
8	0·267	12	−3·26	0·210	−4·15	−3·27	0·266	0·266
9	0·244	11	−3·60	0·269	−3·23	−3·34	0·261	0·264
10	0·244	11	−3·60	0·230	−3·83	−3·34	0·261	0·264
11	0·222	10	−3·95	0·264	−3·30	−3·42	0·256	0·263
12	0·222	10	−3·95	0·256	−3·43	−3·42	0·256	0·263
13	0·222	10	−3·95	0·303	−2·71	−3·42	0·256	0·263
14	0·222	10	−3·95	0·264	−3·30	−3.42	0·256	0·263
15	0·222	10	−3·95	0·226	−3·89	−3·42	0·256	0·263
16	0·200	9	−4·32	0·285	−2·98	−3·49	0·251	0·262
17	0·178	8	−4·70	0·316	−2·53	−3·57	0·246	0·261
18	0·156	7	−5·10	0·200	−4·32	−3·66	0·241	0·259

taken as the parameter values for individual batters, and these are being estimated using the results in columns (2) to (4). Column (7) gives the EBE $\delta^1(x_i)$ of θ_i, column (8) the retransformed $\delta^1(x_i)$, i.e. an EBE of p_i. The values in column (8) differ in the third decimal place from the corresponding results in Table 2 of Efron and Morris; the differences are due to rounding at stages of the calculation.

The 'binomial EBE' values in column (9) were obtained by applying the linear EB technique of section 3.7.2 to the results in column (3). Applying formulae in section 3.7.2(c) to the results in column (3) of Table 8.2, the linear EBE is found as $\bar{p}(x_i) = 0{\cdot}2508 + 0{\cdot}001226z_i$.

The ML estimates of the p_i are the values shown in column (2). The sums of squared differences between actual and estimated p_i values are:

maximum likelihood	: 0·0754
retransformed δ^1	: 0·0214
binomial EBE	: 0·0228

Both EBEs perform notably better than the MLE in this example.

8.2.4 *Teacher expectancy and pupil IQ*

Raudenbush and Bryk (1985) examine the results of $n = 19$ independent studies in each of which the effect of teacher expectancy on pupil IQ was estimated by an 'effect size', d_i, being the mean difference between experimental and control children divided by a standard deviation pooled within groups. Table 8.3 gives values of d_i, a standard error $\hat{\sigma}_i$ of each d_i and other data. In the notation of section 8.1, $\hat{\sigma}_i^2 = \hat{\sigma}^2/m_i$.

In the analysis given by Raudenbush and Bryk the distribution of every d_i given the 'true' δ_i is assumed normal with variance v_i. In the calculations v_i is taken to be equal to the square of the standard error shown in the last column of Table 8.3. The prior distribution from which the δ_i are sampled is assumed normal, $N(\mu_G, \sigma_G^2)$.

Following Rubin (1980) one can estimate μ_G and σ_G^2 by the method of maximum likelihood, using the fact that the marginal distribution of d_i is normal $N(\mu_G, \sigma_G^2 + v_i)$. Then the MLE of σ_G^2 is given by $\hat{\sigma}_G^2$

Table 8.3 *Results of experiments assessing the effect of teacher expectancy on pupil IQ*

Study	*Weeks of prior contact*	c_i	*Effect size* d_i	$\hat{\sigma}_i$
1. Rosenthal et al. (1974)	2	2	0·03	0·125
2. Conn et al. (1968)	21	3	0·12	0·147
3. Jose & Cody (1971)	19	3	−0·14	0·167
4. Pellegrini & Hicks (1972)	0	0	1·18	0·373
5. Pellegrini & Hicks (1972)	0	0	0·26	0·369
6. Evans & Rosenthal (1968)	3	3	−0·06	0·103
7. Fielder et al. (1971)	17	3	−0·02	0·103
8. Claiborn (1969)	24	3	−0·32	0·220
9. Kester (1969)	0	0	0·27	0·164
10. Maxwell (1970)	1	1	0·80	0·251
11. Carter (1970)	0	0	0·54	0·302
12. Flowers (1966)	0	0	0·18	0·223
13. Keshock (1970)	1	1	−0·02	0·289
14. Henrikson (1970)	2	2	0·23	0·290
15. Fine (1972)	17	3	−0·18	0·159
16. Greiger (1970)	5	3	−0·06	0·167
17. Rosenthal & Jacobson (1968)	1	1	0·30	0·139
18. Fleming & Anttonen (1971)	2	2	0·07	0·094
19. Ginsburg (1970)	7	3	−0·07	0·174

satisfying

$$\sum(d_i - \hat{\mu}_G)^2/(\hat{\sigma}_G^2 + v_i)^2 = \sum 1/(\hat{\sigma}_G^2 + v_i)$$

and $\hat{\mu}_G = \sum d_i(\hat{\sigma}_G^2 + v_i)^{-1}/\sum(\hat{\sigma}_G^2 + v_i)^{-1}$. The solution has to be found iteratively, for example by starting with a trial value of $\hat{\sigma}_G^2$, calculating the trial $\hat{\mu}_G$, and continuing in this manner. Alternatively the EM algorithm can be used as described in Rubin (1980).

The results of such calculations using the data in Table 8.3 are $\hat{\mu}_G = 0{\cdot}1011$, $\hat{\sigma}_G^2 = 0{\cdot}0456$. Individual EB estimates are given by $(d_i\hat{\sigma}_G^2 + \hat{\mu}_G v_i)/(\hat{\sigma}_G^2 + v_i)$; for example, the EBE for study 4 in the table becomes 0·3674.

A linear Bayes approach can be adopted, without the assumptions of normality, as outlined in sections 1.12 and 4.6. Since marginally we have $E(d_i - \mu_G)^2/(\sigma_G^2 + v_i) = 1$, we can estimate μ_G and σ_G^2 by solving

$$\sum(d_i - \hat{\mu}_G)^2/(\hat{\sigma}_G^2 + v_i) = n,$$

with $\hat{\mu}_G$ defined as in the case of ML estimation. This alternative approach gives $\hat{\mu}_G = 0{\cdot}0777$, $\hat{\sigma}_G^2 = 0{\cdot}0125$, and the EBE for study 4 as 0·1685.

Table 8.3 also gives values of a concomitant variable C = weeks of prior contact. Raudenbush and Bryk coded c-values $\geqslant 3$ to 3 to give c_i as in Table 8.3. A quick examination suggests that there is an association between D and C which might be taken into account in developing EB estimates of individual δ_i values. From equations (4.7.2) and (4.7.3) we obtain the estimates

$$\hat{\gamma}_0 = 0{\cdot}407, \hat{\gamma}_1 = -0{\cdot}157, \hat{\tau}^2 = 0{\cdot}000.$$

According to Raudenbush and Bryk (1985) the EB estimate of δ_i can be written as

$$\hat{\delta}_{Gi} = \{d_i\hat{\tau}^2 + (\hat{\gamma}_0 + \hat{\gamma}_1 c_i)v_i\}/(v_i + \tau^2).$$

The linear Bayes method gives the following estimates:

$$\hat{\gamma}_0 = 0{\cdot}418, \hat{\gamma}_1 = -0{\cdot}162, \hat{\tau}^2 = 0{\cdot}006.$$

8.3 Examples involving standard discrete distributions

8.3.1 Bacteria in samples of drinking water

Von Mises (1943) refers to a study of water quality in which $n = 5$ samples were taken from batches of water, interest being in whether a sample contains at least one bacterium. Let θ be the probability of a positive result, i.e. at least one bacterium present. For a given batch the probability of x positive results is given by the binomial $(5, \theta)$ distribution. In repetitions of the procedure with successive batches the marginal X distribution is the mixed binomial

$$p_G(x) = \int \binom{5}{x} \theta^x (1-\theta)^{5-x} dG(\theta).$$

Von Mises discusses estimation of the mixing distribution using a sample of $N = 3420$ observations on the marginal distribution. The actual problem considered by von Mises was not one of estimating θ but one of evaluating the decision rule according to which a batch of water is accepted as having θ in the range 0 to θ_1 ($= 0{\cdot}63$) when $x = 0$ is observed. Thus interest centres on the conditional probability

$$P(0 < \Theta \leqslant \theta_1 | X = 0) = \int_0^{\theta_1} \binom{5}{x} \theta(1-\theta) dG(\theta)/p_G(x).$$

After first estimating the first two moments of $G(\theta)$ a lower bound for $P(0 < \Theta \leqslant \theta_1 | X = 0)$ is obtained. A similar calculation is performed for the case $X = 1$. In fact, this procedure is mathematically close to the point estimation discussed in earlier chapters. In both cases an estimate is made from previous data of the ratio of two integrals with respect to $G(\theta)$ of functions of θ indexed by x.

8.3.2 *True scores in psychological testing*

In a typical situation considered by Lord (1969) a subject is required to answer m items in a test. On each item the score is 1 or 0 according as the answer is correct or not, and the test score is the sum, X, of the item scores. A simple model is that the probability of a correct answer is θ on each item. Lord (1969) also considers the more realistic model in which θ varies from item to item but we shall look only at an example in which θ is constant for each subject. The probability θ is also referred to as the true score of the subject. Of course, the EB aspect of this problem is that θ varies from subject to subject. Under these conditions the distribution of X for a given subject is binomial (m, θ), and the marginal X distribution is a mixed binomial.

Cressie (1979) presents a set of data on $N = 12\,990$ subjects on a test with $n = 20$ and calculates EB estimates of θ on the assumption that the marginal X distribution is a mixed binomial. Cressie proposes a

Table 8.4 *Observed frequency distribution of scores and EB estimates of true scores in a psychological test*

Score x	*Frequency* $Nf(x)$	*Simple EB* $\hat{\theta}(x)$	*Smooth EB*
20	63	0·950	0·898
19	141	0·879	0·863
18	220	0·824	0·823
17	319	0·774	0·773
16	424	0·713	0·716
15	622	0·672	0·663
14	776	0·629	0·619
13	1001	0·586	0·583
12	1203	0·546	0·553
11	1443	0·515	0·526
10	1550	0·503	0·500
9	1409	0·483	0·475
8	1235	0·445	0·451
7	1052	0·423	0·426
6	696	0·407	0·402
5	471	0·369	0·379
4	226	0·362	0·356
3	98	0·314	0·334
2	27	0·283	0·314
1	12	0·144	0·296
0	2	0·050	0·280

method of obtaining an approximate simple EB estimate, i.e. an EB estimate which can be calculated without explicitly estimating the mixing distribution. Let $\hat{\theta}(x)$ be the EB estimate of the true score of a subject whose observed score is x. Cressie gives the following formula for $\hat{\theta}(x)$:

$$\theta(\hat{x}) = \begin{cases} 1/m, & x = 1, \\ x/m + (1 - 2x/m)/m + (x/m)(1 - x/m) & \\ \quad \times \{f(x+1) - f(x-1)\}/\{2f(x)\}, & x = 1, 2, \ldots, m-1, \\ 1 - 1/m, & x = m, \end{cases}$$

where $f(x)$ is the observed proportion of subjects with score x.

Table 8.4 gives a frequency distribution of observed scores, the estimate $\hat{\theta}(x)$ and a smooth estimate obtained according to the method of estimating the mixing distribution given in Lord (1969).

8.3.3 *Acceptance sampling*

A common sampling inspection model supposes that batches of items are inspected by sampling a relatively small number of items from each batch, and on the basis of the quality of the sampled items a decision is made about the quality of the batch. In the simplest cases a batch is supposed to contain a proportion λ of 'defective' items, the rest being 'good'. The decision becomes a matter of estimating the value of λ, or of accepting or rejecting the hypothesis that λ is smaller than some critical value. More complicated procedures are also considered, as we indicate below.

What makes a Bayesian approach to this problem quite natural is that it seems reasonable to assume that λ varies randomly from batch to batch, according to a prior distribution function $G(\lambda)$. In this context schemes in which the loss structure takes the cost of sampling into account have been considered by many authors. Hald (1960) takes the loss caused by an accepted defective item as 1, the cost per item associated with rejected batches as k_r, and the cost of sampling as k_s per item. The cost of each rejected batch is $nk_s + (N-n)k_r$, and of each accepted batch is $nk_s + (N-x)$, where x is the number of defective items in a sample of n items drawn at random from the batch.

Let $p(x|\lambda)$ be the probability of getting x defectives in the sample when the proportion of defectives in the batch is λ. Suppose the decision rule is to accept batches when $x \leqslant c$ and to reject when $x > c$. Then the average cost for batches of quality λ is

$$K(n, c, \lambda) = nk_s + \sum_{x=0}^{c} (N\lambda - x)p(x|\lambda) + (N - nk_r) \sum_{x=c+1}^{n} p(x|\lambda)$$

The overall average cost is $K(n, c) = \int_0^1 K(n, c, \lambda) dG(\lambda)$, where $G(\lambda)$ is the prior distribution function of λ expressing the batch to batch random variation in λ. We can write $K(n, c)$ as

$$K(n, c) = nk_s + \sum_{x=0}^{c} [NE(\Lambda|x) - x]p_G(x) + (N - nk_r) \sum_{x=c+1}^{n} p_G(x),$$

where $p_G(x) = \int p(x|\lambda) dG(\lambda)$, as before.

Usually an optimum combination of n and c can be found, i.e. such as to minimize $K(n, c)$. This, and similar mathematical problems have been treated in detail by Hald (1960, 1967), Wetherill and Campling (1966) and others. Hald (1960) refers to an example of Kjaer (1957) giving an empirical prior distribution obtained after full inspection of a number of batches. Such a distribution can be used instead of the true $G(\lambda)$ to devise an empirical Bayes sampling inspection plan. More commonly full inspection will not be carried out, but observations providing an estimate of $p_G(x)$ may be available, thus providing the data for application of the EB methods that have been developed. For a comment on this approach see Bohrer (1966).

A somewhat simpler example is treated by Cressie and Seheult (1985) who consider estimation of the number of defectives, d, in a batch, when a sample of size m is taken without replacement from a batch of size M; in our notation $d = m\lambda$. The distribution of X given d is hypergeometric. A simple EB approach is possible through the relation

$$\theta(x) = E\left(\frac{D - x}{M - m + 1 - D + x}\middle| X = x\right) = \frac{(x+1)p_G(x+1)}{(m-x)p_G(x)},$$
$$x = 0, 1, \ldots, m - 1,$$

where $p_G(x)$ is the marginal X probability function, as in earlier chapters. Cressie and Seheult use an approximation for $E(D \mid X = x)$ of the form

$$x + (M - m)\frac{\theta(x)}{1 + \theta(x)}\left(1 - \frac{\theta(x) - \theta(x-1)}{\theta(x)\{1 + \theta(x-1)\} + \theta(x) + 1}\right).$$

In the numerical example given by Cressie and Seheult the hypergeometric distribution is assumed to be adequately approximated by a binomial distribution, the probabilities $p_G(x)$ are estimated directly by observed frequency ratios, and the estimate of $\theta(x)$ is monotonized according to the method of van Houwelingen (1977). The data reported by Cressie and Seheult are shown in Table 8.5. The coding of answers in question forms used in a household survey was checked by random selection of $m = 42$ forms from each of 91 batches of forms, the observed X being the number of forms with errors in one particular question. The batch sizes varied as shown in Table 8.5.

A linear EB approach to this example is possible in which one can use the formulae

$$E(X \mid \lambda) = m\lambda/M$$
$$E(X^2 \mid \lambda) = m(M - m)\lambda/\{M(M-1)\}$$
$$+ [(m/M)^2 - m(M - m)/\{M^2(M-1)\}]\lambda^2.$$

Table 8.5 *Batch number i, batch size M_i, observed number of defectives x_i in samples of size $m = 42$, and nearest integer to the monotonized simple EBE β_i^**

i	M_i	x_i	β_i^*	i	M_i	x_i	β_i^*	i	M_i	x_i	β_i^*
1	238	0	1	31	201	0	1	61	193	0	1
2	199	0	1	32	140	0	1	62	233	1	3
3	216	0	1	33	232	1	3	63	229	1	3
4	145	0	1	34	141	1	2	64	171	0	1
5	248	3	14	35	208	0	1	65	192	0	1
6	228	1	3	36	215	1	3	66	186	0	1
7	215	0	1	37	143	0	1	67	212	0	1
8	254	0	2	38	180	0	1	68	170	0	1
9	210	0	1	39	240	0	1	69	224	0	1
10	185	0	1	40	223	0	1	70	256	0	2
11	165	2	7	41	258	1	3	71	207	2	9
12	221	0	1	42	175	3	10	72	141	1	2
13	257	1	3	43	250	3	14	73	258	0	2
14	241	4	19	44	257	0	2	74	249	0	1
15	198	0	1	45	226	0	1	75	222	0	1
16	190	0	1	46	243	0	1	76	196	0	1
17	160	0	1	47	259	0	2	77	254	3	15
18	150	0	1	48	249	0	1	78	176	3	10
19	239	0	1	49	249	0	1	79	256	4	20
20	212	1	3	50	204	0	1	80	156	1	2
21	226	0	1	51	248	2	11	81	253	0	2
22	194	3	11	52	195	1	3	82	142	1	2
23	244	0	1	53	175	0	1	83	257	0	2
24	226	0	1	54	147	0	1	84	190	0	1
25	166	0	1	55	245	0	1	85	238	0	1
26	220	0	1	56	180	5	16	86	237	1	3
27	174	0	1	57	187	0	1	87	228	4	18
28	168	0	1	58	256	1	3	88	185	1	3
29	159	0	1	59	189	1	3	89	216	0	1
30	186	0	1	60	228	1	3	90	145	2	7
								91	182	0	1

8.3.4 Oilwell discoveries

Table 8.6 gives data on oilwell discoveries used by Clevenson and Zidek (1975) in a study of simultaneous estimation of the means of independent Poisson laws. The numbers of discoveries, X_i, in certain months are shown. They are taken to be observations on Poisson random variables with means λ_i. From additional data more accurate estimates of the λ values are available. These are also shown in the

Table 8.6 *Observed monthly numbers of oilwell discoveries, X_i and corresponding expected values, λ_i*

i	X_i	λ_i	i	X_i	λ_i
1	0	1·17	19	1	0·50
2	0	0·83	20	2	0·50
3	0	0·50	21	0	1·33
4	1	1·00	22	0	0·83
5	2	0·83	23	0	0·33
6	1	0·83	24	1	1·50
7	0	1·17	25	5	1·33
8	2	0·83	26	0	0·67
9	0	0·67	27	1	0·67
10	0	0·17	28	0	0·33
11	0	0·00	29	0	0·33
12	1	0·33	30	1	0·33
13	3	1·50	31	0	0·50
14	0	0·50	32	1	0·83
15	0	1·17	33	0	0·67
16	3	1·33	34	1	0·33
17	0	0·50	35	0	0·00
18	2	1·17	36	1	0·50

table, and for the purpose of this exercise are assumed to be the true λ values.

The simultaneous estimator of $(\lambda_1, \lambda_2, \ldots, \lambda_n)$ derived by Clevenson and Zidek is

$$\{Z/(Z+\beta+n-1)\}(X_1, X_2, \ldots, X_n),$$

where $Z = X_1 + X_2 + \cdots + X_n$. Their approach is one of compound estimation in which the loss when estimating by $(\delta_1, \delta_2, \ldots, \delta_n)$ is $\sum(\delta_i - \lambda_i)^2/\lambda_i$.

A straightforward Bayes approach to the problem is to use the gamma prior density $g(\lambda) = (1/\Gamma(\beta))\alpha^\beta \lambda^{\beta-1}\exp(-\alpha\lambda)$. Then the loss structure $(\delta-\lambda)^2/\lambda$ leads to the Bayes estimate

$$\delta_G(x) = (\beta + x - 1)/(\alpha + 1).$$

If we put $\beta = 1$ and note that the mean of the marginal X distribution is then $E(X_G) = \int \alpha\lambda \exp(-\alpha\lambda)d\lambda = 1/\alpha$, a natural estimate of $1/\alpha$ from past data is Z/n. This gives the empirical Bayes estimate

$$\hat{\delta}_G(x) = \{Z/(Z+n)\}x,$$

which is exactly of the same form as the compound estimate with $\beta = 1$. This is the version of the estimate used by Clevenson and Zidek in the illustration with the oilwell data.

For the data in Table 8.6 we have $Z = 29$ with $n = 36$, so the EB estimate is $\hat{\delta}_G(x) = (29/65)x$. Also, $\sum(X_i - \lambda_i)^2/\lambda_i = 39{\cdot}12$, and $\sum(\delta_G(X_i) - \lambda_i)^2/\lambda_i = 14{\cdot}26$, showing substantial improvement of the estimation by using EB rather than ML.

8.3.5 *Cancer mortality rates*

Table 8.7 is extracted from Tsutakawa, Shoop and Marienfeld (1985). It shows an estimate of the mid-period number of persons at risk, m_i, and the number of stomach cancer deaths, y_i, in the $i = 1, 2, \ldots, 84$ largest cities in Missouri for the period 1972–81. The data are for males aged 45–64 years.

It is assumed that y_i is a realization of a Poisson random variable Y_i

Table 8.7 *Stomach cancer mortality in Missouri cities, males aged 45–64 years in 1972–1981*

m_i	y_i	m_i	y_i	m_i	y_i	m_i	y_i
98066	99	1185	0	647	1	443	0
53637	54	1104	0	631	1	423	1
46394	80	1083	0	603	0	419	2
12890	17	1025	1	601	0	403	0
10975	11	917	1	600	1	395	0
7436	13	917	0	592	0	395	0
3814	3	877	0	588	3	389	1
3461	2	874	0	583	1	386	0
3349	1	857	1	582	3	383	0
3215	1	855	0	582	0	372	1
2708	2	854	0	581	1	368	1
2530	4	842	0	556	0	350	1
2145	4	799	2	527	0	339	1
1823	2	731	5	524	1	333	1
1668	2	721	0	522	0	325	0
1627	3	709	1	517	1	318	0
1407	1	706	3	512	1	317	0
1356	0	680	1	493	0	312	1
1339	3	676	1	490	1	307	0
1209	1	664	0	481	0	305	0
1208	1	657	0	448	1	305	0

with mean $m_i\lambda_i$. The Y_i are independent of each other, and the λ_i values are regarded as being sampled at random from a prior distribution G. In the analysis given by Tsutakawa *et al.* (1985) the reparametrization $\alpha_i = \ln\{\lambda_i/(1-\lambda_i)\}$ is used, and the distribution from which the α_i are sampled is taken as $N(\mu, \sigma^2)$. Under these assumptions the marginal probability of y_i is

$$p_G(y_i) = \int \exp(-m_i\lambda_i)(m_i\lambda_i)^{y_i}(1/y_i!)(1/\sigma)\phi\{(\alpha_i - \mu)/\sigma\}d\alpha_i$$

where $\phi(\cdot)$ is the standard normal density, and $\lambda_i = 1/\{1+\exp(-\alpha_i)\}$. This $p_G(y_i)$ depends on the two parameters μ and σ which can be estimated by the ML method, i.e. by maximizing the marginal likelihood $p_G(y_1)p_G(y_2)\cdots p_G(y_n)$. The posterior mean of λ_i is

$$E(\Lambda_i|y_i) = \int \{1+\exp(-\alpha_i)\}^{-1}h(\alpha_i|y_i, \mu, \sigma)d\alpha_i,$$

where $h(\alpha_i|y_i, \mu, \sigma)$ is the posterior density of α_i, written down in an obvious way from the given assumptions. Replacing μ and σ by their ML estimates $\hat{\mu}$ and $\hat{\sigma}$ gives the EB estimate λ_{iEB}; the usual MLE of λ_i is $\hat{\lambda}_i = y_i/m_i$.

For city i the crude mortality rate is calculated as $r_i = \hat{\lambda}_i \times 10^5$, and the corresponding EB rate is r_{iEB} obtained by replacing $\hat{\lambda}_i$ by λ_{iEB}. Table 8.8 gives crude and EB rates for a selection of the cities listed in Table 8.7. The large fluctuations exhibited by the crude rates are smoothed out by the EB approach.

An alternative EB approach is the linear EB method as given in sections 3.8.3 and 3.9.1. The linear Bayes estimate of λ_i is

$$\bar{\lambda}_i = \omega_0 + \omega_1 y_i/m_i,$$

Table 8.8 *Crude and EB mortality rates for selected Missouri cities*

m_i	y_i	r_i	r_{iEB}	r_{iLEB}
98066	99	100·9	102·8	118·3
3215	1	31·1	102·2	117·5
731	5	684·4	140·3	125·6
664	0	0·0	110·8	117·1
333	1	300·1	118·6	120·8

where ω_0 and ω_1 are given by

$$\begin{bmatrix} 1 & E(\Lambda) \\ E(\Lambda) & E(\Lambda^2)+E(\Lambda)/m_i \end{bmatrix}\begin{bmatrix} \omega_0 \\ \omega_1 \end{bmatrix}=\begin{bmatrix} E(\Lambda) \\ E(\Lambda^2) \end{bmatrix}.$$

In order to obtain an empirical version of $\bar{\lambda}_i$ we need estimates of $\gamma_1 = E(\Lambda)$ and $\gamma_2 = E(\Lambda^2)$. Possible estimates are

$$\bar{\gamma}_1 = (1/n)\sum_{i=1}^{n} y_i/m_i$$

$$\bar{\gamma}_2 = (1/n)\sum_{i=1}^{n} y_i(y_i - 1)/m_i^2.$$

Note that slightly different forms of estimators are given in section 3.9.1. Applying this method to the data in Table 8.7 gives $\bar{\gamma}_1 = 0{\cdot}00119$ and $\bar{\gamma}_2 = 1{\cdot}448 \times 10^{-6}$, and the linear EB estimates of rates given by these values are shown in Table 8.8 under the heading r_{iLEB}.

8.3.6 *A quality measurement plan (QMP)*

Suppose that a quality control scheme is operated in which numbers of defects, x_i, are observed in audit samples at rating periods $i = 1, 2, \ldots, n$. A traditional method of statistical quality control is by plotting a Shewhart-type chart, in this case a T-rate chart of values of $T_i = (e_i - x_i)/\sqrt{e_i}, i = 1, 2, \ldots, n$ plotted in sequence. In this definition of T_i the symbol e_i is the expected rate according to a set standard. In an attempt to develop an even more useful technique for quality control and measurement Hoadley (1981a, b) proposed an EB approach to the problem, actually what is now known as a Bayes-EB approach, following the FB approach discussed in section 7.2.2.

Let m_i be the audit sample size at the ith period. The observation x_i is taken to be a realization of a r.v. X_i whose mean is $m_i\alpha_i$, where α_i is the true defect rate at the ith period. The rate α_i is regarded as a realization of a random process. The model proposed by Hoadley is specified as follows:

1. The distribution of X_i is Poisson with mean $sm_i\lambda_i$, noting the reparametrization $\lambda_i = \alpha_i/s$, where s is a standard defects rate assumed known and fixed in advance. The sequence $\{\lambda_i\}$, $i = 1, 2, \ldots, n$, is assumed to comprise independent realizations of a r.v. Λ whose distribution is gamma with mean ϕ_1 and variance ϕ_2.

2. The data set available for estimating the λ_i's is the sequence $\{I_i; i = 1, 2, \ldots, n\}$, where $I_i = x_i/e_i$ is an unbiased estimate of λ_i at the ith rating period. The ratio I_i is also called the **quality index**.
3. In a Bayes approach to the problem the parameters ϕ_1 and ϕ_2 are assumed to have a proper prior distribution function $P(\boldsymbol{\phi})$ defined such that the marginal distribution of ϕ_1 is gamma with mean and variance determined from the distribution of quality between different product types. The marginal distribution of $\omega = \sigma^2/(\sigma^2 + \phi_2)$ is gamma with mean and variance again depending on factors such as distribution of quality between product types. The parameter σ^2 is the average sampling variance over the time interval covering n rating periods, and it is estimated independently as discussed below.

The exact posterior distribution of the current defect rate, given the observed number of defectives, can in principle be obtained by substituting appropriately in formula (7.2.1). The distribution $G(\lambda|\boldsymbol{\phi})$ is taken as gamma with scale parameter ϕ_2/ϕ_1 and shape parameter ϕ_1^2/ϕ_2 and a proper prior $P(\boldsymbol{\phi})$ has to be used for $\boldsymbol{\phi}$. Two difficulties generally arise in implementing these ideas. One is that the choice of P is not obvious, the other is that rather complicated numerical integration may be required. These difficulties have led to the approximate methods mentioned in Chapter 7 to obtain the posterior distribution of Λ and its percentiles. The approximate solution given by Hoadley (1981a, b) can be summarized as follows:

1. Calculate the 'Bayes' estimate, $\hat{\phi}_1$, as the mean of the conditional distribution of the process mean, ϕ_1, given $I_i, i = 1, 2, \ldots, n$. This estimate is a weighted average of I_i values obtained by using the form of marginal prior distribution assumed above.
2. Calculate the 'Bayes' estimate, $\hat{\omega}$, of ω as the mean of the conditional distribution of ω using the form of marginal prior assumed above. It is of the form

$$\hat{\omega} = \hat{\sigma}^2/D(S)$$

where D is a known function of $S = v\sum_{i=1}^{n} q_i(I_i - \hat{\phi}_1)^2$, and v and $q_i, i = 1, 2, \ldots, n$, are known constants.
3. At the ith component the sampling variance of I_i is λ_i/e_i. The average sampling variance over all rating periods can then be estimated as $\hat{\sigma}^2 = \sum_{i=1}^{n} q_i(I_i/e_i)$, where the q_i are weights determined so as to obtain an optimal linear estimate.

Table 8.9 *Quality inspection data*

1977 period	3	4	5	6	7	8
Rating period (i)	1	2	3	4	5	6
Number inspected (n_i)	500	500	500	500	500	500
Number of defects (x_i)	17	20	19	12	7	11
Quality index (I_i)	2·4	2·9	2·7	1·7	1·0	1·6

4. Calculate the estimate of the process variance ϕ_2 from the relation $\hat{\omega} = \hat{\sigma}^2/(\hat{\sigma}^2 + \phi^2)$.
5. Calculate the estimate of the current shrinkage factor w_k, where $k = n + 1$, as

$$\hat{w}_k = (\hat{\phi}_1/e_k)/\{(\hat{\phi}_1/e_k) + \hat{\phi}_2\}.$$

6. Calculate the EB estimate of the current quality λ_k as

$$\hat{\lambda}_k = \hat{w}_k \hat{\phi}_k + (1 - \hat{w}_k) I_k.$$

Hoadley (1986) gives a summary of the earlier work and illustrates the QMP technique on the small data set reproduced in Table 8.9.

To obtain a QMP plot one needs to know, for a selected 'current' rating period k, say, the posterior mean and variance of Λ_k. In the following we illustrate the calculation of the posterior mean at $k = 6$. The equality of the sample sizes m_i simplifies formulae by making all weights p_i equal to each other; the same applied to the q_i weights.

The processes average is estimated as

$$\hat{\phi}_1 = \sum_{i=1}^{6} I_i/6 = 2{\cdot}0.$$

The average sampling variance is estimated as

$$\hat{\sigma}^2 = \sum_{i=1}^{6} (I_i/e_i)/6,$$

and since the standard defects per unit is $s = 0{\cdot}014$, we get $\hat{\sigma}^2 = 0{\cdot}28$.

For this example Hoadley (1987) used an approximate solution to obtain an estimate of the process variance ϕ_2 as the difference between the estimated 'total variance' and the estimated average sampling variance. The total variance is estimated by

$$\sum_{i=1}^{6} (I_i - \hat{\phi}_1)^2/5 = 0{\cdot}54.$$

Thus the process variance estimate is 0·26.

The posterior mean of Λ_6 is calculated as follows:

$$\hat{w}_6 = (\hat{\phi}_1/e_6)/\{(\hat{\phi}_1/e_6) + \hat{\phi}_2\} = 0{\cdot}28/0{\cdot}54 = 0{\cdot}52;$$
$$\hat{\lambda}_6 = (0{\cdot}52)(2{\cdot}0) + (0{\cdot}48)(1{\cdot}6) = 1{\cdot}8.$$

Hoadley (1987) goes on to estimate the posterior variance of Λ as $\text{var}(\Lambda_6|I_1, I_2, \ldots, I_6) \simeq 0{\cdot}15$. The shape and scale parameters of the posterior gamma distribution are estimated as

$$\hat{\alpha} = \hat{\lambda}_6^2/\text{var}(\Lambda_6|I_1, I_2, \ldots, I_6) = 21{\cdot}6,$$
$$\hat{\beta} = \text{var}(\Lambda_6|I_1, I_2, \ldots, I_6)/\hat{\lambda}_6.$$

The 1st, 5th, 95th, 99th percentiles are then obtained respectively as 1·01, 1·2, 2·5, 2·8, and it is noted that the posterior 1st percentile is greater than the standard set at 1·0.

By contrast the corresponding T-rate is

$$T_6 = (e_6 - x_6)/\sqrt{e_6} = (7 - 11)/\sqrt{7} = -1{\cdot}5,$$

which is within the control limits $(-2{\cdot}0, +2{\cdot}0)$.

It will be noted that the illustration above is very similar to a linear EB approach for the Poisson case as far as point estimation is concerned. More complicated assumptions about the priors are needed to derive the posterior distribution. These assumptions, and the availability of suitable data for estimating hyper-prior parameters, are crucial for the results to be reliable.

8.4 Miscellaneous EB applications

8.4.1 Calibration

A typical rather simple calibration problem arises in the following way: accurate measurements $x_1, x_2, \ldots, x_n$, are made of a certain characteristic of n objects. A quick, less accurate method of measurement gives results $y_1, y_2, \ldots, y_n$. These results are used to estimate α and β in the relation

$$y_i = \alpha + \beta x_i + e_i,$$

where the e_i are independent identically distributed errors. Then the estimates $\hat{\alpha}$ and $\hat{\beta}$ are used to estimate the true value, x, of a new object which yields the quick result y. An obvious estimate of x is $(y - \hat{\alpha})/\hat{\beta}$. Of course, if α and β are known the natural estimate is $(y - \alpha)/\beta$.

If x is regarded as a realization of a random variable X with distribution function $G(x)$, a Bayes estimate of y can be developed. Suppose for the moment that α and β are known, and to simplify calculations, that the distribution of e is $N(0, \sigma^2)$ and that $G(x)$ is $N(\mu_G, \sigma_G^2)$. Then straightforward application of results of the bivariate normal distribution gives the Bayes estimate of x:

$$E(X \mid y) = \frac{\beta^2 \sigma_G^2 \{(y-\alpha)/\beta\} + \sigma^2 \mu_G}{\beta^2 \sigma_G^2 + \sigma^2}$$

In some calibration experiments the initial set of x values is controlled and so does not provide any information about $G(x)$. On the other hand, if those initial x values are sampled at random, like x, then they can be used to obtain estimates of μ_G and σ_G^2. Replacing these two parameters by their estimates gives an empirical Bayes estimate of x. Usually α and β will not be known and may be replaced by estimates $\hat{\alpha}$ and $\hat{\beta}$ derived by the method of least squares, or some other suitable method. Suppose that S_{xx}, S_{yy}, S_{xy}, are the sample variances of X and Y, and the sample covariance of X and Y, while $\bar{x}$ and $\bar{y}$ are the sample means. Then, if the least squares estimates are used for α and β, and noting that σ_G^2 is estimated by S_{xx}, and $\beta^2\sigma_G^2 + \sigma^2$ by S, the empirical version of the Bayes estimate of y given above is

$$\hat{E}(X \mid y) = \frac{S_{xy}}{S_{yy}}(y - \bar{y}) + \bar{x}, \qquad (8.4.1)$$

also known as the 'inverse estimator'.

A somewhat more general approach is presented in Lwin and Maritz (1980) where the connection between y and x is written as

$$y = m(x, \boldsymbol{\theta}) + e,$$

and the conditional distribution function of Y given $X = x$ is taken to be $F[\{y - m(x, \boldsymbol{\theta})\}/\sigma]$. Then, by the same argument as above the Bayes estimate of x is

$$E(X \mid y) = \int x f[\{y - m(x, \boldsymbol{\theta})\}/\sigma] dG(x) \Big/ \int f[\{y - m(x, \boldsymbol{\theta})\}/\sigma] dG(x).$$

Also, if $\hat{\theta}$ and $\hat{\sigma}$ are estimates of θ and σ, and if the observed x values are sampled at random like the current x to be estimated, an empirical version of the Bayes estimate is

$$\hat{x}_n(y) = \sum_{i=1}^{n} x_i f[\{y - m(x_i, \hat{\boldsymbol{\theta}})\}/\hat{\sigma}] \Big/ \sum_{i=1}^{n} f[\{y - m(x_i, \hat{\boldsymbol{\theta}})\}/\hat{\sigma}]. \quad (8.4.2)$$

It is interesting to note that this estimate is not the same as the estimate given by (8.4.1) even when $m(x, \boldsymbol{\theta}) = \alpha + \beta x$ and F is the standard normal distribution function. It is not linear in y, unlike both the inverse and the 'classical' estimate $(y - \hat{\alpha})/\hat{\beta}$, where $\hat{\alpha}$ and $\hat{\beta}$ are the least squares estimates of α and β.

An illustration of the application of these estimates to data on the water content of soil specimens reported by Aitchison and Dunsmore (1975, p. 182) is given in Lwin and Maritz (1980). These data are shown in Table 8.10. In every (x_i, y_i) pair the x value is an accurate laboratory determination and the y value is an 'on site' measurement. The form of $m(x, \boldsymbol{\theta})$ was taken to be $\alpha + \beta x$, and least squares estimates of α and β were used. A normal probability plot of residuals suggested that F could be taken as the standard normal distribution with p.d.f. $\phi(u)$. Thus the non-linear estimate becomes

$$\hat{x}_n(y) = \sum_{i=1}^{n} x_i \phi\{(y - \hat{\alpha} - \hat{\beta}x)/\hat{\sigma}\} \Big/ \sum_{i=1}^{n} \phi\{(y - \hat{\alpha} - \hat{\beta}x)/\hat{\sigma}\},$$

where $\hat{\alpha}$ and $\hat{\beta}$ are the least squares estimates of α and β.

Table 8.10 *Laboratory measurement x_i and on site measurement y_i of water content of soil samples, and three predictors of x_i: (1) classical (2) inverse (3) non-linear*

i	x_i	y_i	(1)	(2)	(3)
1	35·3	23·7	32·5	32·2	33·2
2	27·6	20·2	28·3	28·2	27·0
3	36·2	24·5	33·5	33·2	33·9
4	21·6	15·8	22·7	22·8	22·9
5	39·8	29·2	40·0	39·2	39·2
6	24·1	17·8	25·3	25·3	25·7
7	16·1	10·1	15·0	15·6	15·0
8	27.5	19·0	26·7	26·7	26·1
9	33·1	24·3	33·7	33·3	35·6
10	12·8	10·6	16·6	17·0	17·4
11	23·1	15·2	21·7	21·9	21·1
12	19·6	11·4	16·5	16.9	17·1
13	26·1	19·7	27·7	27·7	27·3
14	19·3	12·7	18·5	18·9	18·9
15	18·8	12·6	18·4	18·8	19·1
16	39.8	31·8	44·7	43·9	39·8
m.s.e.			4·6	4·3	3·6

The estimated, or predicted, x values in Table 8.10 were computed by treating every x_i, $i = 1, \ldots, 16$, as the unknown x, and the remaining 15 observations as the previous results from the calibration experiment. The last line in the table gives the mean squared error calculated as $(1/n)\sum(x_i - \text{estimated } x_i)^2$.

8.4.2 *Two-way contingency tables*

Let X_{ij} be the frequency in row i column j of a two-way contingency table, $i = 1, 2, \ldots, n$, $j = 1, 2, \ldots, p$. There are different ways in which Bayesian ideas can be used in the analysis of such data. We begin with an account of an approach described by Laird (1978), who refers to several others. Assume that the X_{ij} are multinomially distributed with $E(X_{ij}) = N\pi_{ij}$, where N is the total number of observations. Let

$$\ln \pi_{ij} = u_0 + u_{1i} + u_{2j} + u_{12ij},$$

and let $\mathbf{u}_1 = (u_{11}, u_{12}, \ldots, u_{1n-1})$, $\mathbf{u}_2 = (u_{21}, u_{22}, \ldots, u_{2p-1})$ while $\mathbf{u}_{12}$ is the vector of all u_{12ij}. Also let $\mathbf{X}$ be the vector of the cell frequencies and $\mathbf{u}^{\mathrm{T}} = (\mathbf{u}_1^{\mathrm{T}}, \mathbf{u}_2^{\mathrm{T}}, \mathbf{u}_{12}^{\mathrm{T}})$. Laird assumes that the u_{1i} and the u_{2i} are *a priori* independent with 'flat' distributions while the u_{12ij} are *a priori* independently and identically distributed with common $N(0, \sigma^2)$. The posterior density of $\mathbf{u}$ is then

$$f(\mathbf{u}|x, \sigma^2) = \{m(\mathbf{x}, \sigma^2)\}^{-1} l(\mathbf{x}|\mathbf{u}) f(\mathbf{u}_{12}|\sigma^2),$$

where $l(\mathbf{x}|\mathbf{u})$ is the usual likelihood of the observed frequencies $\mathbf{x}$, $f(\mathbf{u}_{12}|\sigma^2)$ is the multivariate normal prior density of $\mathbf{u}_{12}$, and

$$m(\mathbf{x}, \sigma^2) = \int l(\mathbf{x}|\mathbf{u}) f(\mathbf{u}_{12}|\sigma^2) d\mathbf{u}.$$

We may interpret $m(\mathbf{x}, \sigma^2)$ as the marginal likelihood of $\mathbf{x}$ given σ^2. For computational reasons it is suggested that the posterior mode, $\mathbf{u}^*$ be taken as point estimate of $\mathbf{u}$. A Bayes point estimate of π_{ij} is then

$$\pi_{ij}^* = \exp(u_{1i}^* + u_{2j}^* + u_{12ij}^*) \Big/ \sum_{l,k} \exp(u_{1l}^* + u_{2k}^* + u_{12lk}^*).$$

This Bayes estimate depends on σ^2. Moreover, in the model as formulated estimation of σ^2 is possible, one way of doing it being by using the marginal likelihood $m(\mathbf{x}, \sigma^2)$. Laird (1978) gives several methods of estimation of σ^2, all of them based on the idea of

Table 8.11 *Numbers of deaths according to occupation (row) and cause (column)*

	Cause of death										
1	10	78	7	74	14	28	111	49	46	7	38
2	281	733	70	1129	234	332	1185	783	500	266	608
3	58	267	28	240	39	69	294	259	134	49	295
4	3	55	2	50	7	18	59	39	17	17	13
5	5	54	2	37	6	8	44	28	15	6	7
6	1	34	3	38	4	9	26	23	13	2	1
7	7	60	0	47	3	11	43	39	14	12	15
8	16	71	2	63	7	15	78	56	38	11	17
9	3	15	0	5	0	2	18	14	4	1	4
10	40	183	7	185	31	73	249	136	77	70	91
11	7	32	2	32	4	18	60	29	24	16	10
12	9	54	2	56	6	25	66	37	23	13	16
13	9	101	4	70	13	34	82	42	49	19	18
14	8	71	0	41	1	20	54	34	28	15	10
15	13	116	5	87	10	25	115	88	52	23	14
16	3	30	0	6	0	1	4	22	2	2	3

maximizing $m(\mathbf{x}, \sigma^2)$. Replacing σ^2 by the estimate $\hat{\sigma}^2$ in the formula above gives an empirical version of the Bayes estimate.

Laird gives an application to a data set concerning male deaths classified according to occupation and cause of death; these data are reported in Good (1956). Table 8.11 gives a subset of the original collection of data. Numbers of deaths in 16 occupations (rows) and 11 causes (columns) are shown.

Laird reports an estimate $\hat{\sigma}^2 = 0{\cdot}078$ from these data and gives the empirical Bayes expected frequencies for a selection of cells in Table 8.12 under the heading EB estimate. The cells chosen are those in which the observed frequencies are 0 and 3.

In many instances a more natural Bayes approach would be indicated by the notion that the rows in the contingency table are randomly generated so that the k row probabilities may be supposed to have a k-dimensional joint distribution. In the notation of section 4.5 the row probabilities are θ_i, $i = 1, \ldots, p$, and the n rows are thought of as a random sample of p-dimensional vectors. A tractable model for the prior distribution of $(\theta_1, \theta_2, \ldots, \theta_p)$ is the Dirichlet distribution with parameters $\alpha_1, \alpha_2, \ldots, \alpha_p$. This leads to a different empirical Bayes approach of which details are given in section 4.5. In essence it entails estimation of the parameters α_i, $i = 1, \ldots, p$, using the n observed row vectors.

Recall from section 4.5 that the method of moments gives the following equations, (4.5.12), for estimating the α_i:

$$(1/n) \sum_{i=1}^{n} x_{ij}/m_i = \bar{\alpha}_j/\bar{A} = \bar{\beta}_j \tag{8.4.3}$$

$$(1/n) \sum_{j=1}^{p} \sum_{i=1}^{n} x_{ij}(x_{ij} - 1)/\{m_i(m_i - 1)\}$$

$$= \left\{ \sum_{j=1}^{p} \bar{\alpha}_j^2 + \bar{A} \right\} \Big/ \{\bar{A}(\bar{A} + 1)\}. \tag{8.4.4}$$

These equations can be solved as indicated in section 4.5 giving the result (4.5.8).

Applying this method to the data in Table 8.11 is not entirely satisfactory because we have the rather large number of parameters, $p = 11$, and only $n = 13$ observed vectors. Nevertheless, the estimates are as follows: $\bar{\beta}_1 = 0{\cdot}0291$, $\bar{\beta}_2 = 0{\cdot}2092$, $\bar{\beta}_3 = 0{\cdot}0077$, $\bar{\beta}_4 = 0{\cdot}1588$, $\bar{\beta}_5 = 0{\cdot}0198$, $\bar{\beta}_6 = 0{\cdot}0537$, $\bar{\beta}_7 = 0{\cdot}1984$, $\bar{\beta}_8 = 0{\cdot}1477$, $\bar{\beta}_9 = 0{\cdot}0794$, $\bar{\beta}_{10} = 0{\cdot}0387$, $\bar{\beta}_{11} = 0{\cdot}0576$, $A = 90{\cdot}30$.

The resulting EB estimates for the cells selected in Table 8.12 are shown under the heading (2).

Table 8.12 *Estimated cell frequencies for certain i and j in Table 8.4.2 when* $\hat{\sigma}^2 = 0{\cdot}078$

i	*j*	*Observed frequency*	*EB estimate*	*(2)*	*(3)*
7	3	0	1·88	0·52	0·37
9	3	0	0·55	0·29	0·24
9	5	0	1·29	0·75	—
14	3	0	2·09	0.53	0·37
16	3	0	0·57	0·31	0·25
16	5	0	1·33	0·80	—
4	1	3	6·25	4·26	3·88
6	3	3	1·48	2·33	2·51
7	5	3	4·73	3·52	—
9	1	3	1·99	2·38	2·50
16	1	3	2·05	2·52	2·62
16	11	3	3·61	3·67	—

The number of parameters to be estimated can be reduced by pooling cells in Table 8.11. For example, if we pool causes 4–11 we get $k = 4$ parameters estimated as follows: $\bar{\beta}_1 = 0{\cdot}0291$, $\bar{\beta}_2 = 0{\cdot}2092$, $\bar{\beta}_3 = 0{\cdot}0077$, $A = 57{\cdot}44$.

The resulting EB estimates for the cells from columns 1–3 selected in Table 8.12 are shown under the heading (3).

8.4.3 Stratified sampling

Stratified sampling of populations is performed for various reasons, including convenience, the desire to improve precision, and also because results for individual strata may be of interest. In the latter event empirical Bayes ideas may be useful for smoothing out irregularities in individual estimates, especially if they are based on relatively sparse data. The justification for such an approach can be found in the notion that the stratum parameters are randomly assigned, if this seems appropriate. Alternatively one may appeal to an empirical Bayes justification of a compound decision approach.

Typically we have in stratum i an estimate y_i of a population characteristic λ_i, subject to a variance σ_i^2. This σ_i^2 will depend on the stratum sample size and on the size of the stratum, among other things. We shall assume that it is known, or that a reasonably good estimate of it is available. If the λ_i are taken to be randomly generated by a $N(\mu_G, \sigma_G^2)$ distribution, and if the distribution of $y_i|\lambda_i$ is $N(\lambda_i, \sigma_i^2)$, then the methods of estimating μ_G and σ_G^2 are exactly like those described in section 8.2.1. When concomitant variables x_i are observed, incorporating such information in an EB approach is possible as described in section 8.1.

A good example of this type is described by Fay and Herriot (1979). It has to do with estimating per capita income (PCI) in a large number, approximately 39 000, of local government units. Many of these were places with populations smaller than 500 persons. Original estimates of PCI for 1970 were made on the basis of a 20% sample of the population so that the sampling errors for the small places can be substantial. Certain concomitant information was used, such as 1969 tax-return data and data on housing from a 1970 census. A further point of considerable interest in the Fay and Herriot paper is that special censuses of a selection of the small places conducted in 1973 made it possible to compare various 1972 PCI estimates with the supposedly true values; the authors refer to some problems associated with the special census values. The EB estimates, actually referred to as James–Stein estimates by Fay and Herriot, exhibit superior performance relative to two other, more common, estimates.

8.4.4 *Empirical Bayes estimation of rates in longitudinal studies*

In longitudinal studies aimed at estimating rates of change it is not uncommon for observations over relatively short time intervals to be made on many subjects. A typical study of this sort is the assessment of rate of change in lung function of workers in a certain industry. Over the period of the study workers are tested from time to time and an attempt is made to estimate the rate of change for each worker. Suppose that on subject i responses y_{ij} are measured at times t_{ij}, $j = 1, 2, \ldots, m_i$. Then a reasonable estimate of the subject's rate of change is the regression coefficient b_i obtained by fitting a least squares line to the (y, t) values of the individual subject. If the total time period over which the observations are taken is relatively short the subject's true response curve can be assumed well approximated by a straight line over the observed time interval. The slope b_i can be regarded as an estimate of the subject's rate of change at time $\bar{t}_i$, the mean of the t_{ij} values.

Hui and Berger (1983) describe a study of this sort, its details being briefly as follows. As above, the observed least squares slope of subject i is taken as an estimate of a true slope at a point somewhere within the observed time range. Instead of taking this point as $\bar{t}_i$, however, Hui and Berger adjust it to $t_i = \bar{t}_i + (1/2)M_{3i}/M_{2i}$, where M_{3i} and M_{2i} are the third and second sample moments of the t-values of subject i. Thus b_i is an estimate of $\beta_i(t_i)$ with variance $v_i = \sigma_i^2/\sum_{j=1}^{i}(t_{ij} - \bar{t}_i)^2$. The prior distributional assumptions about the slopes $\beta_i(t_i)$ are that they are generated by a $N(\gamma_0 + \gamma_1 t_i, \tau^2)$ distribution. The specification of the model so far is exactly along the lines of the discussion of section 8.1, t_i here being the concomitant variable. A novel aspect of the treatment by Hui and Berger is that the v_i values are not assumed known, or at least relatively accurately estimated. Instead, let s_i^2 be the usual unbiased estimate of σ_i^2 obtained from the fitting of a straight line by least squares to the data of subject i. Assume that $(m_i - 2)s_i^2$ is distributed like $\sigma_i^2 X^2(m_i - 2)$, and that the σ_i^2 values are generated by a prior distribution of the inverse gamma form with density

$$\pi(\sigma^2) \propto \sigma^{-2(A+1)} \exp\{-C/(2\sigma^2)\},$$

where A and B are parameters which can be estimated from the data. These estimates are incorporated to give a somewhat more comprehensive EB method.

An application of the approach described above to data on bone loss in postmenopausal women is given by Hui and Berger.

References

Aitchison, J. (1964) Bayesian tolerance regions. *J. Roy. Statist. Soc.*, B, **26**, 161–175.

Aitchison, J.A. and Dunsmore, I.R. (1975) *Statistical Prediction Analysis*. Cambridge: Cambridge University Press.

Atwood, C. (1984) Approximate tolerance intervals based on maximum likelihood estimates. *J. Amer. Statist. Assoc.*, **79**, 459–465.

Bartlett, M.S. and MacDonald, P.D.M. (1968) Least squares estimation of distribution mixtures. *Nature*, **217**, 195–6.

Behboodian, J. (1975) Structural properties and statistics of finite mixtures, in *Statistical Distributions in Scientific Work* (eds G.P. Patil *et al.*), Vol 1. Dordrecht: Reidel, pp. 103–112.

Bennet, G.K. and Martz, H.F. (1972) A continuous empirical Bayes Smoothing technique. *Biometrika.*, **59**, 361–9.

Berger, J.O. (1986) *Statistical Decision Theory and Bayesian Analysis*. Springer-Verlag, New York.

Berkey, C.S. (1982) Bayesian approach for a nonlinear growth model. *Biometrics*, **38**, 953–961.

Bohrer, R. (1966) Discussion on the paper by Wetherill and Campling. *J. Roy. Statist. Soc.*, B, **28**, 414.

Bowman, K.O. and Shenton, L.R. (1973) Space of solutions for a normal mixture. *Biometrika*, **60**, 629–636.

Box, G.E.P. and Tiao, G.C. (1973) *Bayesian Inference in Statistical Analysis*. Reading, Massachusetts: Addison-Wesley.

Cacoullos, T. (1966) Estimation of multivarate density. *Ann. Inst. Math. Statist.*, **18**, 179–89.

Choi, K. (1966) Estimates for the parameters of a finite mixture of distributions. *Univ. of Missouri Tech. Report* No. 18.

Clevenson, M.L. and Zidek, J.V. (1975) Simultaneous estimation of the means of independent Poisson laws. *J. Amer. Statist. Assoc.*, **70**, 698–705.

Copas, J.B. (1969) Compound decisions and empirical Bayes. *J. Roy. Statist. Soc.*, B, **31**, 397–425.

Copas, J.B. (1974) On symmetric compound decision rules for dichotomies. *Ann. Statist.*, **2**, 199–204.

Cox, D.R. (1975) Prediction intervals and empirical Bayes confidence intervals, in *Perspectives in Probability and Statistics* (ed. J. Gani). Applied Probability Trust, pp. 47–55.

Cox, D.R. and Hinkley, D.V. (1974) *Theoretical Statistics.* London: Chapman & Hall.

Cressie, N. (1979) A quick and easy empirical Bayes estimate of true scores. *Sankhya*, B, **41**, 101–108.

Cressie, N.A.C. (1982) A useful empirical Bayes identity. *Ann. Statist.*, **10**, 725–729.

Cressie, N. and Seheult, A. (1985) Empirical Bayes estimation in sampling inspection. *Biometrika*, **72**, 451–458.

Day, N.E. (1969) Estimating the components of a mixture of normal distributions. *Biometrika*, **56**, 463–474.

Deely, J.J. and Kruse, R.L. (1968) Construction of sequences estimating the mixing distribution. *Ann. Math. Statist.*, **39**, 286–288.

Deely, J.J. and Lindley, D.V. (1981) Bayes empirical Bayes. *J. Amer. Statist. Assoc.*, **76**, 833–841.

Deely, J.J. and Zimmer, W.J. (1969) Shorter confidence intervals using prior observations. *J. Amer. Statist. Assoc.*, **64**, 378–386.

De Finetti, B. (1964) Foresight, its logical laws, its subjective sources, in *Studies in Subjective Probability* (eds H.E. Kyburg and H.E. Smokler), Wiley, New York.

De Groot, M.H. (1970) *Optimal Statistical Decisions.* McGraw-Hill, New York:

Dempster, A.P., Laird, N.M. and Rubin, D.B. (1977) Maximum likelihood estimation from incomplete data via the EM algorithm (with discussion). *J. Roy. Statist. Soc.*, B, **39**, 1–38.

Der Simonian, R. (1986) Maximum likelihood estimation of a mixing distribution. *Appl. Statist.*, **35**, 301–309.

Efron, B. and Morris, C.N. (1972) Empirical Bayes estimators on vector observations – an extension of Stein's method. *Biometrika*, **59**, 335–347.

Efron, B. and Morris, C. (1975) Data analysis using Stein's estimator and its generalizations. *J. Amer. Statist. Assoc.*, **70**, 311–319.

Evans, I.G. (1964) Bayesian estimation of the variance of a normal distribution. *J. Roy. Statist. Soc.*, B, **26**, 63–68.

Fairfield-Smith, H. (1936) A discriminant function for plant selection. *Ann. Eugenics*, **7**, 240–260.

Fay, R.E. III and Herriot, R.A. (1979) Estimation of income for small places: An application of James–Stein procedures to census data. *J. Amer. Statist. Assoc.*, **74**, 269–77.

Ferguson, T.S. (1967) *Mathematical Statistics: A Decision Theoretic Approach.* Academic Press, New York.

Finney, D.J. (1974) Problems, data and inference: The Address of the President (with Proceedings). *J. Roy. Statist. Soc.*, **A137**, 1–23.

Good, I.J. (1956) On estimation of small frequencies in contingency tables. *J. Roy. Statist. Soc.*, **B18**, 113–24.

Griffin, B.S. and Krutchkoff, R.G. (1971) Optimal Linear estimators: an empirical Bayes version with application to the Binomial distribution. *Biometrika*, **58**, 195–201.

Guttman, I. (1970) *Statistical Tolerance Regions: Classical and Bayesian.* London: Griffin.

Hald, A. (1960) The compound hypergeometric distribution and a system of single sampling inspection plans based on prior distributions and costs. *Technometrics*, **2**, 275–340.

Hald, A. (1967) Asymptotic properties of Bayesian single sampling plans. *J. Roy. Statist. Soc.*, B, **29**, 162–173.

Hannan, J.F. and Robbins, H. (1955) Asymptotic solutions of the compound decision problem for two completely specified distributions. *Ann. Math. Statist.*, **26**, 37–51.

Hannan, J.F. and van Ryzin, J.R. (1965) Rate of convergence in the compound decision problem for two completely specified distributions. *Ann. Math. Statist.*, **26**, 37–51.

Hartigan, J.A. (1969) Linear Bayes methods. *J. Roy. Statist. Soc.*, **B31**, 446–56.

Hoadley, B. (1981a) The quality measurement plan. *Bell System Technical Journal*, **60**, 215–271.

Henderson, C.R., Kempthorne, O., Searle, S.R. and Van Krosigk, C.M. (1959) The estimation of environmental and genetic trend from records subject to culling. *Biometriks*, **15**, 192–218.

Hoadley, B. (1981b) Empirical Bayes analysis and display of failure rates. *Proc. Electronic Components Conference*, pp. 499–505.

Hoadley, B. (1986) Quality measurement plan (QMP), in *Encyclopaedia of Statistics* (eds N.L. Johnson and S. Kotz), Wiley, New York.

Hoerl, A.E. and Kennard, R.W. (1970) Ridge regression; biased estimation for non-orthogonal problems. *Technometrics*, **12**, 55–67.

Hudson, H.M. (1978) A natural identity for exponential families. *Ann. Statist.*, **6**, 473–484.

Hui, S.L. and Berger, J.O. (1983) Empirical Bayes estimation of rates in longitudinal studies. *J. Amer. Statist. Assoc.*, **78**, 753–760.

James, W. and Stein, C. (1960) Estimation with quadratic loss. *Proceedings of the Fourth Berkeley Symposium on Mathematical Statistics and Probability*, Vol. 1, 361–379.

Johns, M.V. Jr (1957) Non-parametric empirical Bayes procedures. *Ann. Math. Statist.*, **28**, 649–669.

Joshi, V.M. (1969) Admissibility of the usual confidence sets for the mean of a univariate or bivariate normal population. *Ann. Math. Statist.*, **40**, 1042–1067.

Kjaer, G.J. (1957) Kontrol med returem ballage. *Nordisk Tidsskrift for Industrial Statistik*, **2**, 15.

Krutchkoff, R.C. (1967) A Supplementary sample non-parametric empirical Bayes approach to some statistical decision problems. *Biometrika* **54**, 451–8.

Kullback, S. (1959) *Information Theory and Statistics*. Wiley, New York.

Laird, N.M. (1978) Empirical Bayes methods for two-way contingency tables. *Biometrika*, **65**, 581–590.

Laird, N.M. and Louis, T.A. (1987) Empirical Bayes confidence intervals based on bootstrap samples. *J. Amer. Statist. Assoc.*, **82**, 739–757.

Lemon, G.F. and Krutchkoff, R.G. (1969) An empirical Bayes smoothing technique. *Biometrika.*, **56**, 361–5.

Lin, Pi-Erh (1975) Rates of convergence in empirical Bayes estimation problems: continuous case. *Ann. Statist.*, **3**, 155–164.

Lindley, D.V. (1961) The use of prior probability distributions in statistical inference and decision. Proc. Fourth Berkeley Symposium on Math. Statist. and Prob., **1**, 453–68.

Lindley, D.V. (1962) Discussion on 'Confidence sets for the mean of a multivariate normal distribution' by C. Stein, *J. Roy. Statist. Soc.*, B, **24**, 265–296.

Lindley, D.V. (1965) *Introduction to Probability and Statistics*, Part 2. Cambridge: Cambridge University Press.

Lindley, D.V. (1971) The estimation of many parameters, in *Foundations of Statistical Inference* (eds V.P. Godambe and D.A. Sprott). Toronto: Holt, Rinehart and Winston.

Lindley, D.V. and Smith, A.F.M. (1972) Bayes estimates for the linear model. *J. Roy. Statist. Soc.*, B, **34**, 1–18.

Lindsay, B.G. (1983a) The geometry of mixture likelihoods: a general theory. *Ann. Statist.*, **11**, 86–94.

Lindsay, B.G. (1983b) The geometry of mixture likelihoods, part II: the exponential family. *Ann. Statist.*, **11**, 783–792.

Lord, F.M. (1969) Estimating true-score distributions in psychological testing (an empirical Bayes estimation problem). *Psychometrika*, **34**, 259–299.

Lord, F.M. and Cressie, N. (1975) An empirical Bayes procedure for finding an interval estimate. *Sankhya*, A, **37**, 1–7.

Lukacs, E. (1960) *Characteristic Functions.* London: Griffin.

Lwin, T. (1976) Optimal linear estimators of location and scale parameters using order statistics and related empirical Bayes estimation. *Scand. Act. J.*, 79–91.

Lwin, T. and Maritz, J.S. (1976) Empirical Bayes interval estimation. *Scand. Act. J.*, 185–196.

Lwin, T. and Maritz, J.S. (1982) A note on the problem of statistical calibration. *Appl. Statist.*, **29**, 135–141.

Lwin, T. and Maritz, J.S. (1982) An analysis of the linear-calibration controversy from the perspective of compound estimation. *Technometrics*, **24**, 235–41.

Martz, H.F. and Krutchkoff, R.G. (1969) Empirical Bayes estimators in a multiple linear regression model. *Biometrika*, **56**, 367–74.

Maritz, J.S. (1966) Smooth empirical Bayes estimation for one-parameter discrete distributions. *Biometrika*, **53**, 417–429.

Maritz, J.S. (1968) On the smooth empirical Bayes approach to testing of hypotheses and the compound decision problem. *Biometrika*, **55**, 83–100.

Maritz, J.S. (1981) *Distribution free statistical methods.* Chapman and Hall, London.

Maritz, J.S. and Lwin, T. (1975) Construction of simple empirical Bayes estimators *J. Roy. Statist. So.*, B, **37**, 421–5.

Maritz, T.S. (1989) Linear emperical Bayes estimation of quantiles. *Stat.* and *Prob. Letters* (in press).

Medgyessy, P. (1961) *Decomposition of Superpositions of Distribution Functions.* Budapest: Hungarian Academy of Sciences.

Miyasawa, K. (1961) An empirical Bayes estimator of the mean of a normal population. *Bull. Inter. Statist. Inst.*, **38**, 181–7.

Morris, C.N. (1983a) Parametric empirical Bayes confidence intervals, in *Scientific Inference, Data Analysis, and Robustness* (ed. G.E.P. Box). Academic Press, New York.

Morris, C.N. (1983b) Parametric empirical Bayes inference: theory and applications. *J. Amer. Statist. Assoc.*, **78**, 47–59.

Nelder, J.A. (1972) Discussion on paper by Lindley and Smith. *J. Roy. Statist. Soc.*, **34**, 1–18.

Nichols, W.G. and Tsokos, C.P. (1972) Empirical Bayes point estimation in a family of probability distributions. *Inter. Statist. Rev.*, **40**, 147–51.

Parzen, E. (1962) On estimation of a probability density function and mode. *Ann. Math. Statist.*, **33**, 1065–76.

Raiffa, H. and Schlaifer, R. (1961) *Applied Statistical Decision Theory.* Harvard University Press, Boston.

Rao, C.R. (1975) Simultaneous estimation of parameters in different linear models and applications to biometric problems. *Biometrics*, **31**, 545–554.

Raudenbush, S.W. and Bryk, A.S. (1985) Empirical Bayes meta-analysis. *J. Educ. Statist.*, **10**, 75–98.

Robbins, H. (1951) Asymptotically subminimax solutions of compound statistical decision problems. *Proc. Berkeley Symposium on Math. Statist. and Prob.*, 131–148.

Robbins, H. (1955) An empirical Bayes approach to statistics. *Proc. Third Berkeley Symposium on Math. Statist. and Prob.*, **1**, 157–64.

Robbins, H. (1964) The empirical Bayes approach to statistical problems. *Ann. Math. Statist.*, **35**, 1–20.

Rolph, J.E. (1968) Bayes estimation of mixing distributions. *Ann. Math. Statist.*, **39**, 1289–1302.

Rubin, H. (1977) Robust Bayesian estimation, in *Statistical Decision Theory and Related Topics II*, (eds S.S. Gupta and D.S. Moore), Academic Press, New York.

Rubin, D.B. (1980) Using empirical Bayes techniques in the law school validity studies. *J. Amer. Statist. Assoc.*, **75**, 801–816.

Rutherford, J.R. and Krutchkoff, R.G. (1967) The empirical Bayes approach: estimating posterior quantiles. *Biometrika*, **55**, 672–675.

Rutherford, J.R. and Krutchkoff, R.G. (1969) ε-asymptotic optimality of empirical Bayes estimators. *Biometrika*, **56**, 220–23.

Samuel, E. (1963a) An empirical Bayes approach to the testing of certian parametric hypotheses. *Ann. Math. Statist.*, **34**, 1370–1385.

Samuel, E. (1963b) Asymptotic solutions of the sequential compound decision problem. *Ann. Math. Statist.*, **34**, 1079–1094.

Samuel, E. (1964) Convergence of the losses of certain decision rules for the sequential compound decision problem. *Ann. Math. Statist.*, **34**, 1606–21.

Samuel, E. (1965) Sequential compound estimators. *Ann. Math. Statist.*, **36**, 879–889.

Samuel, E. (1966) Sequential compound rules for the finite decision problem. *J. Roy. Statist. Soc.*, B, **28**, 63–72.

Serfling, R.J. (1980) *Approximation Theorems of Mathematical Statistics.* Wiley, New York.

Silverman, B. (1986) *Density Estimation for Statistics and Data Analysis.* Chapman and Hall, London.

Simar, L. (1976) Maximum likelihood estimation of a compound Poisson process. *Ann. Statist.*, **4**, 1200–1209.

Smith, A.F.M. (1973) Bayes estimates in one-way and two-way models. *Biometrika*, **60**, 319–329.

Southward, G.M. and van Ryzin, J.R. (1972) Estimating the mean of a random binomial parameter. *Proc. Sixth Berkeley Symp. Math. Statist. Prob.*, **4**, 249–263.

Stein, C. (1955) Inadmissibility of the usual estimator for the mean of a multivariate normal distribution. *Proc. Third Berkeley Symp. Math Stat. Prob.*, **1**, 197–206.

Stein, C. (1962) Confidence sets for the mean of a multivariate normal distribution. *J. Roy. Statist. Soc.*, B, **24**, 265–296.

Tallis, G.M. (1969) This identifiability of mixtures of distributions. *J. Appl. Probab.*, **6**, 389–398.

Tallis, G.M. and Chesson, P. (1982) Identifiability of mixtures. *J. Austral. Math. Soc.*, A, **32**, 339–348.

Teicher, H. (1961) Identifiability of mixtures. *Ann. Math. Statist.*, **32**, 244–248.

Teicher, H. (1963) Identifiability of finite mixtures. *Ann. Math. Statist.*, **34**, 1265–1269.

Titterington, D.M., Smith, A.F.M. and Makov, U.E. (1986) *Statistical Analysis of Finite Mixture Distributions.* Wiley, New York.

Tsutakawa, R.K., Shoop, G.L. and Marienfeld, C.J. (1985) Empirical Bayes estimation of cancer mortality rates. *Statistics in Medicine*, **4**, 201–212.

van Houwelingen, J.C. (1977) Monotonizing empirical Bayes estimators for a class of discrete distributions with monotone likelihood ratio. *Statist. Neerl.*, **31**, 95–104.

van Ryzin, J.R. (1966a) The compound decision problem with $m \times n$ finite loss matrix. *Ann. Math. Statist.*, **37**, 412–424.

van Ryzin, J.R. (1966b) The sequential compound decision problem with $m \times n$ finite loss matrix. *Ann. Math. Statist.*, **37**, 954–975.

von Mises, R. (1943) On the correct use of Bayes' formula. *Ann. Math. Statist.*, **13**, 156–165.

Wald, A. (1942) Setting of tolerance limits when the sample is large. *Ann. Math. Statist.*, **13**, 389–399.

Wetherill, G.B. and Campling G.E.G. (1966) The decision theory approach to sampling inspection. *J. Roy. Statist. Soc.*, B, **28**, 381–416.

Wilks, S.S. (1941) Determination of sample sizes for setting tolerance limits. *Ann. Math. Statist.*, **12**, 91–96.

Winkler, R.L. (1972) A decision theoretic approach to interval estimation. *J. Amer. Statist. Assoc.*, **67**, 137–191.

Yakowitz, S.J. and Spragins, J.D. (1968) On the identifiability of finite mixtures. *Ann. Math. Statist.*, **39**, 209–214.

Author index

Subject index